INORGANIC SYNTHESES

Volume XVI

Future Volumes

XVII Alan G. MacDiarmid
XVIII Bodie Douglas
XIX Duard F. Shriver

Editor-in-Chief

FRED BASOLO

Department of Chemistry
Northwestern University
Evanston, Illinois

INORGANIC SYNTHESES

Volume XVI

McGRAW-HILL BOOK COMPANY

New York St. Louis San Francisco Auckland Düsseldorf
Johannesburg Kuala Lumpur London Mexico Montreal
New Delhi Panama Paris São Paulo Singapore
Sydney Tokyo Toronto

INORGANIC SYNTHESES, VOLUME XVI

 Library of Congress Catalog Card Number 39-23015

ISBN 0-07-004015-X

1234567890 KPKP 785432109876

CONTENTS

Preface ix
Notice to Contributors x

Chapter One METAL NITROSYL COMPLEXES 1

1. Tetranitrosylchromium and Carbonyltrinitrosylmanganese 1
   A. Tetranitrosylchromium 2
   B. Carbonyltrinitrosylmanganese 4
2. Dithiocarbamate Complexes of Iron and Cobalt Nitrosyls 5
   A. Bis(diethyldithiocarbamato)nitrosyliron 5
   B. Bis(dimethyldithiocarbamato)nitrosylcobalt 7
3. Ammine Complexes of Osmium, Including Amminenitrosyls 9
   A. Pentaammine(dinitrogen)osmium(II) Iodide 9
   B. Pentaammineiodoosmium(III) Iodide 10
   C. Hexaammineosmium(III) Iodide 10
   D. Pentaamminenitrosylosmium(3+) Halide Monohydrate 11
   E. Tetraamminehydroxonitrosylosmium(2+) Halide 11
   F. Tetraamminehalonitrosylosmium(2+) Halide 12
4. Nitrosylammineruthenium Complexes 13
   A. Tetraamminechloronitrosylruthenium(2+) Chloride 13
   B. Tetraammineacidonitrosylruthenium(2+) Perchlorates 14
5. Dinitrosylcobalt Complexes 16
   A. (*N*,*N*,*N*′,*N*′-Tetramethylethylenediamine)dinitrosylcobalt(1+) Tetraphenylborate 17
   B. Bis(triphenylphosphine)dinitrosylcobalt(1+) Tetraphenylborate 18
   C. [Ethylenebis(diphenylphosphine)]dinitrosylcobalt(1+) Tetraphenylborate 19
6. Bis(triphenylphosphine)chlorodinitrosylruthenium(1+) Tetrafluoroborate 21
7. Dicarbonyl-η-cyclopentadienylnitrosylmolybdenum and Bis(dihalo-η-cyclopentadienylnitrosylmolybdenum) Derivatives 24
   A. Dicarbonyl-η-cyclopentadienylnitrosylmolybdenum 24
   B. Bis(dichloro-η-cyclopentadienylnitrosylmolybdenum) 26
   C. Bis(dibromo-η-cyclopentadienylnitrosylmolybdenum) 27
   D. Bis(diiodo-η-cyclopentadienylnitrosylmolybdenum) 28
8. Bis(methyldiphenylphosphine)dichloronitrosylcobalt 29
   A. Room-Temperature Form 30
   B. Low-Temperature Form 30
9. Tris(triphenylphosphine)nitrosylcobalt and Tris(triphenylphosphine)nitrosylrhodium 32
   A. Tris(triphenylphosphine)nitrosylcobalt 33
   B. Tris(triphenylphosphine)nitrosylrhodium 33
10. Carbonylchloronitrosylrhenium Compounds 35
   A. Octacarbonyl-di-μ-chlorodirhenium 35
   B. Pentacarbonyl-tri-μ-chloro-nitrosyldirhenium 36
   C. Tetracarbonyl-di-μ-chlorodichlorodinitrosyldirhenium 37
11. Decacarbonyldi-μ-nitrosyl-trimetal Compounds 39

A. Decacarbonyldi-μ-nitrosyl-triruthenium 39
B. Decacarbonyl-di-μ-nitrosyl-triosmium 40
12. *trans*-Bis(triphenylphosphine)chloronitrosyliridium(1+) Tetrafluoroborate 41

Chapter Two METAL CARBONYL COMPOUNDS 45

13. Dodecacarbonyltriruthenium 45
14. Dodecacarbonyltriruthenium 47
15. Hexadecacarbonylhexarhodium 49
16. Di-μ-chloro-bis[tricarbonylchlororuthenium(II)] 51
17. Dicarbonyl(η-cyclopentadienyl)(thiocarbonyl)manganese(I) 53
18. Transition Metal Carbonyl Derivatives of Magnesium 56
A. Bis(dicarbonyl-η-cyclopentadienyliron)bis(tetrahydrofuran)magnesium 56
B. Tetrakis(pyridine)bis(tetracarbonylcobalt)magnesium 58
C. Bis[dicarbonyl-η-cyclopentadienyl(tributylphosphine)molybdenum]tetrakis-(tetrahydrofuran)magnesium 59
19. Tris(pentacarbonylmanganese)thallium(III) 61
20. (Alkylamino)difluorophosphine- and Halodifluorophosphinetetracarbonyliron Complexes 63
A. (Diethylamino)difluorophosphinetetracarbonyliron 64
B. Chlorodifluorophosphinetetracarbonyliron 66
C. Trifluorophosphinetetracarbonyliron 67
21. Tris[bis(2-methoxyethyl)ether]potassium and Tetraphenylarsonium Hexacarbonyl-metallates(1−) of Niobium and Tantalum 68
A. Tris[bis(2-methoxyethyl)ether]potassium Hexacarbonylniobate(1−) 69
B. Tris[bis(2-methoxyethyl)ether]potassium Hexacarbonyltantalate(1−) 71
C. Tetraphenylarsonium Hexacarbonyltantalate(1−) 72
D. Tetraphenylarsonium Hexacarbonylniobate(1−) 72

Chapter Three WERNER TYPE METAL COMPLEXES 75

22. (Dinitrogen oxide)pentaammineruthenium(II) Salts 75
23. Tetrakis(isothiocyanato)bis(2,2′-bipyridine)niobium(IV) 78
24. Malonato Complexes of Chromium(III) 80
A. Potassium Tris(malonato)chromate(III) Trihydrate 80
B. Potassium *cis*- and *trans*-diaquabis(malonato)chromate(III) Trihydrate 81
25. Selenoureametal Complexes 83
A. Tetrakis(selenourea)cobalt(II) Perchlorate 84
B. Sulfatotris(selenourea)cobalt(II) 85
C. Dichlorobis(selenourea)mercury(II) 86
D. Dibromobis(selenourea)mercury(II) 86
E. Di-μ-chloro-dichlorobis(selenourea)dimercury(II) 86
26. Nickel Tetrafluorooxovanadate(IV) Heptahydrate 87
27. Polymeric Chromium(III)-bis(phosphinates) 89
A. Poly[aquahydroxo-bis(μ-R,R′-phosphinato)-chromium(III)] 90
B. Poly[hydroxo-bis(μ-R,R′-phosphinato)-chromium(III)] 91
28. Resolution of the *cis*-Bromoamminebis(ethylenediamine)cobalt(III) Ion 93
A. (+)-*cis*-Bromoamminebis(ethylenediamine)cobalt(III) (+)-α-bromocamphor-π-sulfonate [(+)(+)-diastereoisomer] 93
B. (−)-*cis*-Bromoamminebis(ethylenediamine)cobalt(III) Dithionate 94
C. (+)-*cis*-Bromoamminebis(ethylenediamine)cobalt(III) Chloride 95
D. (−)-*cis*-Bromoamminebis(ethylenediamine)cobalt(III) Chloride 96
29. Bis(alkylphosphine)trichlorotitanium(III) Complexes 97
A. Bis(methylphosphine)trichlorotitanium(III) 98
B. Bis(dimethylphosphine)trichlorotitanium(III) 100
C. Bis(trimethylphosphine)trichlorotitanium(III) 100
D. Bis(triethylphosphine)trichlorotitanium(III) 101

Chapter Four OTHER TRANSITION METAL COMPLEXES 103

30. Reagents for the Synthesis of η-Diene Complexes of Tricarbonyliron and Tricarbonylruthenium 103
   A. (Benzylideneacetone)tricarbonyliron(0) 104
   B. (1,5-Cyclooctadiene)tricarbonylruthenium(0) 105
31. Bis(η-cyclopentadienyl)niobium Complexes 107
   A. Dichlorobis(η-cyclopentadienyl)niobium(IV) 107
   B. Bis(η-cyclopentadienyl)(tetrahydroborato)niobium(III) 109
   C. Bis(η-cyclopentadienyl)hydrido(dimethylphenylphosphine)niobium(III) 110
   D. Bis(η-cyclopentadienyl)dihydrido(dimethylphenylphosphine)niobium(V) Hexafluorophosphate or Tetrafluoroborate 111
   E. Bromobis(η-cyclopentadienyl)dihydrido(dimethylphenylphosphine)niobium(III) 112
32. Dichloro(1,3-propanediyl)platinum and *trans*-Dichlorobis(pyridine)(1,3-propanediyl)platinum Derivatives 113
   A. Dichloro(1,3-propanediyl)platinum 114
   B. *trans*-Dichlorobis(pyridine)(2-phenyl-1,3-propanediyl)platinum 115
33. Olefin(β-diketonato)silver(I) Compounds 117
   A. (1,1,1,5,5,5-hexafluoro-2,4-pentanedionato)(1,5-cyclooctadiene)silver(I) 117
   B. Olefin(β-diketonato)silver(I) 118
34. Trichloromethyltitanium and Tribromomethyltitanium 120
   A. Trichloromethyltitanium 122
   B. Tribromomethyltitanium 124
35. Ethylenebis(triphenylphosphine)palladium(0) and Related Complexes 127
   A. Ethylenebis(triphenylphosphine)palladium(0) 127
   B. Ethylenebis(tricylohexylphosphine)palladium(0) and Ethylenebis(tri-*o*-tolyl phosphite)palladium(0) 129

Chapter Five MAIN-GROUP AND ACTINIDE COMPOUNDS 131

36. Tetraethylammonium, Tetraphenylarsonium, and Ammonium Cyanates and Cyanides 131
   A. Tetraethylammonium Cyanate and Cyanide 131
   B. Tetraphenylarsonium Cyanate and Cyanide 134
   C. Ammonium Cyanate 136
37. Diindenylmagnesium 137
38. Fluoromethylsilanes 139
   A. Trifluoromethylsilane 139
   B. Difluorodimethylsilane and Fluorotrimethylsilane 141
39. Uranium Hexachloride 143
40. Chlorotris(η-cyclopentadienyl) Complexes of Uranium(IV) and Thorium(IV) 147
   A. Chlorotris(η-cyclopentadienyl)uranium(IV) 148
   B. Chlorotris(η-cyclopentadienyl)thorium(IV) 149

Chapter Six LIGANDS USED TO PREPARE METAL COMPLEXES 153

41. Trimethylphosphine 153
42. Tertiary Phosphines 155
   A. Methyldiphenylphosphine 157
   B. Ethyldiphenylphosphine 158
   C. *n*-Butyldiphenylphosphine 158
   D. Cyclohexyldiphenylphosphine 159
   E. Benzyldiphenylphosphine 159
   F. 2-(Diphenylphosphino)triethylamine 160
43. Diphenylphosphine 161

44. Tri(phenyl-d_5)phosphine 164
45. Methyl Difluorophosphite 166
46. Bi-, Tri-, and Tetradentate Phosphorus-Sulfur Ligands 168
A. [*o*-(Methylthio)phenyl]diphenylphosphine 171
B. Bis[*o*-(methylthio)phenyl]phenylphosphine 172
C. Tris[*o*-(methylthio)phenyl]phosphine 173
47. Tetradentate Tripod Ligands Containing Nitrogen, Sulfur, Phosphorus, and Arsenic as Donor Atoms 174
A. Tris[2-(diphenylphosphino)ethyl]amine 176
B. Tris[2-(diphenylarsino)ethyl]amine 177
C. Tris[2-(methylthio)ethyl]amine 177
48. Dimethyl(pentafluorophenyl)phosphine and Dimethyl(pentafluorophenyl)arsine Ligands 180
A. Dimethyl(pentafluorophenyl)phosphine 181
B. Dimethyl(pentafluorophenyl)arsine 183
49. Tris(*o*-dimethylarsinophenyl)arsine and Tris(*o*-dimethylarsinophenyl)stibine 184
A. Tris(*o*-dimethylarsinophenyl)arsine 186
B. Tris(*o*-dimethylarsinophenyl)stibine 187
50. *cis*-2-Diphenylarsinovinyldiphenylphosphine and 2-Diphenylarsinoethyldiphenylphosphine 188
A. *cis*-2-Diphenylarsinovinyldiphenylphosphine 189
B. 2-Diphenylarsinoethyldiphenylphosphine 191
51. [2-(Isopropylphenylphosphino)ethyl]diphenylphosphine 192
52. [(Phenylisopropylphosphino)methyl]diphenylphosphine Sulfide 195
53. *N,N,N',N'*-Tetrakis(diphenylphosphinomethyl)ethylenediamine 198
54. Ethylenebis(nitrilodimethylene)tetrakis(phenylphosphinic acid) 199
55. [2-(Phenylphosphino)ethyl]diphenylphosphine 202
A. Isopropylphenylvinylphosphinate 203
B. [2-(Phenylphosphino)ethyl]diphenylphosphine 204
56. 1,1,1-Trifluoro-4-mercapto-4-(2-thienyl)-3-buten-2-one 206

Chapter Seven COMPOUNDS OF BIOLOGICAL INTEREST 213

57. Metalloporphyrins 213
A. *meso*-Tetraphenylporphyrincopper(II) 214
B. Hematoporphyrin IX dimethylesterchloroiron(III) 216
58. (1,4,8,11-Tetraazacyclotetradecane)nickel(II) Perchlorate and 1,4,8,11-Tetraazacyclotetradecane 220
A. (1,4,8,11-Tetraazacyclotetradecane)nickel(II) Perchlorate 221
B. 1,4,8,11-Tetraazacyclotetradecane 223
59. *N,N'*-Ethylenebis(monothioacetylacetoniminato)cobalt(II) and Related Metal Complexes 225
A. *N,N'*-Ethylenebis(monothioacetylacetonimine) 226
B. *N,N'*-Ethylenebis(monothioacetylacetoniminato)cobalt(II) 227
60. Metallatranes 229
A. 1-Ethyl-2,8,9-trioxa-5-aza-germatricyclo[3.3.3.0]undecane 229
B. 1-Ethyl-2,8,9-trioxa-5-aza-stannatricyclo[3.3.3.0]undecane 230
C. 2,8-Dioxa-5-aza-1-stannobicyclo[2.2.0]octane 232
61. *cis*-Bis(diethyldithiocarbamato)dinitrosylmolybdenum 235
62. Some η-Cyclopentadienyl Complexes of Titanium(III) 237
A. Dichloro(η-cyclopentadienyl)titanium(III) Polymer 238
B. Dichloro(η-cyclopentadienyl)bis(dimethylphenylphosphine)titanium(III) 239

Index of Contributors 243
Subject Index 247
Formula Index 253

PREFACE

This volume contains checked synthetic procedures for 130 compounds. An attempt is made to group syntheses in different chapters in accordance with specific areas of interest. For example, the first chapter presents the syntheses of some metal nitrosyls and the last chapter is devoted to the preparation of a few compounds of biological interest. There is also a chapter on the syntheses of ligands, which might be more appropriate for a volume of *Organic Syntheses*. My justification for including this chapter is that the compounds are often prepared by inorganic chemists and the ligands are certainly largely used by coordination and/or organometallic chemists. Other syntheses are grouped in different chapters with more traditional headings.

Inorganic Syntheses, Inc., is a nonprofit organization dedicated to the selection and presentation of tested procedures for the synthesis of compounds of more than routine interest. The Editorial Board seeks the cooperation and help of the entire scientific community in its attempt to realize these goals. Persons familiar with the syntheses of important compounds should take the initiative to prepare manuscripts suitable for INORGANIC SYNTHESES and to encourage others in the field to do likewise. Equally important are the persons willing to devote the time to check these procedures and to work out with the submitter any significant alterations that may improve the synthesis. My experience in working with submitters and checkers has been most gratifying, and I am pleased to acknowledge their cooperation and help in preparing Volume XVI.

Directions for submitting syntheses for INORGANIC SYNTHESES follow this Preface. The editor for Volume XVII is A. G. MacDiarmid, for Volume XVIII is B. E. Douglas, and for Volume XIX is D. F. Shriver.

Finally I thank those members of Inorganic Syntheses, Inc., who read the original manuscripts and gave me valuable advice and help. I am especially thankful to Professors Robert D. Feltham and Devon W. Meek for contacting prospective authors for syntheses of metal nitrosyls and of ligands, respectively. I also thank S. Kirschner and W. H. Powell for their help.

Fred Basolo

NOTICE TO CONTRIBUTORS

The INORGANIC SYNTHESES series is published to provide all users of inorganic substances with detailed and foolproof procedures for the preparation of important and timely compounds. Thus the series is the concern of the entire scientific community. The Editorial Board hopes that all chemists will share in the responsibility of producing INORGANIC SYNTHESES by offering their advice and assistance in both the formulation of and the laboratory evaluation of outstanding syntheses. Help of this kind will be invaluable in achieving excellence and pertinence to current scientific interests.

There is no rigid definition of what constitutes a suitable synthesis. The major criterion by which syntheses are judged is the potential value to the scientific community. An ideal synthesis is one which presents a new or revised experimental procedure applicable to a variety of related compounds, at least one of which is critically important in current research. However, syntheses of individual compounds that are of interest or importance are also acceptable.

The Editorial Board lists the following criteria of content for submitted manuscripts. Style should conform with that of previous volumes of INORGANIC SYNTHESES. The *Introduction* should include a concise and critical summary of the available procedures for synthesis of the product in question. It should also include an estimate of the time required for the synthesis, an indication of the importance and utility of the product, and an admonition if any potential hazards are associated with the procedure. The *Procedure* should present detailed and unambiguous laboratory directions and be written so that it anticipates possible mistakes and misunderstandings on the part of the person who attempts to duplicate the procedure. Any unusual equipment or procedure should be clearly described. Line drawings should be included when they can be helpful. All safety measures should be stated clearly. Sources of unusual starting materials must be given, and, if possible, minimal standards of purity of reagents and solvents should be stated. The scale should be reasonable for normal laboratory operation, and any problems involved in scaling the procedure either up or down should be discussed. The criteria for judging the purity of the final product should be delineated clearly. The section on *Properties* should list and discuss those physical and chemical characteristics that are relevant to judging the purity of the product and to permitting its handling and use in an intelligent manner. Under *References,* all pertinent literature citations should be listed in order.

The Editorial Board determines whether submitted syntheses meet the general specifications outlined above. Every synthesis must be satisfactorily reproduced in a different laboratory from that from which it was submitted.

Each manuscript should be submitted in duplicate to the Secretary of the Editorial Board, Professor Jay H. Worrell, Department of Chemistry, University of South Florida, Tampa, FL 33620. The manuscript should be typewritten in English. Nomenclature should be consistent and should follow the recommendations presented in "The Definitive Rules for Nomenclature of Inorganic Chemistry," *J. Am. Chem. Soc.,* **82,** 5523 (1960). Abbreviations should conform to those used in publications of the American Chemical Society, particularly *Inorganic Chemistry*.

Chapter One

METAL NITROSYL COMPLEXES

Assembled through the cooperation of Professor Robert D. Feltham.

1. TETRANITROSYLCHROMIUM AND CARBONYLTRINITROSYLMANGANESE

Submitted by S. K. SATIJA* and B. I. SWANSON*
Checked by C. E. STROUSE†

Three of the four-coordinate metal nitrosyl complexes of the isoelectronic and isostructural $Ni(CO)_4$ series, $Mn(NO)_3(CO)$, $Fe(NO)_2(CO)_2$, and $Cr(NO)_4$, are formed when respective binary metal carbonyl solutions (*n*-pentane, cyclohexane) are irradiated in the presence of a slow stream of NO.[1,2] The fourth member of the series, $Co(NO)(CO)_3$, can be prepared in a quantitative yield by bubbling a slow stream of NO through a *n*-pentane solution of Co_2-$(CO)_8$ for ca. 15 min. Passage of NO through the solution for a longer time results in the very slow conversion of $Co(NO)(CO)_3$ to the complex $Co(NO)_3$.[3] Earlier methods for the synthesis of $Mn(NO)_3(CO)$ give low yields and require several steps.[4,5] The new photolytic procedure described here gives a good yield of $Mn(NO)_3(CO)$ in only one step starting from $Mn_2(CO)_{10}$. Unfortunately, the yield of $Fe(NO)_2(CO)_2$ in the photochemical reaction of $Fe(CO)_5$ and NO is very low. The low yield results from continuous conversion of Fe-$(NO)_2(CO)_2$ to $Fe(NO)_4$[6] and polymerization of $Fe(CO)_5$ to insoluble poly-

*Department of Chemistry, University of Texas at Austin, Austin, Tex. 78712.
†Department of Chemistry, University of California at Los Angeles, Los Angeles, Calif. 90024.

nuclear carbonyls of iron. This procedure can conveniently be employed to prepare isotopically substituted species (^{15}NO, $N^{18}O$, and $^{15}N^{18}O$) of Mn-$(NO)_3(CO)$ and $Cr(NO)_4$ by using a closed gas-circulating system and occasional removal of CO.

The high solubility of these complexes in stopcock grease requires the use of apparatus made with O-ring joints and grease-free Teflon needle stopcocks.[7] Because of the sensitivity of these compounds to air and moisture, all manipulations must be carried out in an inert atmosphere or in a vacuum system.

■ **Caution.** *Since NO is easily oxidized by air to give poisonous N_2O_4 gas, the reaction should be done in a hood, and proper ventilation should be maintained in the room. Furthermore, carbon monoxide and volatile metal carbonyls and nitrosyls are all highly toxic.*

A. TETRANITROSYLCHROMIUM

$$Cr(CO)_6 + 4NO(g) \xrightarrow{h\nu} Cr(NO)_4 + 6CO(g)$$

Procedure

The reaction vessel (Fig. 1) is a Pyrex tube connected to an O-ring joint which is equipped with a capillary gas inlet and grease-free needle stopcocks. A quartz immersion well fitted with an O-ring joint can be inserted into the reaction vessel. A 500-mg sample of $Cr(CO)_6$* is dissolved in 300 ml of pentane† directly in the reaction vessel, and the immersion well is inserted into it. The solution is flushed with a slow stream of nitrogen for 30 min to ensure removal of oxygen. While the flow of nitrogen is maintained, the Teflon plug of the outlet stopcock of the reaction vessel is quickly replaced by a serum cap. After removal of oxygen, a slow stream of NO‡ is bubbled through the solution.§ The reaction mixture is then irradiated with a 450-W mercury lamp (Hanovia type) using a Corex filter. A quick flow of cold water is maintained through the water jacket of the immersion well to prevent overheating of the reaction mixture and of the lamp during irradiation. The solution begins to turn red within 10 min, indicating the formation of $Cr(NO)_4$. The reaction is monitored by re-

*Alfa, Ventron Corporation, Chemical Division, Beverly, Mass. 01915.

†Pentane is purified by washing with conc. H_2SO_4, sodium bicarbonate solution, and distilled water; it is then dried over Drierite and subjected to fractional distillation (36°) through a 4-ft-long column.

‡C.P. grade NO from Matheson Gas Products, East Rutherford, N.J. 07073.

§Before being passed through the reaction mixture, both N_2 and NO from the tank are passed through a glass-bead-embedded trap (−78°) to remove moisture and other condensable impurities.

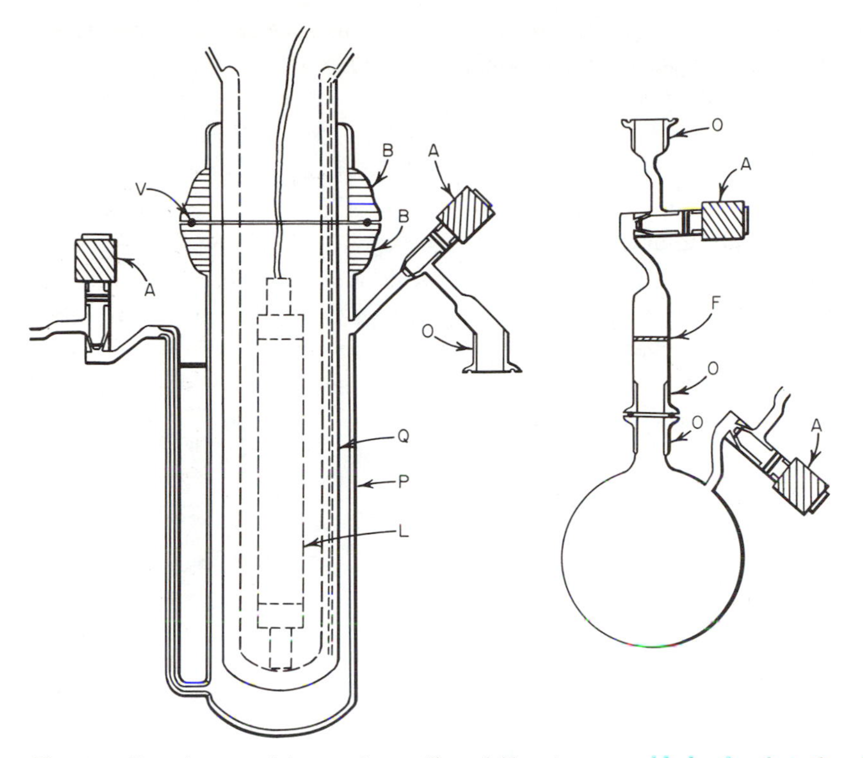

Fig. 1. Reaction vessel, immersion well, and filtration assembly for the photochemical preparation of metal nitrosyls.[7] *A = Teflon needle stopcock, 0–5 mm; B = O-ring connector, tube i.d. 75 mm; F = fritted glass disk, o.d. 15 mm; L = mercury lamp; O = O-ring connector, tube i.d. 15 mm; P = Pyrex tubing, o.d. 80 mm, length 10 in.; Q = water-cooled quartz immersion well, o.d. 65 mm; V = Viton O ring.*

moving small aliquots from the reaction vessel at 20-min intervals and following the disappearance of the 1986 cm^{-1} $Cr(CO)_6$ peak in the infrared. The reaction is complete after $1\frac{1}{2}$ hr.* On completion of the reaction the serum cap is quickly replaced by the Teflon plug (see Caution above), and the flow of NO is stopped. The deep-red solution of $Cr(NO)_4$ is vacuum-filtered through a fritted-glass disk into a 500-ml flask (Fig. 1).[7] The flask is maintained at room temperature, and the solvent is vacuum-distilled into a 0° trap until the red solution is reduced to a volume of 20 ml. The remaining solvent is distilled in a high vacuum from the flask at −78° to a liquid-nitrogen trap. The product is purified by sublimation at 15° and 0.5 torr pressure. Final yield is about 200–210 mg (50–55% of the theory). It has not been possible to obtain an elemental analysis of $Cr(NO)_4$ because of its ease of decomposition. The identity

*The irradiation period depends on how long the lamp has been in service.

of the compound is established by its mass spectrum, which gives a parent ion peak at 172 and peaks for the fragments $Cr(NO)_x$; $x = 1, 3$.

Properties

Tetranitrosylchromium is a reddish-black volatile solid which melts at 38–39°.[2] It is highly soluble in common organic solvents, giving an intense red color in solution. A pentane solution of the complex absorbs very strongly at 1716 cm^{-1} in the infrared. The $Cr(NO)_4$ shows signs of decomposition under vacuum at room temperature, giving metallic chromium and NO.[1] The compound can be stored under dry nitrogen at 0°, but extended storage should preferably be in a cyclohexane solution at 0°.

B. CARBONYLTRINITROSYLMANGANESE

$$Mn_2(CO)_{10} + 2NO(g) \xrightarrow{h\nu} 2Mn(NO)(CO)_4 + 2CO(g)$$

$$Mn(NO)(CO)_4 + 2NO(g) \xrightarrow{h\nu} Mn(NO)_3(CO) + 3CO(g)$$

Procedure

The procedure for the synthesis of $Mn(NO)_3(CO)$ is the same as that for $Cr(NO)_4$, except for the following changes. The time of irradiation in this case is 30–45 min. Irradiation for a longer time than this results in conversion of $Mn(NO)_3(CO)$ to a light-brown solid which has not yet been identified. A Pyrex filter is used for irradiating the $Mn_2(CO)_{10}$ solution. The reaction goes through the intermediate $Mn(NO)(CO)_4$ as shown by the formation of its infrared bands at 2106 (w), 2034 (s), 1979 (s), and 1763 (s) cm^{-1}.[8] The final yield is about 210–220 mg (45–50% of the theory). *Anal.* Calcd. for $Mn(NO)_3(CO)$: Mn, 31.77; C, 6.94; N, 24.27. Found: Mn, 31.70; C, 6.77; N, 24.03.

Properties

Carbonyltrinitrosylmanganese is a dark-green volatile solid (m.p. 27°) which is highly soluble in common organic solvents. The infrared spectrum of the compound, in pentane, shows strong bands at 2091, 1824, and 1733 cm^{-1}. The CO group in $Mn(NO)_3(CO)$ can easily be replaced by a variety of ligands such as phosphines, arsines, stibines, and amines to give derivatives of the type $Mn(NO)_3(R)$.[9] The complex $Mn(NO)_3(CO)$ decomposes under vacuum at 40°, but it can be stored under nitrogen at 0°.

References

1. B. I. Swanson and S. K. Satija, *Chem. Commun.*, **1973,** 40.
2. M. Herberhold and A. Razavi, *Angew. Chem., Int. Ed.*, **11,** 1092 (1972).
3. A. Burg and I. Sabherwal, *Chem. Commun.*, **1970,** 1001.
4. H. Wawersik and F. Basolo, *Inorg. Chem.*, **6,** 1066 (1966).
5. C. G. Barraclough and J. Lewis, *J. Chem. Soc.*, **1960,** 4842.
6. W. P. Griffith, J. Lewis, and G. Wilkinson, *J. Chem. Soc.*, **1958,** 3993.
7. D. F. Shriver, "The Manipulation of Air-Sensitive Compounds," McGraw-Hill Book Company, New York, 1969.
8. P. M. Treichel, E. Pitcher, R. B. King, and F. G. A. Stone, *J. Am. Chem. Soc.*, **83,** 2593 (1961).
9. W. Heiber and H. Tengler, *Z. Anorg. Allg. Chem.*, **318,** 136 (1962).

2. DITHIOCARBAMATE COMPLEXES OF IRON AND COBALT NITROSYLS

Submitted by O. A. Ileperuma* and R. D. Feltham*
Checked by E. D. Johnson†

A. PREPARATION OF BIS(DIETHYLDITHIOCARBAMATO)-NITROSYLIRON

$$Fe^{2+} + NO \rightarrow Fe(NO)^{2+}$$

$$Fe(NO)^{2+} + 2S_2CN(C_2H_5)_2^- \rightarrow Fe(NO)[S_2CN(C_2H_5)_2]_2$$

The preparation of this compound has been reported,[1] but this reference is relatively unavailable and the compound is not always obtained in pure form. The preparation described below[2] gives consistent and reliable results.

Procedure

The reaction is carried out in a 500-ml, three-necked, round-bottomed flask equipped with a magnetic stirrer and a 100-ml dropping funnel. All connections are made with polyvinyl tubing, and all joints should be lubricated with high-vacuum grease. Electronic-grade nitrogen should be used. The exit gases

*Department of Chemistry, University of Arizona, Tucson, Ariz. 85721.
†Department of Chemistry, Northwestern University, Evanston, Ill. 60201.

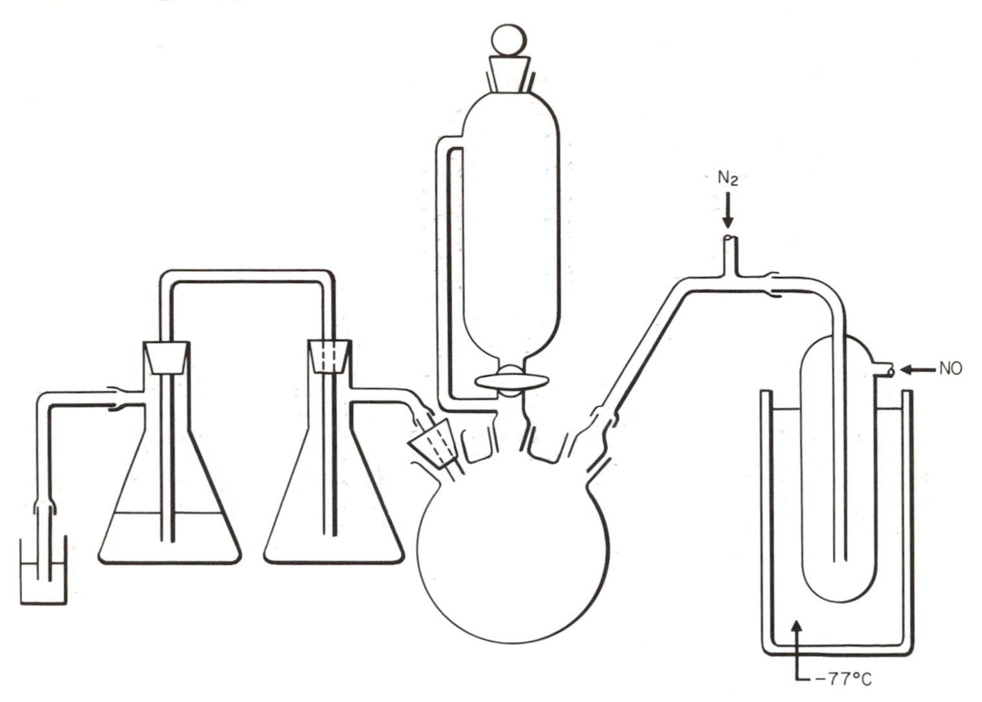

Fig. 2. Apparatus for the preparation of $Fe(NO)[S_2CN(C_2H_5)_2]_2$.

are allowed to pass through a bubble trap into water. Care must be taken that air does not enter the flask during the reaction (see Fig. 2). (■ **Caution.** *Because nitrogen(II) oxide is toxic, the entire reaction should be carried out in an efficient hood.*)

Iron(II) sulfate heptahydrate (5.56 g, 20.0 mmoles) is dissolved in 100 ml of carefully degassed water, about 0.1 g of finely powdered zinc is added [to convert any iron(III) to iron(II)], and the mixture is then cooled in an ice bath to a temperature of 0–10°. Nitrogen is bubbled through this solution for about 10 min. Nitrogen(II) oxide from the cylinder is purified by passing through three gas traps maintained at −78° using Dry Ice and acetone. A moderate flow of purified nitrogen(II) oxide is passed through the iron(II) solution for about 30 min. A dark-green solution is obtained at this stage. A solution of 9.01 g of sodium diethyldithiocarbamate (40.0 mmoles) in 50 ml of degassed water is added through the dropping funnel over a 10-min period, while the passage of nitrogen(II) oxide is continued. Efficient stirring should be maintained throughout the reaction. The passage of nitrogen(II) oxide is continued for another 15 min, and a dark-green precipitate is obtained at this stage. The completion of the reaction can be determined when there is no backward flow of liquid in the exit tube from the bubble trap.

When the reaction is complete, the nitrogen(II) oxide is replaced by a strong flow of nitrogen. Extreme care must be taken in handling the iron nitrosyl formed in solution because it is highly sensitive to air, and especially so to nitrogen(IV) oxide, which can arise by the air oxidation of nitrogen(II) oxide. The flow of nitrogen is stopped when no more brown fumes are observed at the exit tube (about 10 min). The green precipitate is removed by filtration under nitrogen, washed three times with water and finally with a little methanol, and dried under vacuum. The yield is about 6.50 g (85% based on iron added).

The compound can be purified [from any residual tris(diethyldithiocarbamato) iron(III) which is always present in these samples] and recrystallized simultaneously by doing a Soxhlet extraction with *n*-heptane. The compound (6.5 g) in the Soxhlet extractor is extracted with 150 ml of heptane for 24 hr. Dark-green crystals separate from the heptane extract and can be collected on a filter. Yield: about 4.5 g.

The product can be identified by its infrared spectrum and elemental analysis. *Anal.* Calcd. for $Fe(NO)[S_2CN(C_2H_5)_2]_2$: C, 31.40; H, 5.26; N, 10.99. Found: C, 31.96; H, 5.10; N, 10.73. Infrared spectrum observed, $\nu(NO) = 1670\ cm^{-1}$.

B. PREPARATION OF BIS(DIMETHYLDITHIOCARBAMATO)NITROSYLCOBALT

$$Co(H_2O)_6(ClO_4)_2 + NO + 2H_2NCH_2CH_2NH_2 \rightarrow Co(NO)(C_2H_4N_2H_4)_2(ClO_4)_2$$

$$Co(NO)(C_2H_4N_2H_4)_2(ClO_4)_2 + 2NaS_2CN(CH_3)_2 \rightarrow Co(NO)[S_2CN(CH_3)_2]_2 + 2NaClO_4$$

The cobalt complex, $Co(NO)[S_2CN(CH_3)_2]_2$, is well characterized.[3,4] However, direct synthesis of it from the reaction between NO and $Co[S_2CN(CH_3)_2]_2$ leads to impure materials which are very difficult to purify. The ethylenediamine complex, $Co(NO)(C_2H_4N_2H_4)_2(ClO_4)_2$, can readily be prepared in good yield[5] and is easily converted into the dithiocarbamate derivative. The products from these reactions are obtained in a relatively pure form and can easily be further purified.

■ **Caution.** *Perchlorate salts of metal complexes can be explosive and should be handled with care.*

Procedure

Cobalt(II) perchlorate hexahydrate (10.02 g, 27.4 mmoles) is dissolved in 150 ml of methanol. After the solution has been purged with nitrogen, ethylenediamine (5.00 ml, 74.9 mmoles) is added to the deoxygenated solution, and NO is bubbled through the solution until the uptake of NO ceases. After excess NO has been removed by purging again with nitrogen, the red crystalline precipitate which has formed is removed by filtration, using a frit of medium porosity, and washed with a small amount of methanol, 1:1 methanol-ether, and finally ether. The red complex is then dried in a desiccator over Linde 4A molecular sieve. The red solid is stable in air and can be filtered in air; however, solutions of the complex are readily oxidized. Yield: 9.20 g (85% based on cobalt).

The product is identified by its infrared spectrum and its elemental analysis. *Anal.* Calcd. for $Co(NO)(C_2H_4N_2H_4)_2(ClO_4)_2$: C, 11.77; H, 3.95; N, 17.16. Found: C, 12.53; H, 4.31; N, 17.54. Infrared spectrum, $\nu(NO) = 1663$ cm^{-1}.

The dimethyldithiocarbamate derivative of the cobalt nitrosyl is prepared in the following manner. A solution of sodium dimethyldithiocarbamate (1.87 g, 15.1 mmoles) in 100 ml of methanol is added dropwise to 3.08 g (7.6 mmoles) of solid $Co(NO)(C_2H_4N_2H_4)_2(ClO_4)_2$ over 10 min. Since these solutions are air-sensitive, all operations are carried out under nitrogen. The mixture is stirred for 2 hr. At the end of that time, the solvent is carefully removed under a vacuum. The crude product is treated with 50 ml of oxygen-free water. After the mixture has been stirred for 2 hr, the solid is removed by filtration and dried for 12 hr in vacuum. The solid is then dissolved in a minimum amount of dichloromethane under nitrogen, cooled to $-10°$, and allowed to stand overnight. The dark, black-brown crystals which form are removed by filtration and dried in vacuum. Yield: 1.2 g (50% based on cobalt). Further quantities of material can also be obtained from the filtrate, giving a maximum yield* of 85%. The compound is identified by its infrared spectrum and its elemental analysis. *Anal.* Calcd. for $Co(NO)[S_2CN(CH_3)_2]_2$: C, 21.9; H, 3.7; N, 12.8; S, 38.3. Found: C, 21.9; H, 3.8; N, 12.6; S, 38.7. Infrared spectrum, $\nu(NO) = 1630$ cm^{-1}.

References

1. L. Cambi and A. Cagnasso, *Atti Accad. Naz. Lincei,* **13,** 254, 809 (1931).
2. H. Buttner and R. D. Feltham, *Inorg. Chem.,* **11,** 971 (1972).
3. L. Malatesta, *Gazz. Chem.,* **70,** 734 (1940).
4. J. H. Enemark and R. D. Feltham, *J. Chem. Soc., Dalton Trans.,* **1972,** 718.
5. R. D. Feltham and R. S. Nyholm, *Inorg. Chem.,* **4,** 1334 (1965).

*The checker reports a 35% yield but suggests that a yield of 80% is obtained if acetone is used as the solvent.

3. AMMINE COMPLEXES OF OSMIUM, INCLUDING AMMINENITROSYLS

Submitted by F. BOTTOMLEY* and S. B. TONG*
Checked by R. O. HARRIS† and N. K. HOTA†

Until recently ammine complexes of osmium have been little studied compared with their ruthenium analogs. This appears to have been caused by the lack of suitable synthetic routes to them. The discovery of pentaammine(dinitrogen)osmium (II)[1] opened convenient routes to pentaammines of osmium (III), and a convenient synthesis of hexaammineosmium(III)[2] gave new routes to the previously unknown nitrosyls of osmium(II). Here are given the synthesis of $[Os(NH_3)_5(N_2)]I_2$ and its conversion to $[Os(NH_3)_5I]I_2$; the synthesis of $[Os(NH_3)_6]I_3$ and its conversion to $[Os(NH_3)_5(NO)]X_3 \cdot H_2O (X = Cl,$ Br, I); and the preparation of $[OsX(NH_3)_4(NO)]^{2+} (X = OH, Cl, Br, I)$ from $[Os(NH_3)_5(NO)]^{3+}$.

A. PENTAAMMINE(DINITROGEN)OSMIUM(II) IODIDE

$$[OsCl_6]^{2-} + N_2H_4 \cdot H_2O \rightarrow [Os(NH_3)_5(N_2)]^{2+} + \cdots$$

Procedure

Commercial grade $(NH_4)_2[OsCl_6]$‡ (2.37 g) is added in small portions to ice-cold hydrazine hydrate (85%, 23 ml). After the vigorous effervescence subsides, the solution is refluxed, with stirring, for 15 hr. A yellow precipitate of $[Os(NH_3)_5(N_2)]Cl_2$ is formed and is filtered from the cooled mixture, washed with ethanol and ether, and air-dried. To the filtrate is added solid potassium iodide until precipitation of $[Os(NH_3)_5(N_2)]I_2$ is complete. The product is collected on a filter and washed and dried as above. The chloride salt is converted to the iodide by metathesis using solid potassium iodide until precipitation is complete. Total yield of $[Os(NH_3)_5(N_2)]I_2$: 2.36 g (78%). *Anal.* Calcd. for $OsN_7H_{15}I_2$: N, 17.6. Found: N, 17.0. Infrared spectrum, $\nu(N_2)$ (iodide salt) = 2033 (vs) 2043 (sh).

*Chemistry Department, University of New Brunswick, Fredericton, E3B 5A3 New Brunswick, Canada.

†Chemistry Department, Scarborough College, University of Toronto, West Hill, Ontario, Canada.

‡The salt $(NH_4)_2[OsCl_6]$ can be purchased from the J.M.M. group of companies, e.g., Matthey Bishop, Malvern, Pa. 19355, or Johnson, Matthey, and Mallory, 814 Sun Life Building, Montreal, Quebec, Canada.

B. PENTAAMMINEIODOOSMIUM(III) IODIDE

$$[Os(NH_3)_5(N_2)]^{2+} + 1/2I_2 \rightarrow [Os(NH_3)_5I]^{2+} + N_2$$

Procedure

A solution of $[Os(NH_3)_5(N_2)]I_2$ (2.36 g) in hot water (90 ml) is treated with 3 drops of hydriodic acid (57%) and heated for 5 min. Iodine (0.55–1.10 g) is added to the hot solution until gas evolution ceases. The excess iodine is removed on a filter, the solution is cooled in ice, and hydriodic acid (57%) is added until the precipitation of $[Os(NH_3)_5I]I_2$ is complete. The orange-yellow precipitate is filtered off, washed with ethanol and ether, and air-dried. A second crop is obtained by evaporating the filtrate to 20 ml and cooling. Total yield: 2.67 g (96%). The product is recrystallized by dissolving in a minimum amount of hot water, filtering, and reprecipitating with hydriodic acid. The sample is dried *in vacuo* (P_4O_{10}). *Anal.* Calcd. for $OsN_5H_{15}I_3$: H, 2.30; N, 10.68. Found: H, 2.21; N, 10.88.

C. HEXAAMMINEOSMIUM(III) IODIDE

$$2[OsCl_6]^{2-} + 12NH_3 + Zn \rightarrow 2[Os(NH_3)_6]^{3+} + Zn^{2+} + 12Cl^-$$

Procedure

A mixture of $(NH_4)_2[OsCl_6]$ (0.50 g), zinc dust (0.5 g), and aqueous ammonia (d. 0.880 g/ml, 15 ml) is refluxed, with stirring, for 5 hr under argon.* The resultant mixture of excess zinc and metallic osmium in a very pale yellow solution is treated with 3 ml of aqueous ammonia and filtered; potassium iodide (2 g) is added to the filtrate.† The yellow precipitate of $[Os(NH_3)_6]I_3$ is removed by filtration in air, washed with ethanol and ether, and air-dried. Yield: 0.261 g (33%). The crude product is purified by dissolving in a minimum amount of cold water, filtering, and reprecipitating with hydriodic acid.‡ The sample is dried *in vacuo* (P_4O_{10}). A further crop (approximately 8%) can be obtained by evaporating the combined filtrates and recycling. *Anal.* Calcd. for $OsN_6H_{18}I_3$: H, 2.70; N, 12.49; I, 56.56. Found: H, 2.57; N, 12.45; I, 56.91.

*Sometimes an intractable black solid forms when $(NH_4)_2[OCl_6]$ is treated with aqueous ammonia. It is advisable to activate the commercial material by prior refluxing for 15 min in a small amount of 1*M* HCl.

†Coprecipitation of $[Zn(NH_3)_4]I_2$ may occur. This white complex redissolves on further addition of aqueous ammonia.

‡Solutions of $[Os(NH_3)_6]^{3+}$ are unstable. Isolation of the complex should be completed as rapidly as possible.

D. PENTAAMMINENITROSYLOSMIUM(3+) HALIDE MONOHYDRATE

$$[Os(NH_3)_6]^{3+} + NO + H^+ \rightarrow [Os(NH_3)_5(NO)]^{3+} + NH_4^+$$

Procedure

To a deoxygenated slurry of $[Os(NH_3)_6]I_3$ (0.096 g) in water (9 ml) is added 5 drops of 57% hydriodic acid; then nitric oxide is passed through the solution for 1 day. Argon is admitted to remove the excess nitric oxide, and the orange solution is filtered. The filtrate is treated with hydriodic acid (57%, 4 ml) and ice-cooled. The resultant precipitate of $[Os(NH_3)_5(NO)]I_3 \cdot H_2O$ is filtered off, washed with ethanol and ether, and air-dried. Yield: 0.0788 g (79%) (26% based on $(NH_4)_2[OsCl_6]$). The product is recrystallized by dissolving in warm water, filtering, and reprecipitating with several drops of 57% hydriodic acid. The crystals are collected on a filter, washed as above, and dried *in vacuo* (P_4O_{10}).

$[Os(NH_3)_5(NO)]Br_3 \cdot H_2O$ and $[Os(NH_3)_5(NO)]Cl_3 \cdot H_2O$ are prepared metathetically from the iodide salt by using concentrated HBr or HCl. *Anal.* Calcd. for $OsN_6H_{17}O_2I_3$: N, 11.94; I, 54.1. Found: N, 11.90; I, 55.0. Infrared spectrum, $\nu(NO) = 1892$ (s) and 1879 (vs). Calcd. for $OsN_6H_{17}O_2Br_3$: N, 14.93; Br, 42.6. Found: N, 15.02; Br, 42.7. Infrared spectrum, $\nu(NO) = 1893$(s) and 1876 (vs). Calcd. for $OsN_6H_{17}O_2Cl_3$: N, 19.56; Cl, 24.7. Found: N, 19.40; Cl, 24.7. Infrared spectrum, $\nu(NO) = 1893$ (s) and 1871 (vs).

E. TETRAAMMINEHYDROXONITROSYLOSMIUM(2+) HALIDE

$$[Os(NH_3)_5(NO)]^{3+} + OH^- \rightarrow [Os(OH)(NH_3)_4(NO)]^{2+} + NH_3$$

Procedure

A solution of $[Os(NH_3)_5(NO)]I_3 \cdot H_2O$ (0.358 g) in ammonia (d. 0.880 g/ml, 20 ml) is refluxed with stirring for 24 hr. The solution is filtered hot, and the filtrate is evaporated to half volume, ice-cooled, and treated with potassium iodide until precipitation of the product is complete. The yellow crystals of $[Os(OH)(NH_3)_4(NO)]I_2$ are collected on a filter, washed with ethanol and ether, and air-dried. The crude product is recrystallized from 1:1 aqueous ammonia and dried *in vacuo* (P_4O_{10}). Yield: 0.247 g (87%). *Anal.* Calcd. for $OsN_5H_{13}O_2I_2$: N, 12.53; I, 45.5. Found: N, 12.50; I, 46.3. Infrared spectrum, $\nu(NO) = 1817$ (vs). $[Os(OH)(NH_3)_4(NO)]Br_2$ is similarly prepared from $[Os(NH_3)_5(NO)]Br_3 \cdot H_2O$ (75% yield) and $[Os(OH)(NH_3)_4(NO)]Cl_2$ from

$[Os(NH_3)_5(NO)]Cl_3 \cdot H_2O$ (60%), using the appropriate potassium halide as precipitant.

F. TETRAAMMINEHALONITROSYLOSMIUM(2+) HALIDE

$$[Os(OH)(NH_3)_4(NO)]^{2+} + HX \rightarrow [OsX(NH_3)_4(NO)]^{2+} + H_2O$$
$$(X = Cl, Br, I)$$

Procedure

A solution of $[Os(OH)(NH_3)_4(NO)]I_2$ (0.213 g) in hydriodic acid (57%, 10 ml) and water (3 ml) is refluxed for 23 hr. The resultant orange-yellow precipitate of $[OsI(NH_3)_4(NO)]I_2$ is removed by filtration of the hot solution,* washed with ethanol and ether, and air-dried. Yield 0.152 g (60%). The product is purified by dissolving in hot water (15 ml), filtering, and reprecipitating the orange crystals with several drops of hydriodic acid. The sample is dried *in vacuo* (P_4O_{10}). *Anal.* Calcd. for $OsN_5H_{12}OI_3$: N, 10.47; I, 56.9. Found: N, 10.32; I, 57.1. Infrared spectrum, $\nu(NO) = 1844$ (vs). $[OsBr(NH_3)_4(NO)]Br_2$ is prepared similarly by refluxing $[Os(OH)(NH_3)_4(NO)]Br_2$ (0.238 g) in HBr (11 ml) and water (5 ml) for 2 days,* and $[OsCl(NH_3)_4(NO)]Cl_2$ by refluxing $[Os(OH)(NH_3)_4(NO)]Cl_2$ (0.132 g) in HCl (15 ml) for 4 days.* Yields: $[OsBr(NH_3)_4(NO)]Br_2$, 0.20 g (74%); $[OsCl(NH_3)_4(NO)]Cl_2$, 0.12 g (84%). *Anal.* Calcd. for $OsN_5H_{12}OBr_3$: N, 13.26; Br, 45.4. Found: N, 13.58; Br, 45.8. Infrared spectrum, $\nu(NO) = 1842$ (s) and 1830 (sh). Calcd. for $OsN_5H_{12}OCl_3$: N, 17.74; Cl, 26.9. Found: N, 18.00; Cl, 25.1. Infrared spectrum, $\nu(NO) = 1832$ (s).

Properties

$[Os(NH_3)_5(N_2)]I_2$ is air-stable and inert to replacement of N_2, except on oxidation. It can be diazotized with HNO_2, giving $[Os(NH_3)_4(N_2)_2]^{2+}$. Other chemical and physical properties have been reported.[1,3,4] The salt $[Os(NH_3)_6]I_3$ appears to be similar to its ruthenium analog but has been investigated only briefly.[2,5] $[Os(NH_3)_5(NO)]^{3+}$ is reversibly deprotonated to form $[Os(NH_2)(NH_3)_4(NO)]^{2+}$, and may be reduced with zinc to $[Os(NH_3)_6]^{3+}$ or with hydrazine to $[Os(NH_3)_5(N_2)]^{2+}$. None of the nitrosyls synthesized here is susceptible to nucleophilic attack at the NO^+ group. All are diamagnetic, and their other

*All $[OsX(NH_3)_4(NO)]X_2$ complexes must be obtained by precipitation from hot solution. On cooling or if a precipitating anion is added, unknown impurities coprecipitate. If precipitation does not occur during refluxing, the solution may be evaporated until precipitation occurs.

properties are recorded in the literature.[6] The frequencies of the ν(NO) bands in the infrared spectra are given with the analyses above.

References

1. A. D. Allen and J. R. Stevens, *Chem. Commun.*, **1967,** 1147.
2. F. Bottomley and S. B. Tong, *Inorg. Chem.*, **13,** 243 (1974).
3. A. D. Allen and J. R. Stevens, *Can. J. Chem.*, **50,** 3093 (1972).
4. H. A. Scheidegger, J. N. Armor, and H. Taube, *J. Am. Chem. Soc.*, **92,** 5580 (1970).
5. G. W. Watt, E. M. Potrafke, and D. S. Klett, *Inorg. Chem.*, **2,** 868 (1963).
6. F. Bottomley and S. B. Tong, *J. Chem. Soc., Dalton Trans.*, **1973,** 217.

4. NITROSYLAMMINERUTHENIUM COMPLEXES

Submitted by A. F. SCHREINER* and S. W. LIN*
Checked by F. BOTTOMLEY† and M. MANGAT†

One of the most prominent features of ruthenium chemistry is that the metal forms numerous compounds containing the $(RuNO)^{3+}$ fragment.[1] Furthermore, the metal-nitrosyl fragment has in general been of interest to many because of the potential π-acceptor ability of NO and its possible existence as NO^+, NO^-, or NO. Explicit details of how to synthesize the ruthenium-NO complexes are virtually nonexistent, and procedures, when available, are laborious. Here we give the preparative descriptions for such compounds by newly developed methods which are explicit and convenient. The information is also valuable in view of the cost of ruthenium.

The starting material for each preparation is the commercially available "$RuCl_3 \cdot nH_2O$." This chloride salt is used to prepare a solution of the hexaammine, $[Ru(NH_3)_6]^{2+}$, which upon reaction with the appropriate reagent leads to complex ions of the type $[Ru(NH_3)_4(NO)X]^{2+}$.

A. TETRAAMMINECHLORONITROSYLRUTHENIUM(2+) CHLORIDE

$$2RuCl_3 \cdot nH_2O + 12NH_3(aq) + Zn \rightarrow 2[Ru(NH_3)_6]Cl_2 + Zn^{2+} + 2Cl^- + nH_2O$$

$$[Ru(NH_3)_6]^{2+}(aq) \xrightarrow{Cl_2} [Ru(NH_3)_4(NO)Cl]^{2+}$$

*Chemistry Department, North Carolina State University, Raleigh, N.C. 27607.
†Chemistry Department, University of New Brunswick, Fredericton, E3B 53A New Brunswick, Canada.

Procedure

One gram of hydrated ruthenium trichloride (Engelhard Industries, "$RuCl_3 \cdot nH_2O$") is added to a reaction flask containing aqueous ammonia (25 ml, 8*M*), and flushing with N_2 is begun. Following the addition of zinc dust (3 g), the mixture is stirred for 24 hr, during which time the original red-violet solution turns a deep-yellow color. This solution, containing $[Ru(NH_3)_6]^{2+}$, is then filtered free of excess zinc. The deep-yellow filtrate is treated with gaseous Cl_2 (ca. 0.25 cc/sec) until the solution becomes acidic. During the chlorination step a small quantity of a fawn-colored precipitate forms, which dissolves when the next step is carried out. At this next step, concentrated NH_4OH (20 ml) is added, and the solution is boiled 20 min (solution I). Solution I should be used within 5 hr of its preparation. After addition of HCl (20 ml, 6*M*), refluxing (2 hr), and cooling (0°), a dark-yellow precipitate forms after the solution goes through a color change from yellow, to blue, to green, and, finally, to yellow. Following recrystallization (0.1*M* HCl), 0.7 g (60% yield) of orange crystalline compound is obtained. The complex is fully identifiable as $[Ru(NH_3)_4(NO)Cl]Cl_2$ from the number and positions of its infrared bands[2] (see Table III); no bands are present other than the ones previously given in an independent characterization.[2]

B. TETRAAMMINEACIDONITROSYLRUTHENIUM(2+) PERCHLORATES

$$[Ru(NH_3)_4(NO)Cl]Cl_2 + X^- + 2ClO_4^- \rightarrow [Ru(NH_3)_4(NO)X](ClO_4)_2 + 3Cl^-$$

Procedure

The preparation of tetraammineacetatonitrosylruthenium(2+) perchlorate, $[Ru(NH_3)_4(NO)(C_2H_3O_2)](ClO_4)_2$, is carried out in the following manner. $[Ru(NH_3)_4(NO)Cl]Cl_2$ (0.2 g) is dissolved in hot water (20 ml, 90°), sodium acetate (1 g) is added, and this solution is acidified with acetic acid and heated (90°, 30 min). The hot yellow solution is filtered into 5 ml of saturated $NaClO_4$ solution, and the mixture is then cooled (0°, 30 min). A yellow precipitate forms; this is retained on a filter, washed (ethanol, ether), redissolved in cold water (15 ml), and filtered into 5 ml of saturated $NaClO_4$ solution. The mixture is then cooled (0°, 30 min); a yellow precipitate which forms is retained on a filter and washed with ethyl alcohol and ether. The yellow compound (0.10 g, 33% yield) is identified from its infrared spectrum and elemental analysis.

Anal. Calcd. for $C_2H_{15}N_5O_{11}Cl_2Ru$: C, 5.23; H, 3.27; N, 15.28. Found: C, 4.77; H, 3.93; N, 15.06.

The preparation of the analogous cyanato complex, $[Ru(NH_3)_4(NO)NCO]$-$(ClO_4)_2$, is very similar: KNCO is used in place of $NaC_2H_3O_2$, and the complex cation is initially precipitated as the iodide salt, $[Ru(NH_3)_4(NO)NCO]I_2$, salt *A*, from 2*M* KI solution. This iodide ion is displaced by filtering a solution of *A* into a saturated solution of $NaClO_4$. Salt *A* is identified from its infrared spectrum and elemental analysis. *Anal.* Calcd. for $CH_{12}N_6O_2I_2Ru$: N, 16.95; I, 51.20. Found: N, 16.84; I, 50.32.

Properties

Several conveniently measurable properties further characterize the compounds, namely, molar conductances, electronic spectra, and infrared band positions in the mid-infrared region. We give the molar-conductance values in Table I for the compounds of $L^- = C_2H_3O_2^-$ and NCO^-. The data are firmly indicative of 2:1 type complexes.

TABLE I Molar Conductance Values* of $[Ru(NH_3)_4(NO)(L)]X_2$

L^-	X^-	Concentration, *M*	Molar conductance, Ω^{-1}	Type compound
$C_2H_3O_2^-$	ClO_4^-	0.000508	222	2:1
NCO^-	I^-	0.000420	250	2:1

*Aqueous solution.

Several electronic absorption-band positions of these complexes are given in Table II.

TABLE II Electronic Bands of $[Ru(NH_3)_4(NO)L]^{2+}$ in Aqueous Solution[3]

Complex ion	Counterion	ν_{max}, nm	ε_{max}
$[Ru(NH_3)_4(NO)Cl]^{2+}$	Cl^-	331	268
$[Ru(NH_3)_4(NO)(C_2H_3O_2)]^{2+}$	ClO_4^-	328	130
		300	240
		(225)	(4930)*
$[Ru(NH_3)_4(NO)(NCO)]^{2+}$	ClO_4^-	450	433
		344	

*Charge transfer.

A number of strong, distinct mid-infrared band positions for these and similar compounds are given in Table III. These were taken from spectra obtained here in KBr pellets. The N—O stretching frequency is especially diagnostic, since this band is sharp, strong, and isolated from other modes. Several other band positions diagnostic of the coordinated ligands NCO^- and $C_2H_3O_2^-$ are also given in this table. The colors of these solid compounds depend on crystal size and vary from light yellow (powders) to orange (large crystals).

TABLE III Useful Infrared Bands of $[Ru(NH_3)_4(NO)L]^{2+}$

Complex	$\nu(NO)$, cm^{-1}	Other selected useful bands
$[Ru(NH_3)_4(NO)Cl]Cl_2$*,†	1880	1300 (s, sh), 855 (m), 609 (m, sh), 482 (m, sh)
$[Ru(NH_3)_4(NO)(C_2H_3O_2)]I_2$†	1882	$\nu(COO)$ 1365 (s)
$[Ru(NH_3)_4(NO)(NCO)]I_2$†	1890	$\nu(NC)$ 2250 (s)

*This work and reference 2.
†This work.

References

1. W. P. Griffith, "The Chemistry of the Rarer Platinum Metals," Interscience Publishers, London, 1967.
2. J. R. Durig, E. E. Mercer, and W. A. McAllister, *Inorg. Chem.*, **5**, 1881 (1966).
3. A. F. Schreiner, S. W. Lin, P. J. Hauser, E. A. Hopcus, D. J. Hamm, and J. D. Gunter, *Inorg. Chem.*, **11**, 880 (1972).

5. DINITROSYLCOBALT COMPLEXES

Submitted by D. GWOST* and K. G. CAULTON*
Checked by J. R. NAPPIER,† O. A. ILEPERUMA,† and R. D. FELTHAM†

Complexes of nitric oxide are receiving increasing attention as potential catalysts.[1] The origin of this interest lies in the proposal,[2] since verified,[3] that NO might exhibit a bent mode of coordination in which the nitrogen possesses a nonbonding lone pair. Since "bent NO" donates two fewer electrons to a metal than the linear isomer does, linear-bent tautomerism raises the possibility of coordinative unsaturation and catalysis.

Syntheses of nitrosyl complexes ("nitrosylation") have, in the past, focused on the metal-containing product, there being little concern for stoichiometry

*Chemistry Department, Indiana University, Bloomington, Ind. 47401.
†Chemistry Department, University of Arizona, Tucson, Ariz. 85721.

with respect to NO itself. A recent study[4] of the role of NO in nitrosylation of iron and cobalt has demonstrated that this molecule can function as a reducing agent.

$$M^{n+} + NO + ROH \rightarrow M^{(n-1)+} + RONO + H^{+}$$
$$M^{(n-1)+} + xNO \rightarrow M(NO)_{x}^{(n-1)+}$$

The complexes reported below were made earlier[5] by bridge-splitting reactions on $[Co(NO)_2X]_2$. Although this dimer is an isolable intermediate in the procedures reported here, it is best prepared by reduction of CoX_2 in the presence of excess Co.[6] However, the following preparations represent the most convenient route to a variety of base adducts of the form $Co(NO)_2L_2^+$. Since all these compounds and intermediates are more or less air-sensitive, operations should be carried out using standard inert-atmosphere techniques.[7]

■ **Caution.** *Because of the toxicity of* NO, *the following reactions must be performed in a well-ventilated hood.*

A. (*N,N,N′,N′*-TETRAMETHYLETHYLENEDIAMINE) DINITROSYLCOBALT(1+) TETRAPHENYLBORATE

$$CoCl_2 + 2TMEDA^* + 3NO \xrightarrow{CH_3OH} \xrightarrow[CH_3OH]{Na[BPh_4]}$$
$$Co(NO)_2(TMEDA)[BPh_4] + [H(TMEDA)][BPh_4] + 2NaCl + CH_3ONO\uparrow$$

Procedure

A 250-ml, three-necked flask, equipped with a magnetic stirring bar and a nitrogen inlet, is charged with 2.86 g of $CoCl_2 \cdot 6H_2O$ (12.0 mmoles) and heated to 110° under vacuum (<0.1 torr) for 6–8 hr (the color will change from red to blue). After the flask cools to room temperature, 75 ml of methanol (dried by distillation from magnesium alkoxide),[8] followed by 3.6 ml of TMEDA (24.0 mmoles) (distilled from BaO), is added to form a deep-blue solution. A reflux condenser is placed in the middle neck of the reaction flask, and the whole reaction system is flushed with nitrogen. A gas bubbler is connected to the top of the condenser, and nitrogen is replaced with nitric oxide.† Nitric oxide is passed through the reaction solution for 8–10 hr.‡ At the end of this time, dissolved NO is removed by degassing under vacuum.[7]

*TMEDA is *N,N,N′,N′*-tetramethylethylenediamine.

†NO can be sufficiently purified by passing the gas through a −78° trap before it enters the reaction flask.

‡The exit bubbling rate has been established as approximately 150 cc/hr; this corresponds to ca. 15 bubbles/min through a Nujol bubbler.

A solution of 9.0 g of $Na[BPh_4]$ (26.0 mmoles) in 50 ml of methanol is slowly* added dropwise, using an addition funnel, to the deep-green reaction solution. After the addition of $Na[BPh_4]$ is complete, stiriring is continued for 6–10 hr. The pale-green solid is filtered, washed with methanol, and dried under vacuum. The dried solid is extracted with 50 ml of dichloromethane, using a Soxhlet extractor, until the extracts are colorless. The filtrate is concentrated to 25 ml, and 100 ml of ethanol is slowly added to the solution. The black and white crystals, $[Co(NO)_2(TMEDA)][BPh_4]$ and $[H(TMEDA)][BPh_4]$, respectively, are removed by filtration and washed with 30-ml portions of ethanol warmed to 65° until the $[H(TMEDA)][BPh_4]$ is removed. The remaining black solid is recrystallized from ethanol-chloroform (20:4) to give black crystals. After it is filtered and washed with ethanol and ethyl ether, the solid is dried under vacuum. Yield: 6.12 g of $[Co(NO)_2(TMEDA)][BPh_4]$ (11.2 mmoles or 94%). *Anal.* Calcd. for $C_{30}H_{36}BCoN_4O_2$: C, 65.0; H, 6.6; Co, 10.6; N, 10.1. Found: C, 65.0; H, 6.7; Co, 10.4; N, 10.0.

B. BIS(TRIPHENYLPHOSPHINE)DINITROSYLCOBALT(1+) TETRAPHENYLBORATE

$$CoCl_2 + NEt_3 + 3NO \xrightarrow{\text{methanol}} \tfrac{1}{2}[Co(NO)_2Cl]_2 + [HNEt_3]Cl + CH_3ONO\uparrow$$

$$\xrightarrow[\text{2PPh}_3]{\text{ethanol}} \xrightarrow[\text{2Na[BPh}_4]]{\text{ethanol}} [Co(NO)_2(PPh_3)_2][BPh_4] + [HNEt_3][BPh_4] + 2NaCl$$

Procedure

Anhydrous $CoCl_2$ (12.0 mmoles) is prepared as in Sec. A. After the addition of 75 ml of methanol, 8.3 ml (60.0 mmoles) of NEt_3 (distilled from KOH) is slowly† added to the solution of $CoCl_2$. As in Sec. A, NO is bubbled through the reaction solution for 8–10 hr, and the solution is degassed.

A solution of 7.86 g of PPh_3‡ (30.0 mmoles) in 75 ml of hot ethanol, under a strong flow of N_2 using a funnel inserted in the unused neck of the flask, is quickly added to the red-brown reaction solution of $[Co(NO)_2Cl]_2$;§ the resulting mixture is stirred for 3 hr. A solution of 9.0 g of $Na[BPh_4]$ (26.6 mmoles)

*If the $Na[BPh_4]$ solution is added too fast, the stirred solution becomes a solid mass and must be stirred manually.

†If the NEt_3 is not carefully distilled or is added too fast, a small amount of precipitate is formed.

‡The solid phosphine is degassed before being dissolved in ethanol.

§Care is required because $[Co(NO)_2Cl]_2$ is very air-sensitive, especially in solution.

in 50 ml of ethanol is slowly added dropwise, using an addition funnel, to the reaction solution; stirring is continued for 8–12 hr. The resulting precipitate is filtered, washed with ethanol, and dried under vacuum. The $[Co(NO)_2(PPh_3)_2][BPh_4]$ is washed away from $[HNEt_3][BPh_4]$ with a minimum of chloroform (ca. 100 ml). The filtrate is concentrated to a syrup or slurry (ca. 25 ml); the walls of the flask are washed clean with 5–10 ml of chloroform. A measured volume of methanol, equal to that of the chloroform solution, is slowly added to cause initial precipitation of $[Co(NO)_2(PPh_3)_2][BPh_4]$; the slow addition of 150–200 ml of ethyl ether causes quantitative precipitation of the product. The dark-brown crystals are filtered, washed with ethyl ether, and dried under vacuum. Yield: 9.10 g of $[Co(NO)_2(PPh_3)_2][BPh_4]$ (9.5 mmole) or 79% based on Co. *Anal.* Calcd. for $C_{60}H_{50}BCoN_2O_2P_2$: C, 74.8; H, 5.2; Co, 6.1; N, 2.9. Found: C, 74.3; H, 5.2; Co, 6.3; N, 2.7.

C. [ETHYLENEBIS(DIPHENYLPHOSPHINE)]DINITROSYLCOBALT(1+) TETRAPHENYLBORATE

$$CoCl_2 + NEt_3 + 3NO \xrightarrow{CH_3OH} \tfrac{1}{2}[Co(NO)_2Cl]_2 + HNEt_3Cl + CH_3ONO\uparrow$$

$$\xrightarrow[\text{disphos}]{\text{ethanolic}} \xrightarrow[2Na[BPh_4]]{\text{ethanolic}} [Co(NO)_2(diphos)][BPh_4] + [HNEt_3][BPh_4] + 2NaCl$$

Procedure

The product is prepared as in Sec. B. However, 4.8 g of diphos* (12.0 mmoles) is added as a slurry in 100 ml of ethanol and stirred until all the diphos dissolves in the methanol-ethanol solution. The addition of $Na[BPh_4]$ results in the formation of an orange precipitate. The precipitate is filtered, washed with ethanol, and dried under vacuum. Because the solubility of $[Co(NO)_2(diphos)][BPh_4]$ in chloroform is limited, the product is separated from the white $[HNEt_3][BPh_4]$ by Soxhlet extraction of the dried precipitate with 100 ml of chloroform. The chloroform solution is concentrated to ca. 25 ml; 25 ml of methanol, followed by 100 ml of ethyl ether, is then slowly added. The brick-red crystalline product is filtered, washed with ethyl ether, and dried under vacuum. Yield: 8.46 g of $[Co(NO)_2(diphos)][BPh_4]$ (10.1 mmoles or 84% based on Co). *Anal.* Calcd. for $C_{50}H_{44}BCoN_2O_2P_2$: C, 71.8; H, 5.3; Co, 7.0; N, 3.4. Found: C, 70.0; H, 5.5; Co, 7.0; N, 3.2.

*Diphos is ethylenebis(diphenylphosphine) or 1,2-bis(diphenylphosphino)ethane.

Properties

The dinitrosyl cations prepared here are air-stable as dry solids but sensitive to air and water in solution; all reactions and subsequent work-ups should be carried out in appropriate[7] Schlenk glassware under prepurified nitrogen. Solubilities and NO stretching frequencies appear in Table I.

TABLE I

$Co(NO)_2L_2^+$	ν(NO), cm^{-1}(KBr)	Solubility		
		$CHCl_3$	Acetone	Ethanol
$[Co(NO)_2(TMEDA)][BPh_4]$	1869, 1792	Insol.	Very sol.	Insol.
$[Co(NO)_2(PPh_3)_2][BPh_4]$	1855, 1795	Very sol.	Very sol.	Slightly sol.
$[Co(NO)_2(diphos)][BPh_4]$	1838, 1789	Sol.	Very sol.	Insol.

The characteristic color of these compounds varies with particle size; thus, $[Co(NO)_2(TMEDA)][BPh_4]$ is a pale-green powder or shiny-black crystals. Although the crystal structure of none of the three $[Co(NO)_2L_2]^+$ compounds prepared here has been determined, the x-ray structure of $[Ir(NO)_2(PPh_3)_2]$-$[ClO_4]$ has been reported.[9] The Ir—N—O bond angle is intermediate (164°) between linear (180°) and bent (120°).

References

1. J. P. Collman, N. Hoffman, and D. Morris, *J. Am. Chem. Soc.*, **91,** 5659 (1969); S. Wilson and J. Osborn, *ibid.*, **93,** 3068 (1971); E. Zuech, W. Hughes, D. Kubicek, and E. Kettleman, *ibid.*, **92,** 528, 532 (1970).
2. E. Thorsteinson and F. Basolo, *J. Am. Chem. Soc.*, **88,** 3929 (1966).
3. D. Hodgson, N. Payne, J. McGinnety, R. Pearson, and J. Ibers, *J. Am. Chem. Soc.*, **90,** 4486 (1968).
4. D. Gwost and K. G. Caulton, *Inorg. Chem.*, **12,** 2095 (1973).
5. T. Bianco, M. Rossi, and L. Uva, *Inorg. Chim. Acta*, **3,** 443 (1969); W. Hieber and K. Kaiser, *Z. Anorg. Allg. Chem.*, **362,** 169 (1968).
6. B. Haymore and R. D. Feltham, *Inorganic Syntheses*, **14,** 81 (1973).
7. D. F. Shriver, "The Manipulation of Air-Sensitive Compounds," Chap. 7, McGraw-Hill Book Co., New York, 1969.
8. H. Lund and J. Bjerrum, *Chem. Ber.*, **64,** 210 (1931).
9. D. Mingos and J. Ibers, *Inorg. Chem.*, **9,** 1105 (1970).

6. BIS(TRIPHENYLPHOSPHINE)CHLORODINITROSYL-RUTHENIUM(1+) TETRAFLUOROBORATE

$$RuCl_3(NO)[P(C_6H_5)_3]_2 \xrightarrow[C_6H_6]{Zn/Cu} [RuCl(NO)\{P(C_6H_5)_3\}_2] \xrightarrow[C_2H_5OH]{NO[BF_4]} [RuCl(NO)_2\{P(C_6H_5)_3\}_2][BF_4]$$

Submitted by J. REED,* C. G. PIERPONT,† and R. EISENBERG*
Checked by E. D. JOHNSON†

The cationic complex bis(triphenylphosphine)chlorodinitrosylruthenium(1+) is a striking example of the two predominant modes of nitrosyl coordination within the same molecule.[1] The geometry of $[RuCl(NO)_2\{P(C_6H_5)_3\}_2]^+$ is square pyramidal with a linearly coordinated NO group in the basal plane bonding as formally NO^+ and a bent or angularly coordinated NO group in the apical position bonding as NO^-.

The complex is best prepared by the reaction of nitrosonium salts with the complex $RuCl(NO)[P(C_6H_5)_3]_2$, first reported by Stiddard and Townsend.[2] Because of the extreme reactivity of $RuCl(NO)[P(C_6H_5)_3]_2$, this complex is not isolated in the sequence outlined below but is prepared *in situ* following a minor modification of Stiddard and Townsend's procedure involving the zinc reduction of $RuCl_3(NO)[P(C_6H_5)_3]_2$.

Procedure

A Schlenk apparatus, illustrated in Fig. 3 and similar to that employed by Collman et al.,[4] is used for the reaction sequence. Vacuum and argon sources are provided by a double manifold, to which the stopcocks of the apparatus are attached. Prepurified nitrogen can be used in place of argon.

The compound $RuCl_3(NO)[P(C_6H_5)_3]_2$ (1.52 g, 2.00 mmoles) and 100 ml of freshly distilled benzene§ are introduced into a Schlenk tube equipped with a magnetic stirrer and degassed through several pump and flush cycles. Approxi-

*Brown University, Providence, R.I. 02912; Present address, Department of Chemistry, University of Rochester, Rochester, N.Y. 14627.
†Chemistry Department, West Virginia University, Morgantown, W. Va. 26506.
†Department of Chemistry, Northwestern University, Evanston, Ill. 60201.

§Solvents are dried and distilled under N_2. Benzene is distilled over $LiAlH_4$ and ethanol over Mg.

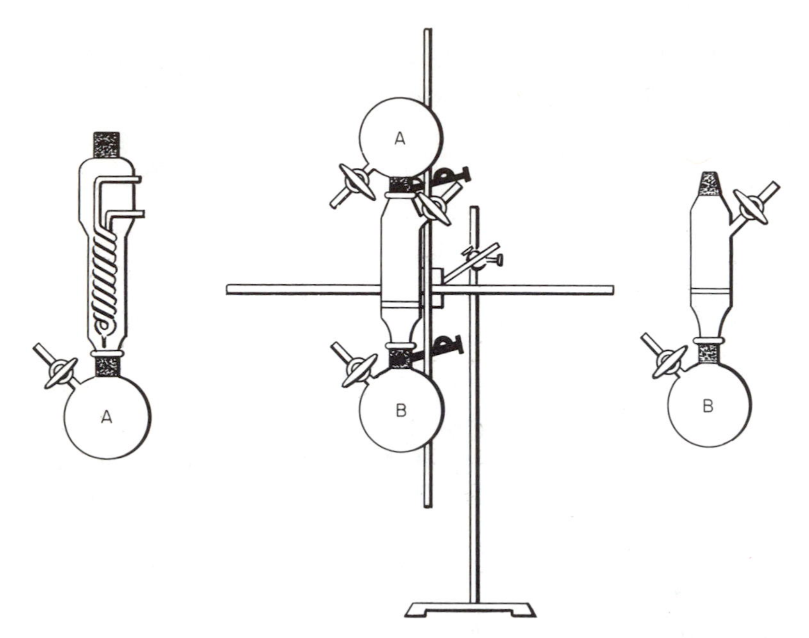

Fig. 3. Schlenk apparatus for the preparation of $[RuCl(NO_2)\{P(C_6H_5)_3\}_2][BF_4]$.

mately 2 g of a Zn/Cu couple* is then added to the system, and the stirred suspension is refluxed for 1 hr. During refluxing, the Schlenk tube is attached to a condenser, which in turn is connected to a mineral-oil bubbler. A color change in the solution from mustard gold to emerald green indicates the successful reduction of the Ru(II) starting material to $RuCl(NO)[P(C_6H_5)_3]_2$. After 1 hr the reaction solution is allowed to cool to room temperature, and the condenser is replaced by a filter assembly consisting of a fritted-funnel tube and an empty Schlenk tube, which are connected through standard-taper ground-glass joints (see Fig. 3). The apparatus is degassed and then inverted to filter off the remaining Zn/Cu couple and insoluble materials. With argon flowing through the stopcock of the Schlenk tube now holding the green solu-

*Since the Zn reduction of $RuCl_3(NO)[P(C_6H_5)_3]_2$ is very dependent on the quality of the zinc, a Zn/Cu couple is used for this purpose, following a suggestion of Stiddard's.[5] The couple is prepared as follows: a solution of 2.0 g of $CuCl_2$ in distilled water is added dropwise to 2.0 g of Zn dust in 40 ml of H_2O with vigorous stirring. Addition is stopped when the now black powder settles to the bottom of the beaker even with continued stirring. Concentrated HCl is then added dropwise (ca. 8–10 drops) until H_2 gas is evolved. After the cessation of gas evolution (ca. 20 min), the black Zn/Cu couple is filtered under N_2, washed with absolute alcohol and ether, and vacuum-dried. The couple does not remain active long and should be used just after preparation. It is oxidatively unstable and when placed on a piece of filter paper after preparation, it causes the paper to burst into flame within 1 min.

tion, the filter assembly is removed, cleaned, and returned to the same position.

Nitrosyl tetrafluoroborate, $NO[BF_4]$* (0.35 g, 2.98 mmoles) is then dissolved in 15 ml of freshly distilled ethanol. The solution is poured onto the fritted disk of the filter and degassed by backflushing with argon, during which argon is passed from the reaction vessel through the frit and the ethanol solution above it. The solution of $NO[BF_4]$ is then added to the reaction vessel, and an immediate color change from green to orange is observed. The solution is stirred for 10 min and filtered to remove the orange product. Note that if the filter assembly is of the design shown in Fig. 3, the filter tube must be removed and inverted *before* the final filtration.† The yield is 1.11 g (70%). *Anal.* Calcd. for $C_{36}H_{30}ClN_2O_2P_2Ru \cdot C_2H_6O$: C, 53.46; H, 4.22; N, 3.28; P, 7.15; Cl, 4.10. Found: C, 53.14; H, 4.32; N, 3.87; P, 7.81; Cl, 4.18. Recrystallization is accomplished using methanol-benzene.

Properties

Bis(triphenylphosphine)chlorodinitrosylruthenium(1+)tetrafluoroborate is an orange crystalline solid (m.p. 196°), which is moderately stable in air (decomposition over a period of several weeks is noted). The complex is very soluble in methanol, methylene chloride (dichloromethane), acetone, dimethylformamide, and acetonitrile. Its solutions are unstable to the air. It is insoluble in ether and hexane and only slightly soluble in benzene.

The complex exhibits two nitrosyl stretching frequencies, 1850 and 1685 cm^{-1}, corresponding to the linear and bent nitrosyls, respectively.[1] Collman et al.[6] have shown that the two nitrosyls interconvert by using labeled NO and monitoring the nitrosyl stretching region. This interconversion presumably goes through a trigonal bipyramidal intermediate. The complex shows no reaction with 1-hexene under 1 atm of H_2 following conditions used in other catalytic studies.[7]

References

1. C. G. Pierpont, D. G. VanDerveer, W. Durland, and R. Eisenberg, *J. Am. Chem. Soc.*, **92,** 4760 (1970); C. G. Pierpont and R. Eisenberg, *Inorg. Chem.*, **11,** 1088 (1972).
2. M. H. B. Stiddard and R. E. Townsend, *Chem. Commun.*, **1969,** 1372.

*Available from Alfa Inorganics, Beverly, Mass. 01915.

†The filter tube of Fig. 3 is of an unsymmetric design and is available commercially, e.g., Kontes K-215100. The checker suggests that a similar commercially available filter tube of symmetric design, e.g., Kontes K-215500, will make some of the operations easier. Both filter designs have been tried, and each works well.

3. M. B. Fairy and R. J. Irving, *J. Chem. Soc. (A)*, **1966**, 475.
4. J. P. Collman, N. W. Hoffman, and J. W. Hosking, *Inorganic Syntheses*, **12,** 8 (1970).
5. M. H. B. Stiddard, private communication.
6. J. P. Collman, P. Farnham, and G. Dolcetti, *J. Am. Chem. Soc.*, **93,** 1788 (1971).
7. J. A. Osborn, F. H. Jardine, J. F. Young, and G. Wilkinson, *J. Chem. Soc. (A)*, **1966,** 1711.

7. DICARBONYL-η-CYCLOPENTADIENYLNITROSYLMOLYBDENUM AND BIS(DIHALO-η-CYCLOPENTADIENYLNITROSYLMOLYBDENUM) DERIVATIVES

Submitted by D. SEDDON,* W. G. KITA,* J. BRAY,* and J. A. McCLEVERTY*
Checked by S. P. ANAND† and R. B. KING†

The synthesis of $(\eta\text{-}C_5H_5)Mo(NO)(CO)_2$ was first reported in 1956,[1] and a revised procedure has been outlined.[2] The compound is a particularly useful precursor for the preparation of a variety of sulfur complexes,[3] carbene derivatives,[4] and halides $[(\eta\text{-}C_5H_5)Mo(NO)X_2]_2$ (X = Cl,[5] Br,[5] I[2,5]). The halides, especially when X = I, are important intermediates in the formation of the stereochemically nonrigid $(C_5H_5)_2Mo(NO)I$ and $(C_5H_5)_3Mo(NO)$ compounds.[6]

A. DICARBONYL-η-CYCLOPENTADIENYLNITROSYLMOLYBDENUM

$$Mo(CO)_6 + NaC_5H_5 \xrightarrow{THF} Na[Mo(CO)_3(\eta\text{-}C_5H_5)] + 3CO$$

$$Na[Mo(CO)_3(\eta\text{-}C_5H_5)] + CH_3CO_2H \xrightarrow{THF} [(\eta\text{-}C_5H_5)Mo(CO)_3H] + NaO_2CCH_3$$

$$(\eta\text{-}C_5H_5)Mo(CO)_3H + p\text{-}CH_3C_6H_4SO_2N(NO)CH_3 \xrightarrow{THF} (\eta\text{-}C_5H_5)Mo(NO)(CO)_2 + CO + p\text{-}CH_3C_6H_4SO_2NHCH_3$$

Procedure

■ **Caution.** *All reactions and operations must be carried out under nitrogen,*

*Department of Chemistry, Sheffield University, Sheffield S3 7HF, England.
†Department of Chemistry, University of Georgia, Athens, Ga. 30602.

unless otherwise stated, and all solvents must be saturated with nitrogen before use.

Sodium tricarbonyl-η-cyclopentadienylmolybdenum is prepared by the recommended procedures,[7] using finely divided sodium (3.0 g), freshly cracked cyclopentadiene (20.0 ml), hexacarbonylmolybdenum (24.0 g), and dry, purified tetrahydrofuran (THF).* Glacial acetic acid (5.2 ml, 5.45 g) in THF (15.0 ml) is added dropwise and with stirring at room temperature to the yellow solution of the tricarbonyl-η-cyclopentadienylmolybdenum ion. The solution becomes orange-red, due to partial oxidation of $(\eta\text{-}C_5H_5)Mo(CO)_3H$ to $[(\eta\text{-}C_5H_5)Mo(CO)_3]_2$. After 10 min it is treated with a solution of *N*-methyl-*N*-nitroso-*p*-toluenesulfonamide (21.0 g) in THF (70 ml). Vigorous effervescence occurs during the addition (slowly over 30 min) of the nitrosylating agent, and after this is finished, the mixture is stirred at room temperature for 4 hr more. The mixture is then filtered through kieselguhr, the residue is washed with THF until the washings are colorless, and the combined filtrate and washings are evaporated (rotary evaporator) to dryness. The residue, a brown tar, is extracted with ether (ca. 400 ml), and the extract is filtered and then evaporated, leaving a deep-red tar, which gradually solidifies to a waxy solid on standing under reduced pressure. This solid is then Soxhlet-extracted with petroleum ether (b.p. 40–60°; 250 ml), and the product forms as orange-red crystals. The yield is 16–19 g [71–85% based on $Mo(CO)_6$].† *Anal.* Calcd. for $C_7H_5NO_3Mo$: C, 34.0; H, 2.0; N, 5.7. Found: C, 33.8; H, 2.0; N, 5.7. The material so prepared is useful for most, but not all, synthetic purposes but may be further purified either by recrystallization from *n*-hexane or, better, by sublimation *in vacuo*. The stability of the compound, when stored without final purification, may vary from batch to batch, but prolonged storage should be attempted only under nitrogen, in the absence of light, at low temperature.

Properties

Dicarbonyl-η-cyclopentadienylnitrosylmolybdenum is a volatile, orange-red, crystalline solid, readily soluble in common organic solvents. The infrared spectrum consists of three strong absorptions: $\nu(CO) = 2021$ and 1948 and $\nu(NO) = 1688$ (in *n*-hexane solution).

*THF was purified by refluxing for 24 hr under nitrogen and over sodium and benzophenone, and was then distilled under nitrogen.

†By preparing the NaC_5H_5 using NaH rather than sodium and omitting the acetic acid treatment, the checkers obtained slightly better yields (19–21 g, 85–94%) of a more readily purified product.

B. BIS(DICHLORO-η-CYCLOPENTADIENYLNITROSYLMOLYBDENUM)

$$2(\eta\text{-}C_5H_5)Mo(NO)(CO)_2 + 2Cl_2 \xrightarrow{CCl_4} [(\eta\text{-}C_5H_5)Mo(NO)Cl_2]_2 + 4CO$$

Procedure

■ **Caution.** *All reactions and operations are carried out under nitrogen.* The apparatus for the preparation of this compound is illustrated in Fig. 4.

The apparatus is initially purged with nitrogen, and the Dreschel bottle *A*

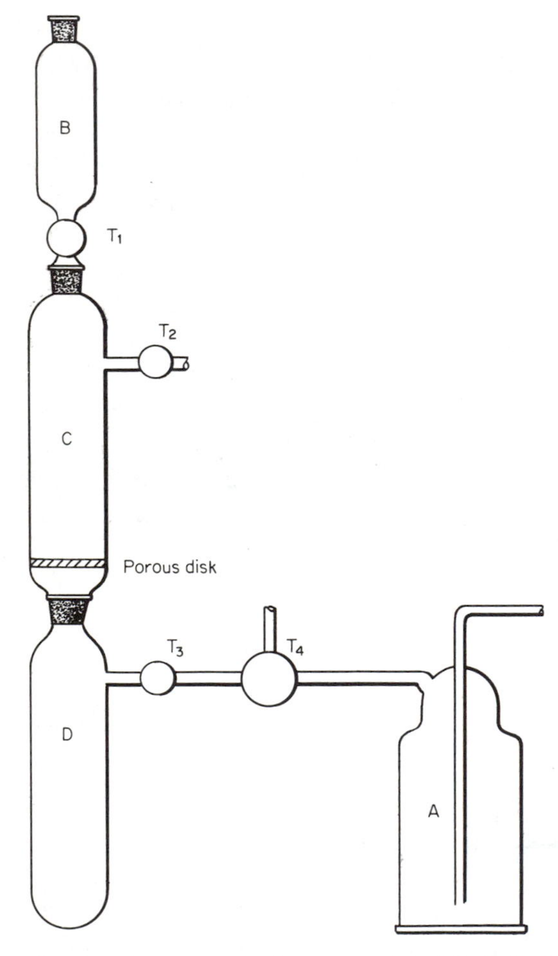

Fig. 4. Apparatus for the preparation of $[(\eta\text{-}C_5H_5)Mo(NO)Cl_2]_2$.

(250-ml capacity) is filled with chlorine at atmospheric pressure.* A solution of freshly sublimed $(\eta\text{-}C_5H_5)Mo(NO)(CO)_2$ (2.5 g)† in carbon tetrachloride (30 ml) is placed in the tap funnel *B*. By appropriate regulation of the taps T_1 and T_2, it is degassed with nitrogen. The solution is then allowed to drip onto a porous disk in the vessel *C*, diffusion through the disk being prevented by maintenance of a slight positive pressure of N_2 in the vessel *D*. By regulation of taps T_3 and T_4, the chlorine in *A* (250 ml), in a slow stream of nitrogen,‡ is passed through *D*, the porous disk, and the solution. The reaction is complete after 10 min, and *n*-pentane (30 ml) is added to the reaction mixture from *B*. The tap funnel is then removed, tap T_3 is closed, the Dreschel bottle *A* is disconnected, and the nitrogen supply is attached to T_2. A water pump is then connected to T_3, and the reaction mixture filtered by opening T_2 and T_3. The product, which precipitates after chlorination and addition of *n*-pentane, is retained on the porous disk, washed with *n*-pentane (50 ml), and dried *in vacuo*. The yield is 2.0 g [76% based on $(\eta\text{-}C_5H_5)Mo(NO)(CO)_2$]. *Anal.* Calcd. for $C_{10}H_{10}N_2O_2Cl_4Mo_2$: C, 22.9; H, 1.9; N, 5.3; Cl, 27.1. Found: C, 22.6; H, 1.9; N, 5.5; Cl, 27.2.

C. BIS(DIBROMO-η-CYCLOPENTADIENYLNITROSYLMOLYBDENUM)

$$2(\eta\text{-}C_5H_5)Mo(NO)(CO)_2 + 2Br_2 \xrightarrow{CCl_4} [(\eta\text{-}C_5H_5)Mo(NO)Br_2]_2 + 4CO$$

Procedure

■ **Caution.** *All reactions are carried out under nitrogen, using nitrogen-saturated solvents.*

Freshly sublimed $(\eta\text{-}C_5H_5)Mo(NO)(CO)_2$ (5.0 g)§ dissolved in carbon tetrachloride (80 ml) is filtered into a conical flask containing a magnetic stirring bar and fitted with a nitrogen inlet and dropping funnel. A solution of bromine (3.2 g) in carbon tetrachloride (20 ml) is then added dropwise to the nitrosyl compound with stirring at room temperature. Effervescence occurs, and after the addition is complete (ca. 15 min), *n*-pentane (30 ml) is added and

*The chlorine gas is used from the cylinder without further purification.

†Use of impure $(\eta\text{-}C_5H_5)Mo(NO)(CO)_2$ results in the formation of substantial amounts of $[Mo(NO)_2Cl_2]_n$, and the desired compound is isolated in yields as low as 10%.

‡A slow stream of chlorine-nitrogen is essential in order to minimize frothing above the porous disk.

§Use of impure $(\eta\text{-}C_5H_5)Mo(NO)(CO)_2$ results in the formation of $[Mo(NO)_2Br_2]_n$, the yield of the desired complex dropping to between 50 and 60%.

stirring is continued for a further 30 min. The product, which precipitates on bromination and addition of the *n*-pentane, is then filtered off, washed with carbon tetrachloride and *n*-pentane, and dried *in vacuo*. The yield is 6.1 g [86% based on (η-C_5H_5)Mo(NO)$(CO)_2$]. *Anal.* Calcd. for $C_{10}H_{10}N_2O_2Br_4Mo_2$: C, 17.1; H, 1.4; N, 4.0; Br, 45.6. Found: C, 17.7; H, 2.0; N, 3.6; Br, 45.4.

D. BIS(DIIODO-η-CYCLOPENTADIENYLNITROSYLMOLYBDENUM)

$$2(\eta\text{-}C_5H_5)Mo(NO)(CO)_2 + 2I_2 \xrightarrow{CH_2Cl_2} [(\eta\text{-}C_5H_5)Mo(NO)I_2]_2 + 4CO$$

Procedure

Finely divided iodine (30.0 g) is Soxhlet-extracted into a solution of (η-C_5H_5) Mo(NO)$(CO)_2$ (30.0 g)* in dichloromethane (300 ml) until the extracting solution from the Soxhlet thimble is colorless (ca. 15 hr). The reaction mixture is then cooled. The deep-purple, almost black, product is filtered off, washed with dichloromethane (until the washings are colorless), and dried *in vacuo*. The yield is 49.5 g [92% based on (η-C_5H_5)Mo(NO)$(CO)_2$]. *Anal.* Calcd. for $C_{10}H_{10}N_2O_2I_2Mo_2$: C, 18.9; H, 1.6; N, 4.4; I, 40.0. Found: C, 18.9; H, 1.5; N, 4.3; I, 40.2.

Properties

The purple diiodide, $[(\eta\text{-}C_5H_5)Mo(NO)I_2]_2$, is stable in air for very long periods, but the brown dibromide and, especially, the orange dichloride have very limited stability toward aerial decomposition. The dibromide can be stored for several months under nitrogen at low temperatures.

The complexes are soluble in polar solvents such as acetonitrile and tetrahydrofuran but are only slightly soluble in chloroform or dichloromethane and are insoluble in carbon tetrachloride and hydrocarbons. All the complexes exhibit one NO stretching frequency (KBr pellet) at 1670 cm^{-1}.

The compounds are believed to contain [(η-C_5H_5)Mo(NO)X] groups bridged by two additional X atoms. Cleavage of the X bridges is readily effected by Lewis bases and by halide ions.[2,5]

*Crude (η-C_5H_5)Mo(NO)$(CO)_2$ is perfectly satisfactory for this preparation.

References

1. T. S. Piper and G. Wilkinson, *J. Inorg. Nucl. Chem.*, **3**, 104 (1956).
2. R. B. King, *Inorg. Chem.*, **6**, 30 (1967).
3. T. A. James and J. A. McCleverty, *J. Chem. Soc. (A)*, **1970**, 3308; *ibid.*, **1971**, 1068; R. B. King and M. B. Bisnette, *Inorg. Chem.*, **6**, 469 (1967).
4. E. O. Fischer and H.-J. Beck, *Chem. Ber.*, **104**, 3101 (1971).
5. J. A. McCleverty and D. Seddon, *J. Chem. Soc., Dalton Trans.*, **1972**, 2526.
6. J. L. Calderon, F. A. Cotton, and P. Legdzins, *J. Am. Chem. Soc.*, **91**, 2528 (1969); R. B. King, *Inorg. Chem.*, **7**, 90 (1968).
7. E. O. Fischer, *Inorganic Syntheses*, **7**, 136 (1963); R. B. King and F. G. A. Stone, *ibid.*, **8**, 99 (1963).

8. BIS(METHYLDIPHENYLPHOSPHINE)DICHLORO-NITROSYLCOBALT

$$CoCl_2[P(CH_3)(C_6H_5)_2]_2 + NO \rightarrow CoCl_2(NO)[P(CH_3)(C_6H_5)_2]_2$$

Submitted by G. DOLCETTI,* M. GHEDIM,* and C. A. REED†
Checked by T. E. NAPPIER‡ and R. D. FELTHAM‡

Compounds of the type $CoX_2(NO)(PR_3)_2$ were briefly described by Booth and Chatt,[1] who noted the presence of two NO bands in the infrared spectrum. Collman et al.[2] suggested that these two NO bands were due to the presence of two different hybridization isomers. For the complex $PR_3 =$ methyldiphenylphosphine, two different crystalline forms were obtained at different temperatures: a low-temperature form with one NO band (1750 cm^{-1}) and a room-temperature form with two NO bands [1753 (s), and 1640 (m) cm^{-1}].[3] The preparation of both forms of $CoCl_2(NO)[P(CH_3)(C_6H_5)_2]_2$ is described below.

General Procedure

Because the nitrosyl complexes are extremely air-sensitive in solution, air must be rigorously excluded. The Schlenk-tube techniques described by Shriver[4] are suitable for the synthesis of these compounds. Air can be excluded by flushing the apparatus with nitrogen and introducing the solvents with a syringe. All solvents must be deoxygenated with nitrogen, followed by freezing at 77 K (liquid nitrogen) and alternately evacuating the Schlenk tube and

*Department of Chemistry, Università della Calabria, 87100 Cosenza, Italy.
†Department of Chemistry, Stanford University, Stanford, Calif. 94305.
‡Department of Chemistry, University of Arizona, Tucson, Ariz. 85721.

refilling it with nitrogen. All the solvents are then transferred using gastight Hamilton syringes previously flushed with nitrogen.

A. ROOM-TEMPERATURE FORM

Procedure

The starting material, $CoCl_2[P(CH_3)(C_6H_5)_2]_2$, can be prepared[3] by mixing a saturated ethanol solution of cobalt dichloride with methyldiphenylphosphine (Co/ligand, 1:2). The blue solid which forms is separated by filtration, washed with ethanol, and dried under vacuum. This complex (1 g, or 2 mmoles) is placed in a Schlenk tube along with a magnetic stirring bar. After flushing the Schlenk tube by alternately applying vacuum and refilling with nitrogen, the exact amount of dichloromethane (20 ml) to give a saturated solution is injected through the stopcock of the Schlenk tube with a syringe. Stirring is begun, and nitric oxide (Matheson 98% purified by passing over a column of silica gel at $-77°$) is passed through the solution for a maximum of 2 min. The flow of nitric oxide is then stopped and replaced by a flow of nitrogen. Deoxygenated hexane (60 ml) is added slowly from a syringe. After several hours, the dark-brown crystals are removed by filtration, washed with ethanol-hexane, and finally dried *in vacuo* (950 mg or 90% yield). *Anal.* Calcd. for $CoCl_2NO[P(CH_3)(C_6H_5)_2]_2$: C, 55.75; H, 4.7; N, 2.5; Cl, 12.7. Found: C, 55.45; H, 4.6; N, 2.5; Cl, 12.3. The corresponding triphenylphosphine, tri(*p*-tolyl) phosphine, and triethylphosphine complexes can be prepared in a similar manner.[5]

Properties

Dichlorobis(methyldiphenylphosphine)nitrosylcobalt is a dark-brown crystalline solid[3] which is moderately stable in air, but in solution it reacts rapidly with oxygen. The complex is soluble in dichloromethane, chloroform, and tetrahydrofuran and is insoluble in hexane, ethanol, methanol, and ether. The compound has the infrared spectrum, $\nu(NO) = 1750$ (s), 1650 (s) cm^{-1} in CH_2Cl_2, and 1735(s) and 1640(m) cm^{-1} in KBr. The ^1H nmr spectrum is τ 7.5 CH_3 J_{PH}(4 Hz) and τ 2.6 (C_6H_5) in $CDCl_3$.

B. LOW-TEMPERATURE FORM

Procedure

A Schlenk-tube apparatus (Fig. 5) with a frit of coarse porosity is flushed

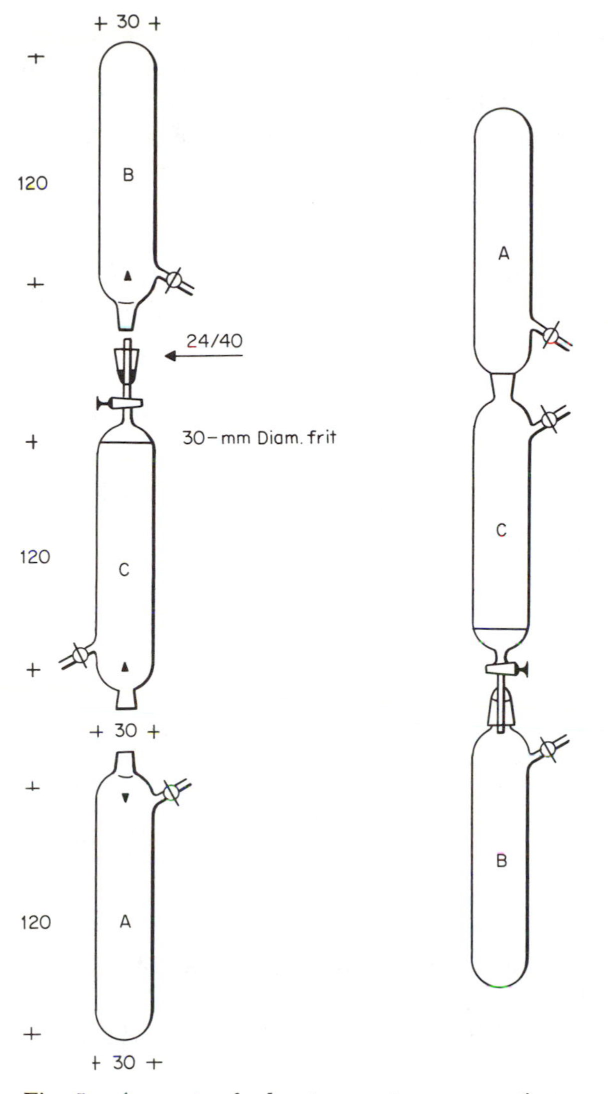

Fig. 5. Apparatus for low-temperature preparations.

with nitrogen by alternately applying a vacuum and refilling with nitrogen. A stirring bar and $CoCl_2[P(CH_3)(C_6H_5)_2]_2$ (1 g, 2 mmoles) are introduced into Schlenk tube *A*. Next, 35 ml of deoxygenated ethanol is added by means of a syringe inserted through a rubber septum placed on the stopcock. Sufficient pure, deoxygenated dichloromethane is then similarly added to the suspension until all the blue crystals *just dissolve*. Nitric oxide is added to the solution via a needle inserted into the septum (1–2 min), and the solution is rapidly cooled to $-77°$. After 2 hr, 60 ml of cold deoxygenated hexane ($-77°$) is added. After several hours, the light-brown crystals which form are filtered. Care must be taken to maintain the wet solid at $-77°$ during all operations (filtering, washing, and drying) to prevent conversion to the room-temperature form. Filtration is accomplished by inverting the apparatus (position 2) after the frit has been cooled by packing it in Dry Ice. Finally, the brown solid is washed with cold deoxygenated ethanol and hexane ($-77°$) and rapidly vacuum-dried first at $-77°$ and finally at room temperature. Yield: 990 mg or 90%. *Anal.* Calcd. for $CoCl_2(NO)[P(CH_3)(C_6H_5)_2]_2$: C, 55.75; H, 4.7; N, 2.5; Cl, 12.7. Found: C, 55.55; H, 4.6; N, 2.5; Cl, 12.6.

Properties

The low-temperature form is a light-brown crystalline solid which is moderately stable in air but is unstable in solution if exposed to air.[3] Its infrared spectrum shows $\nu(NO) = 1737\ cm^{-1}$ (Nujol).

References

1. G. Booth and J. Chatt, *J. Chem. Soc.*, **1962,** 2099.
2. J. P. Collman, P. Farnham, and G. Dolcetti, *J. Am. Chem. Soc.*, **93,** 1788 (1971).
3. J. P. Collman, G. Dolcetti, P. H. Farnham, J. A. Ibers, J. E. Lester, C. S. Pratt, and C. A. Reed, *Inorg. Chem.*, **12,** 1304 (1973).
4. D. F. Shriver, "The Manipulation of Air-Sensitive Compounds," McGraw-Hill Book Company, New York, 1969.
5. G. Dolcetti, N. W. Hoffman, and J. P. Collman, *Inorg. Chem. Acta,* **6,** 531 (1972).

9. TRIS(TRIPHENYLPHOSPHINE)NITROSYLCOBALT AND TRIS(TRIPHENYLPHOSPHINE)NITROSYLRHODIUM

Submitted by G. DOLCETTI,* O. GANDOLFI,* M. GHEDINI,* and N. W. HOFFMAN†
Checked by O. A. ILEPERUMA,‡ T. E. NAPPIER,‡ and R. D. FELTHAM‡

The title compounds are part of a series of four-coordinate complexes of the $(MNO)^0$ group (M = Co, Rh, and Ir).[1–3] Until recently, these complexes

*Department of Chemistry, Università della Calabria, 87100 Cosenza, Italy.
†Department of Chemistry, Stanford University, Stanford, Calif. 94305
‡Department of Chemistry, University of Arizona, Tucson, Ariz. 85721.

have been accessible only with difficulty. The syntheses of these complexes from readily available intermediates are described below.

A. TRIS(TRIPHENYLPHOSPHINE)NITROSYLCOBALT

$$[Co(NO)_2I]_x + (C_6H_5)_3P + Na/Hg \xrightarrow{\text{tetrahydrofuran}} CoNO[P(C_6H_5)_3]_3 + \cdots$$

Procedure

Since the products from these reactions are air-sensitive, all operations must be carried out under nitrogen. Reagent-grade tetrahydrofuran (40 ml) purified by distillation from $LiAlH_4$ under nitrogen is added to a 150-ml Schlenk tube.[4] An excess of 1% sodium amalgam (0.80 g of Na in 80 g of Hg) is added to the tetrahydrofuran, followed by 1.76 g (3.58 mmoles) of $[Co(NO)_2I]_2$[5] and a six-fold excess (5.63 g) of triphenylphosphine. The mixture is stirred for 2 hr. At the end of that time, the deep-violet solution is filtered through a frit[4] packed with powdered silica* in order to avoid direct contact of the amalgam with the frit and to simplify filtration. The violet filtrate is concentrated to about one-half its original volume (20 ml), and approximately 60 ml of deoxygenated methanol is added with a hypodermic syringe.

The dark-violet crystals which form are removed by filtration and washed with methanol, a 4:1 mixture of methanol-ether, and finally dried *in vacuo* for 16 hr over $CaCl_2$. The yield is 60% (based on cobalt). *Anal.* Calcd. for $Co(NO)[P(C_6H_5)_3]_3$: C, 74.00; H, 5.18; N, 1.60. Found: C, 74.0; H, 5.4; N, 1.69. Infrared spectrum, $\nu(NO) = 1640\ cm^{-1}$.

B. TRIS(TRIPHENYLPHOSPHINE)NITROSYLRHODIUM

$$RhCl_3 \cdot 3H_2O + P(C_6H_5)_3 + NO + Zn \xrightarrow{\text{tetrahydrofuran}} Rh(NO)[P(C_6H_5)_3]_3 + \cdots$$

Procedure

Tetrahydrofuran (150 ml), purified by distillation from $LiAlH_4$ and purged with nitrogen for 30 min, is added under a nitrogen flow to 1.0 g of $RhCl_3 \cdot 3H_2O$ contained in a 250-ml-round-bottomed flask equipped with a gas inlet adapter and a reflux condenser. Next, 7.8 g of triphenylphosphine and 6.8 g

*Fluorosil ® is satisfactory for this purpose.

of powdered zinc (zinc dust)* are added, and the slurry is stirred with a magnetic stirrer and is heated at reflux for 30 min. Nitric oxide† is bubbled slowly through the solution at reflux temperature for 12 hr. After the nitric oxide is removed from the solution by purging with nitrogen, the volume is reduced to approximately 40 ml, and the solution is filtered under nitrogen using 5-mm glass tubing to which a piece of filter paper of medium porosity has been attached. Deoxygenated methanol (400 ml) is slowly added under nitrogen flow, precipitating red crystals of $RhNO[P(C_6H_5)_3]_3$. Yield: 1.5 g (40% based on rhodium). *Anal.* Calcd. for $Rh(NO)[P(C_6H_5)_3]_3$: C, 70.9; H, 3.27; P, 10.1; N, 1.52. Found: C, 70.5; H, 3.25; P, 9.8; N, 1.50. Infrared spectrum, $\nu(NO) =$ 1610 (vs) cm^{-1} (KBr).

Properties

The cobalt nitrosyl complex is a deep-violet crystalline solid which is stable in air but decomposes rapidly in solution in the presence of oxygen. It is soluble in chloroform and dichloromethane but insoluble in ether and alcohols. Its chemistry has been studied recently.[1]

The rhodium nitrosyl complex is a deep-red crystalline solid which is moderately stable in the air but decomposes rapidly in solution in the presence of oxygen. It is soluble in chloroform, dichloromethane, and benzene and is insoluble in ether, methanol, and ethanol. The compound is a homogeneous hydrogenation catalyst for unsaturated organic compounds.[1,2,6]

With the procedure in Sec. A it is possible to prepare various $CoNOP_3$ complexes (P = methyldiphenylphosphine, tri-*p*-fluorophenylphosphine, triphenylarsine) whose properties are similar to the triphenylphosphine complex.

By reacting $[Rh(NO_2)Cl]_x$ (prepared by bubbling nitric oxide in a deoxygenated carbon tetrachloride solution of $[Rh(CO)_2Cl]_2$ made by Wilkinson's method[7]) with sodium amalgam in the presence of several phosphines, various $RhNOP_3$ complexes can be synthesized (P = methyldiphenylphosphine, tri-*p*-tolylphosphine, tri-*p*-fluorophenylphosphine, triphenylarsine). All these complexes are homogeneous hydrogenation catalysts for unsaturated organic compounds.[1,2]

The nitrosyltris(triphenylphosphine)iridium complex, $IrNO[P(C_6H_5)_3]_3$, can be prepared from $IrCl_3$ using the procedure in Sec. B.

*The reactivity of the zinc is critical for this synthesis, and care should be exercised to use a form of zinc which is active. The checkers found it essential to use zinc powder.

†Nitric oxide is the Matheson Gas commercial product from which NO_2 is removed by passage through a silica-gel trap at −78° [E. E. Hughes, *J. Chem. Phys.*, **35,** 1531 (1961)].

References

1. G. Dolcetti, N. W. Hoffman, and J. P. Collman, *Inorg. Chim. Acta,* **6,** 531 (1972).
2. G. Dolcetti, *J. Inorg. Nucl. Chem. Lett.*, **9,** 705 (1973).
3. V. G. Albano, P. L. Bellon, and M. Sansoni, *J. Chem. Soc. (A),* **1971,** 2420.
4. D. F. Shriver, "The Manipulation of Air-Sensitive Compounds," McGraw-Hill Book Company, New York, 1969.
5. B. Haymore and R. D. Feltham, *Inorganic Syntheses,* **14,** 81 (1973).
6. J. P. Collman, N. W. Hoffman, and D. E. Morris, *J. Am. Chem. Soc.,* **91,** 5659 (1972).
7. J. A. McClaverty and G. Wilkinson, *Inorganic Syntheses,* **8,** 211 (1966).

10. CARBONYLCHLORONITROSYLRHENIUM COMPOUNDS

Submitted by G. DOLCETTI* and J. R. NORTON†
Checked by M. C. FREDETTE‡ and C. J. L. LOCK‡

Knowledge of polynuclear transition-metal complexes has been obtained largely by x-ray structure determinations, but the systematic exploration of the chemistry of these compounds is only beginning. Convenient syntheses of some polynuclear carbonylchloronitrosylrhenium compounds are reported here, starting from commercially available $Re_2(CO)_{10}$. Some of these compounds are valuable intermediates in the synthesis of other mononitrosyl rhenium complexes.[1,2]

A. OCTACARBONYL-DI-μ-CHLORODIRHENIUM

$$Re_2(CO)_{10} \xrightarrow{Cl_2} [Re(CO)_4Cl]_2$$

Explicit accounts of the preparation of this compound in the literature involve the previous preparation of pure $Re(CO)_5Cl$, which requires the use of high-pressure equipment. The following preparation, the feasibility of which was inferred from the method of Hileman et al.,[3] produces a product of analytical purity with a minimum of effort.

*Department of Chemistry, Università della Calabria, 87100 Cosenza, Italy.
†Department of Chemistry, Princeton University, Princeton, N.J. 08540.
‡Department of Chemistry, McMaster University, Hamilton, Ontario, Canada. L8S 4M1

Procedure

Decacarbonyldirhenium (3 g)* is dissolved in 100 ml of carbon tetrachloride and added to 100 ml of chlorine-saturated carbon tetrachloride. The mixture is stirred magnetically for $1\frac{1}{2}$ hr, during which time white crystals [probably a mixture of $Re(CO)_5Cl$ and product] are deposited. Chlorine and solvent are then removed on the rotary evaporator. The residue is stirred under nitrogen in boiling hexane (350 ml) for 12 hr to effect complete conversion to the desired dimer. After cooling and filtration, the product is washed several times with ethanol to remove small quantities of colored impurity, washed with hexane, and vacuum-dried. The yield is 2.90 g (94%) of small off-white crystals. *Anal.* Calcd. for $Re_2C_8O_8Cl_2$: C, 14.39; H, 0.0; Cl, 10.62. Found: C, 14.4; H, 0.0; Cl, 10.3. The infrared spectrum (CCl_4 solution) in the carbonyl region is 2114 (w), 2032 (s), 2000 (m), and 1959 (m) cm^{-1}.

B. PENTACARBONYL-TRI-μ-CHLORO-NITROSYLDIRHENIUM

$$[Re(CO)_4Cl]_2 \xrightarrow[CCl_4, 77^\circ]{NO} Re_2(NO)(CO)_5Cl_3$$

Procedure

The standard apparatus of Fig. 6 is used. Octacarbonyl-di-μ-chloro-dirhenium (500 mg), produced as described in Sec. A, is dissolved in 700 ml of boiling carbon tetrachloride under nitrogen; nitric oxide is then passed through the hot solution for 11 hr. (Both the rate of gas bubbling and the reflux rate must be kept as low as possible.) Infrared examination of a sample shows when the reaction is complete by the disappearance of the starting-material peak at 2032 cm^{-1} and the appearance of the product peak at 2047 cm^{-1}. The deep-yellow solution is cooled, purged with nitrogen, and filtered in air. (Although the product is quite air-stable, accompanying minor impurities are readily oxidized to dark, insoluble material.) About 50 ml of heptane is added, and the solution is evaporated until crystals separate. Impure green plates (330 mg) separate and are collected on a filter; they are recrystallized from dichloromethane (50 ml) with the addition of hexane, to give 245 mg (51%) of light-orange crystals of pure products. *Anal.* Calcd. for $Re_2C_5NO_6Cl_3$: C, 9.26; N, 2.16; Cl, 16.39; mol. wt., 649. Found: C, 9.2; N, 2.26; Cl, 16.1; mol. wt., 649 (by mass spectroscopy).

*Strem Chemicals, Inc., Danvers, Mass. 01923.

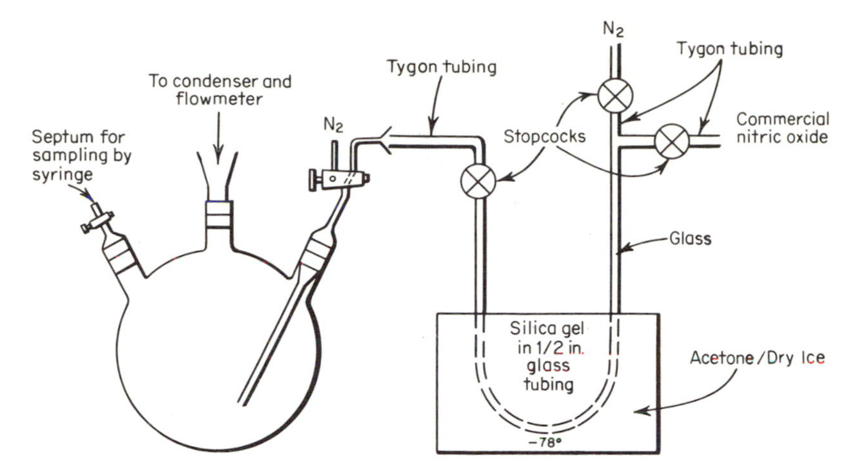

Fig. 6. Apparatus for the preparation of the pentacarbonyl-tri-μ-chloro-nitrosyldirhenium complex.

Properties

Pentacarbonyl-tri-μ-chloro-nitrosylrhenium is a light-orange crystalline solid, quite air-stable both as a solid and in solution. It is soluble in carbon tetrachloride, benzene, and methylene chloride; the compound reacts with ethanol. Its infrared spectrum is as follows: in carbon tetrachloride $\nu(CO) =$ 2113 (m), 2047 (vs), and 1946 (s) cm^{-1}; $\nu(NO) =$ 1813 (m) cm^{-1}. In KBr: 652 (m), 630 (s), 617 (m), 590 (m), 570 (m), 499 (s), 470 (s), 452 (w), 443 (m), and 421 (s) cm^{-1}; $\nu(ReCl)$ at 321 (m), and 276 (s, br) cm^{-1}. Raman spectrum (solid, 6328-A excitation): 2122 (m), 2045 (m), 1932 (sh), 1913 (m), 1815 (w, br), 635 (m), 594 (s), 518 (s), 450 (w), 326 (m), 282 (sh), 272 (m), 225 (m), 140 (s), and 118 (vs) cm^{-1}.

C. TETRACARBONYL-DI-μ-CHLORODICHLORODINITROSYLDI-RHENIUM

$$Re_2(NO)(CO)_5Cl_3 \xrightarrow[\text{benzene},25°]{\text{NO(ethanolic HCl)}} [Re(NO)(CO)_2Cl_2]_2$$

Procedure

Pentacarbonyl-tri-μ-chloro-nitrosyldirhenium (1.1 g; 1.7 mmoles) is dissolved in 600 ml of benzene under nitrogen.* A saturated ethanolic hydrogen

*All solvents used should be deoxygenated by passing a stream of nitrogen through the solvent with a fritted bubbling tube for ½ hr.

chloride solution (50 ml) is then added (the color changes from orange to pale yellow-green), and the solution is purged again with nitrogen. Nitric oxide is bubbled through the mixture at room temperature for $2\frac{1}{2}$ days. Periodically a solution aliquot is evaporated to dryness and redissolved in carbon tetrachloride so that its infrared spectrum can be checked for disappearance of the starting-material peak at 2047 cm^{-1}; numerous peaks attributed to intermediates are observed. (Smaller-scale reactions go to completion much more quickly.)

When the reaction is complete (by infrared analysis), the solution is purged with nitrogen and boiled for 15 min to remove hydrogen chloride. Cooling and removal of solvent on the rotary evaporator give an oil which solidifies after pumping overnight under high vacuum to remove residual ethanol.

Recrystallization from carbon tetrachloride (250 ml), with the addition of heptane, gives 805 mg (1.19 mmoles, 70% yield) of greenish-yellow crystals of product. An additional recrystallization from dichloromethane-hexane produces 759 mg of crystals of the appropriate yellow color. *Anal.* Calcd. for $ReC_2O_3NCl_2$: C, 6.99; N, 4.08; O, 13.98; Cl, 20.99; mol. wt., 686. Found: C, 7.0; N, 4.1; O, 14.0; Cl, 20.5; mol. wt., 686 (mass spectroscopy).

Properties

Tetracarbonyl-di-μ-chlorodichlorodinitrosyldirhenium is a bright-yellow crystalline solid. Its infrared spectrum is as follows: ν(CO)(CCl_4) = 2106 (vs), 2047 (vs); ν(NO) (CCl_4) = 1803 (vs); ν(ReCl) (Nujol) 331 (s), 290 (s), 252 (m) cm^{-1}. The compound is soluble in acetone, benzene, nitrobenzene, 1,2-dichloroethane, carbon tetrachloride, and chloroform and insoluble in aliphatic hydrocarbons. It is very stable both as a solid and in solution.

References

1. F. Zingales, A. Trovati, F. Cariati, and P. Uguagliati, *Inorg. Chem.*, **10**, 507 (1971).
2. J. R. Norton and G. Dolcetti, *Inorg. Chem.*, **12**, 485 (1973).
3. J. C. Hileman, D. K. Huggins, and H. D. Kaesz, *Inorg. Chem.*, **1**, 933 (1962).

11. DECACARBONYL-DI-μ-NITROSYL-TRIMETAL COMPOUNDS

Submitted by J. R. NORTON* and G. DOLCETTI†
Checked by B. F. G. JOHNSON‡ and A. FORSTER‡

Nitrosyl complexes are of considerable structural and chemical interest. Direct syntheses from the corresponding carbonyls are reported here for two nitrosyl-carbonyl clusters containing double nitrosyl bridges. (■ **Caution.** *Considerable amounts of toxic nitric oxide are used in these reactions, which should therefore be carried out in an efficient hood.*)

A. DECACARBONYL-DI-μ-NITROSYL-TRIRUTHENIUM

$$Ru_3CO_{12} \xrightarrow[C_6H_6]{NO(80^\circ)} Ru_3(CO)_{10}(NO)_2$$

Procedure

The preparation is carried out in the general apparatus for nitrosyl complex preparation (Fig. 6).

Five hundred milligrams (0.78 mmole) of dodecacarbonyltriruthenium§ is dissolved in 100 ml of boiling benzene (distilled under nitrogen from Na/K alloy), with a flat glass frit slowly bubbling nitrogen through the solution. As soon as the system is thoroughly flushed and all the carbonyl has gone into solution, the nitrogen flush is replaced by a stream of NO purified by passage through silica gel at −78°.[1]

Rapid NO flow (about 80 ml/min) is continued for 15 min, by which time the reaction mixture has become very dark. Nitrogen is then readmitted and the solution is cooled.¶ Reaction of $Ru_3(CO)_{10}(NO)_2$ with nitric oxide under the conditions required for its formation leads to the production of a brown insoluble polymeric material. Hence the formation and destruction of $Ru_3(CO)_{10}$-

*Department of Chemistry, Princeton University, Princeton, N.J. 08540.
†Department of Chemistry, Università della Calabria, 87100 Cosenza, Italy.
‡University Chemical Laboratory, Lensfield Road, Cambridge CB2 1EW, England.

§Purchased from Strem Chemicals, Inc., Danvers., Mass. 01923, or prepared; see Syntheses 13 and 14.

¶After all the NO has been removed from the system, subsequent operations can be carried out in air as the solutions are fairly stable. However, the chromatography solvent is deoxygenated by bubbling nitrogen through it, the column is flushed with it before use, and the fiitration products are dried by blowing a stream of dry nitrogen through them.

$(NO)_2$ are consecutive, and competitive reactions and maximum yields are obtained by stopping the reaction short of the disappearance of starting material, with chromatographic separation of it from the product mixture.[2]

The brown polymeric by-product (85 mg) is removed on a filter. The filtrate is reduced in volume on a rotary evaporator (bath temperature not above 40°); heptane is added, and then the mixture is placed in the freezer for crystallization. Filtration yields 263 mg of combined product and starting material.

The mixture to be separated is dissolved in several milliliters of a solvent mixture of 3:1 hexane–carbon tetrachloride, applied to a 50- by 3.5-cm chromatographic column constructed with activated J. T. Baker silica gel (150° overnight) as absorbent and developed with 3:1 hexane–carbon tetrachloride. The ruthenium carbonyl moves at almost the speed of the solvent front, whereas the green band containing the product moves with an R_f of about 0.20. After addition of heptane to the green band, concentration of the solution, cooling, and washing with cold pentane, 145 mg (0.23 mmole) of $Ru_3(CO)_{10}(NO)_2$ is obtained. *Anal.* Calcd. for $Ru_3C_{10}N_2O_{12}$: C, 18.67; N, 4.35; mol. wt., 643. Found: C, 18.9; N, 4.6; mol. wt., 643 (mass spectroscopy).

Properties

Decacarbonyl-di-μ-nitrosyl-triruthenium is a golden-green crystalline solid which is very slightly air-sensitive and can be stored for months at room temperature. Infrared spectra: ν(CO) (cyclohexane) = 2110 (w), 2077 (s), 2068 (s), 2061 (sh), 2038 (s), 2030 (s), 2026 (sh), 2015 (w), and 2000 (m) cm^{-1}; ν(NO) (KBr) = 1517 (m) and 1500 (s) cm^{-1}.

References

1. F. E. Hughes, *J. Chem. Phys.*, **35**, 1531 (1961).
2. J. R. Norton, J. P. Collman, G. Dolcetti, and W. T. Robinson, *Inorg. Chem.*, **11**, 382 (1972).

B. DECACARBONYL-DI-μ-NITROSYL-TRIOSMIUM

$$Os_3(CO)_{12} \xrightarrow[100°,\ 24\ hr]{NO(60\ psi),\ benzene} Os_3(CO)_{10}(NO)_2$$

Procedure

Dodecacarbonyltriosmium (prepared from $OsCl_3$[1]) (150 mg) and 10 ml benzene are placed in a 3-oz Fischer-Porter pressure vessel which is then thoroughly purged with argon and charged with 60 psi of nitric oxide.* The sealed reaction mixture is then heated to 100° for 24 hr with magnetic stirring. Upon cooling, a solid [largely unreacted $Os_3(CO)_{12}$] precipitates. After flushing with argon, the mixture is filtered to give a pale-green solution. Addition of heptane and removal of benzene on the rotary evaporator give 20 mg of light-green crystals of $Os_3(CO)_{10}(NO)_2$. Purification for analysis and spectroscopy is effected by chromatography on silica gel under nitrogen with 3:1 hexane–carbon tetrachloride, as described above. *Anal.* Calcd. for $Os_3C_{10}N_2O_{12}$: C, 13.19; H, 0.0; N, 3.07; mol. wt., 911. Found: C, 13.17; H, 0.2; N, 2.83; mol. wt., 911 (mass spectroscopy).

Properties

The osmium-nitrosyl-carbonyl cluster forms green platelike crystals which are air-stable indefinitely. Infrared spectra: ν(CO) (tetrachloroethylene) = 2108 (w), 2068 (s), 2063 (s), 2504 (sh), 2025 (s), 2017 (w), 2008 (s), and 1996 (m) cm^{-1}; ν(NO) (KBr) = 1503 (m), 1484 (s), and 739 (s) cm^{-1}.[1]

References

1. J. R. Norton, J. P. Collman, G. Dolcetti, and W. T. Robinson, *Inorg. Chem.*, **11,** 382 (1972).
2. F. E. Hughes, *J. Chem. Phys.*, **35,** 1531 (1961).

12. *trans*-BIS-(TRIPHENYLPHOSPHINE)CHLORONITROSYL-IRIDIUM(1+) TETRAFLUOROBORATE

$$trans\text{-}IrX(CO)[P(C_6H_5)_3]_2 \xrightarrow[\text{under } N_2]{O_2NC_6H_4CON_3} IrX(N_2)[P(C_6H_5)_3]_2$$

$$\xrightarrow{NOBF_4} [IrX(NO)[P(C_6H_5)_3]_2]BF_4 \qquad (X = Cl, Br)$$

Submitted by R. J. FITZGERALD† and H. -M. W. LIN†
Checked by B. L. HAYMORE‡

The procedure reported by Angoletta and Caglio[1] for the synthesis of the complex *trans*-chloronitrosylbis(triphenylphosphine)iridium(I) cation is long and

*Nitric oxide is the Matheson Gas commercial product from which NO_2 is removed by passage through a silica-gel trap at −78°.[2]

†Department of Chemistry, Illinois Institute of Technology, Chicago, Ill. 60616.
‡Department of Chemistry, Northwestern University, Evanston, Ill. 60201.

leads to small yields. The synthetic procedure suggested by Reed and Roper[2,3] offers substantial improvement over Angoletta's method. The synthesis described here gives the complex in good yield and involves only two steps, beginning with the readily available complexes $IrX(CO)[P(C_6H_5)_3]_2$ (X = Cl, Br).

Procedure

All solvents, methanol, chloroform, absolute ethanol, and *n*-hexane are dried over 4A molecular sieves and degassed by passing a stream of nitrogen through the solvent with a filtered bubbling tube for $\frac{1}{2}$ hr and then cooling to 0° before use.

The compound *trans*-$IrCl(N_2)[P(C_6H_5)_3]_2$ is prepared by a slight modification[4] of the standard procedure.[5] Solid *trans*-$IrCl(CO)[P(C_6H_5)_3]_2$ (0.40 g) is cooled to 0° under nitrogen in a Schlenk-type reaction tube. Chloroform (8.0 ml at 0°) and absolute ethanol (0.15 ml) are added. The standard procedure for the preparation of the dinitrogen complex[5] depends on the small amount of ethanol present in commercial chloroform to esterify the isocyanate intermediate. The addition of ethanol mentioned here assures that it will be present in sufficient amount for the esterification reaction. This same method[5] can be used to prepare *trans*-$IrBr(N_2)[P(C_6H_5)_3]_2$, starting with *trans*-$IrBr(CO)$-$[P(C_6H_5)_3]_2$.* The yellow crystalline product is obtained in 70% yield, and it has a N–N stretching frequency at 2095 cm^{-1}.

The compounds $IrX(N_2)[P(C_6H_5)_3]_2$ are then used to prepare the corresponding salts $[IrX(NO)[P(C_6H_5)_3]_2]BF_4$. Solid 95% $NOBF_4$† (0.016 g, 0.132 mmole) is added to an ice-cold solution of 0.10 g‡ (0.129 mmole) of $IrCl(N_2)$-$[P(C_6H_5)_3]_2$ in 10 ml of chloroform. The color of the solution changes from yellow to red in 10 min. The solution is filtered to remove any remaining solid. The red solution is evaporated in a vacuum desiccator, and the product is recrystallized from a chloroform–*n*-hexane mixture. The red crystalline product weighs 0.10 g (90% yield). *Anal.* Calcd. for $IrNOClP_2BF_4C_{36}H_{30}$: C, 49.74; H, 3.45; N, 1.61. Found: C, 49.1; H, 3.83; N, 1.39. The infrared spectrum is identical to that previously reported[2,3] for $[IrCl(NO)\{P(C_6H_5)_3\}_2]ClO_4$ prepared by other methods, with $\nu(NO) = 1902\ cm^{-1}$.

The bromo compound $[IrBr(NO)[P(C_6H_5)_3]_2]BF_4$ is prepared in a manner

*Starting materials available from Strem Chemicals, Inc., Danvers, Mass. 01923.

†The compound $NOBF_4$ is available from Alfa Inorganics Ventron, Beverly, Mass. 01915. The checker reports equally good yields may be obtained using $NOPF_6$.

‡The checker scaled up the proportion 10 times and obtained the same percentage yield of product.

analogous to that used for the chloro complex except that $IrBr(N_2)[P(C_6H_5)_3]_2$ is used as a starting material. *Anal.* Calcd. for $IrNOBrP_2BF_4C_{36}H_{30}$: C, 47.32; H, 3.28; N, 1.53. Found: C, 47.0; H, 3.53; N, 1.25. The infrared spectrum yielded an intense band attributed to $\nu(NO) = 1902\ cm^{-1}$.

Properties

The complexes *trans*-$[IrX(NO)\{P(C_6H_5)_3\}_2]BF_4$(X = Cl, Br) are dark red crystalline solids with $\nu(NO) = 1902\ cm^{-1}$. The chloro complex exhibits limited transition metal basicity[6] and oxidative-addition reactions;[3] it does not bind with O_2[3] but adds various donor molecules, especially coordinating anions.[3] The complex hydrolyzes in water to give the hydroxy complex and undergoes an exchange reaction in alcohols which yields the alkoxy derivatives.[3] The complexes are presumably square-planar with NO coordinated as NO^+.[6]

References

1. M. Angoletta and G. Caglio, *Gazz. Chim. Ital.*, **93**, 159 (1963).
2. C. A. Reed and W. R. Roper, *J. Chem. Soc. (D)*, **1969**, 155.
3. C. A. Reed and W. R. Roper, *ibid.*, 1459.
4. R. J. Fitzgerald, N. Y. Sakkab, R. S. Strange, and V. Narruttis, *Inorg. Chem.*, **12**, 1081 (1973).
5. J. P. Collman, N. W. Hoffman, and J. W. Hosking, *Inorganic Syntheses*, **12**, 8 (1970).
6. R. J. Fitzgerald and H. M. Wu Lin, *Inorg. Chem.*, **11**, 2270 (1972).

Chapter Two

METAL CARBONYL COMPOUNDS

13. DODECACARBONYLTRIRUTHENIUM

$$[Ru_3O(O_2CCH_3)_6(H_2O)_3]O_2CCH_3 \xrightarrow{CO} Ru_3(CO)_{12}$$

Submitted by B. R. JAMES,* G. L. REMPEL,† and W. K. TEO†
Checked by C. EADY,‡ A. FORSTER,‡ and B. F. G. JOHNSON‡

Dodecacarbonyltriruthenium has been synthesized by a number of methods, the majority of which require high pressures (100–300 atm) of carbon monoxide and elevated temperatures.[1–5] Optimum yields are obtained by a method in which ruthenium trichloride and sodium acetylacetonate (sodium 2,4-pentanedionate) in methanol are treated with equimolar mixtures of hydrogen and carbon monoxide (200–300 atm total pressure) at 140–160°.[6] However, this method also requires high-pressure apparatus, which is not always readily available. Methods for a high-yield synthesis of dodecacarbonyltriruthenium which require only ambient pressures of carbon monoxide have recently been reported.[7,8]

On bubbling carbon monoxide through green 1-propanol solutions of hexakis(μ-acetato)trisaquooxotriruthenium(III) acetate, $[Ru_3O(O_2CCH_3)_6(H_2O)_3]$-$O_2CCH_3$,§ at 80°C the color slowly becomes orange, and after approximately 10 hr, dodecacarbonyltriruthenium begins to form and can be readily isolated. The synthesis provided here gives the details of this method.

*Department of Chemistry, University of British Columbia, Vancouver V6T 1W5, Canada.
†Department of Chemical Engineering, University of Waterloo, Waterloo, Ontario, N2L 3G1, Canada.
‡Chemistry Department, Cambridge University, Cambridge CB2 1EW, England.
§Ruthenium acetate is available from Johnson Matthey, London.

Procedure

To crude hexakis(μ-acetato)trisaquooxotriruthenium(III) acetate[9] (0.50 g) are added 1-propanol (50 ml) and triethylamine (0.90 g).* Carbon monoxide is bubbled slowly through the stirred solution, which is maintained at 75–80° for 15 hr. The solution is cooled under a CO atmosphere at −20° for about 20 hr. The crude, orange crystalline dodecacarbonyltriruthenium is then separated by filtration. The filtrate is treated with CO again at 75–80° for another 15 hr and subsequently cooled under CO at −20° to yield a further crop of crystals. The combined products are then recrystallized from hot hexane. Yield: 0.23 g (59%).† *Anal.* Calcd. for $C_{12}O_{12}Ru_3$: C, 22.49; Ru, 47.46. Found: C, 23.0; Ru, 47.75.

Properties

Dodecacarbonyltriruthenium is an orange-colored crystalline solid, which is stable to air and light. It is soluble in most organic solvents but completely insoluble in water. The infrared spectrum of the dodecacarbonyl, which has sharp bands at 2060 (s), 2030 (s), and 2010 (m) cm^{-1} in *n*-hexane due to terminal CO groups, can be used in identifying and assessing the purity of the compound. No bridging CO bands are observed either in the solid state or in solution.

The dodecacarbonyl has been used as a catalyst for a variety of carbonylation reactions.[11,12]

References

1. W. Manchot and W. J. Manchot, *Z. Anorg. Chem.*, **226,** 385 (1936).
2. W. Hieber and H. Fischer, Ger. Pat. 695589 (1940); *Chem. Abstr.*, **35,** 5657 (1941).
3. M. I. Bruce and F. G. A. Stone, *Angew. Chem., Int. Ed.*, **7,** 427 (1968).
4. J. P. Candlin, Br. Pat. 1,160,765; *Chem. Abstr.*, **71,** 83137 (1969).
5. G. Braca, G. Sbrana, and P. Pino, *Chim. Ind. (Milan)*, **46,** 206 (1964); Ger. Pat. 1216276 (1966); *Chem. Abstr.*, **65,** 8409 (1966).
6. B. F. G. Johnson and J. Lewis, *Inorganic Syntheses*, **13,** 92 (1972).
7. B. R. James and G. L. Rempel, *Chem. Ind.*, **1971,** 1036.
8. J. L. Dawes and J. D. Holmes, *Inorg. Nucl. Chem. Lett.*, **7,** 847 (1971).
9. F. A. Cotton, J. G. Norman, A. Spencer, and G. Wilkinson, *Chem, Commun.*, **1971,** 967.

*Triethylamine neutralizes acetic acid present in the crude acetate salt of ruthenium and acetic acid formed during the reductive hydrolysis of the ruthenium. Acetic acid slowly converts $Ru_3(CO)_{12}$ into $[Ru(CO)_3(O_2CCH_3)]_2$, consequently lowering the yield of $Ru_3(CO)_{12}$.

†When the preparation is carried out in a 100-ml shaking autoclave at carbon monoxide pressures of 3–4 atm. and a temperature of 76°C, the yield of $Ru_3(CO)_{12}$ obtained is between 80 and 85%.

10. G. R. Crooks, G. Gamlen, B. F. G. Johnson, J. Lewis, and I. G. Williams, *J. Chem. Soc. (A)*, **1969,** 2761.
11. B. R. James, *Inorg. Chim. Acta Rev.*, **4,** 73 (1970).
12. J. J. Byerley, G. L. Rempel, N. Takebe, and B. R. James, *Chem. Commun.*, **1971,** 1482.

14. DODECACARBONYLTRIRUTHENIUM

$$RuCl_3 \cdot 3H_2O \xrightarrow{CO} Ru(CO)_nCl_m$$

$$Ru(CO)_nCl_m \xrightarrow[CO]{Zn} Ru_3(CO)_{12} + ZnCl_2$$

Submitted by A. MANTOVANI* and S. CENINI†
Checked by B. R. JAMES‡ and D. V. PLACKETT‡

Dodecacarbonyltriruthenium, $Ru_3(CO)_{12}$, can be prepared in good yields by the usual methods, but these require the use of high-pressure equipment.[1] Recently, however, several syntheses of this carbonyl at atmospheric pressure have been reported.[2–4] The carbonylation of $[Ru(CO)_3Cl_2]_2$ in the presence of zinc does not give satisfactory yields of $Ru_3(CO)_{12}$.[3] Also it requires the previous isolation of $[Ru(CO)_3Cl_2]_2$. This compound can easily be obtained from $Ru_3(CO)_{12}$,§ but other methods of synthesis are not very convenient.

We give here the details of an easy and direct synthesis of $Ru_3(CO)_{12}$ from $RuCl_3 \cdot 3H_2O$ at atmospheric pressure, according to the method reported by Dawes and Holmes.[4] The over-all procedure requires about 15 hr and gives 50–60% yields of $Ru_3(CO)_{12}$.

Procedure

A mixture of $RuCl_3 \cdot 3H_2O$ (obtained from RuO_4 and HCl, and dried under vacuum; Ru: 37–39%) (3.5 g) and 2-ethoxyethanol (60 ml), previously dried over $MgSO_4$ and distilled, is vigorously refluxed (135°) and magnetically stirred, while a fast stream of carbon monoxide is passed through the solution. After ca. 6 hr the starting red solution becomes lemon-yellow, forming an unknown carbonylchlororuthenium complex.[4,5] The solution is left to cool under carbon monoxide at room temperature.

*Cattedra di Chimica, Facoltà di Ingegneria, Via Marzolo 9, 35100 Padova, Italy.

†Centro di Studio per la sintesi e la struttura dei composti dei metalli di transizione nei bassi stati di ossidazione del C.N.R. and Istituto di Chimica Generale dell'Università, Via G. Venezian 21, 20133 Milano, Italy.

‡Department of Chemistry, University of British Columbia, Vancouver V6T 1M5, Canada.

§See Synthesis 16.

Dry ethanol (50 ml) and granular Zn (4 g), previously purified with HCl 10%, washed with distilled water, and dried with acetone, are added.* The granular zinc should be of 30-mesh size, to avoid a partial reduction to ruthenium metal. The solution is then heated at 85°,† while maintaining vigorous magnetic stirring and a fast stream of carbon monoxide. After 6–7 hr, the heating is stopped, and the red-orange suspension is decanted from the unreacted zinc. The crystalline orange $Ru_3(CO)_{12}$ is collected on a filter, washed with methanol, and dried under vacuum. In this way almost pure $Ru_3(CO)_{12}$ is obtained (1.6–1.9 g). Additional $Ru_3(CO)_{12}$, left in the mixture with the zinc, can be obtained by adding the methanol washings from the previous filtration and decanting the solution away from the zinc.

The product can then be recrystallized from hot benzene‡ (150 ml) by evaporating the solution to a small volume (yield: 1.4–1.7 g; 50–60%).§ In this way some brick-red impurities (probably carbonylhydridoruthenium cluster derivatives),[3] which are insoluble in benzene, are eliminated. *Anal.* Calcd. for $Ru_3(CO)_{12}$: C, 22.5. Found: C, 23.3.

After further treatment with CO and Zn under the same conditions as described above, the mother liquor of the carbonylation reaction did not give any additional yield of $Ru_3(CO)_{12}$.

Properties

The properties of $Ru_3(CO)_{12}$ are described on page 46.

References

1. B. F. G. Johnson and J. Lewis, *Inorganic Syntheses,* **13,** 92 (1972) and references therein.
2. B. R. James and G. L. Rempel, *Chem. Ind.,* **37,** 1036 (1971); B. R. James, G. L. Rempel, and W. K. Teo, Synthesis 13 in this book.
3. R. B. King and P. N. Kapoor, *Inorg. Chem.,* **11,** 336 (1972).
4. J. L. Dawes and J. D. Holmes, *Inorg. Nucl. Chem. Lett.,* **7,** 847 (1971).
5. F. A. Cotton and G. Wilkinson, "Advanced Inorganic Chemistry," 2d ed., p. 997, Interscience Publishers, a division of John Wiley & Sons, Inc., New York, 1972.

*The cooled yellow solution can be left overnight under a small flux of carbon monoxide and then allowed to react with Zn without any apparent variation in the yields of the product. However, the checkers note that the solution becomes cloudy on standing overnight under CO.

†The checkers report that the temperature for this zinc reduction should not rise above 85° since decomposition then takes place.

‡The checkers recommend using toluene or *n*-hexane as solvents for recrystallization. With these solvents they obtained a yield of 45% of the pure product.

§The yields of the product appear to be affected by the origin of the "$RuCl_3 \cdot 3H_2O$" employed in the synthesis. The checkers used a Johnson Matthey, London, supply containing 40.6% Ru.

15. HEXADECACARBONYLHEXARHODIUM

$$3Rh_2(O_2CCH_3)_4 + 22CO + 6H_2O \rightarrow Rh_6(CO)_{16} + 6CO_2 + 12CH_3CO_2H$$

Submitted by B. R. JAMES,* G. L. REMPEL,† and W. K. TEO†
Checked by G. CAGLIO‡

The hexanuclear carbonyl $Rh_6(CO)_{16}$ was first prepared by treating anhydrous rhodium trichloride with carbon monoxide at 200 atm for several hours in the presence of a halogen acceptor such as cadmium, copper powder, silver, or zinc at temperatures of 80–230°.[1] At temperatures of 50–80°, the main product was $Rh_4(CO)_{12}$. Optimum yields (80–90%) of $Rh_6(CO)_{16}$ are obtained by allowing methanolic solutions of rhodium trichloride trihydrate to react with carbon monoxide at 40 atm and 60°.[2] Recently, however, there have been reports[3,4] of new high-yield syntheses of rhodium cluster carbonyls which require only ambient pressures of carbon monoxide. Chini and Martinengo[3] have obtained $Rh_6(CO)_{16}$ and $Rh_4(CO)_{12}$ in high yield from the reaction of $Rh_2(CO)_4Cl_2$ with carbon monoxide at atmospheric pressure and room temperature.

Although suspensions of rhodium(II) acetate, $Rh_2(C_2H_3O_2)_4$, in 1-propanol are unaffected by mild carbonylation, protonation of such solutions by aqueous fluoroboric acid (hydrogen tetrafluoroborate) gives Rh_2^{4+} species which react with carbon monoxide at atmospheric pressure and 75° to give $Rh_6(CO)_{16}$ in about 85% yield.[4] The synthesis provided here gives the details of this method.

■ **Caution.** *Carbon monoxide and metal carbonyls are highly toxic, and the reaction should be performed in an efficient fume hood.*

Procedure§

To tetrakis(acetato)dirhodium(II), $Rh_2(O_2CCH_3)_4$,[5,6] (1.0 g) in a 100-ml,

*Department of Chemistry, University of British Columbia, Vancouver V6T 1W5, Canada.

†Department of Chemical Engineering, University of Waterloo, Waterloo, Ontario, Canada.

‡Istituto di Chimica Generale, Università di Milano, 20133 Milano N2L 3G1, Italy.

§The checker makes the following observations: If the temperature exceeds 80°, there is extensive decomposition. The reaction time can be reduced to 9 hr without affecting the yield and quality of the product. The reaction does not require the use of degassed solvents, and the product can be worked in air.

two-necked flask is added 1-propanol (60 ml) and 40% fluoroboric acid (2.0 ml). The flask is equipped with a reflux condenser, a magnetic stirring bar is introduced, and the side neck is attached to a source of carbon monoxide. The gas leaves the flask via the condenser, which is connected to a bubbler. Carbon monoxide is bubbled slowly through the stirred solution, which is maintained at 75–80° for 20 hr. During this time, violet-brown microcrystals of $Rh_6(CO)_{16}$ are deposited. The final orange-brown solution is cooled to room temperature under a carbon monoxide atmosphere. The product is then collected on a filter and washed several times with cold ethanol before drying *in vacuo* at 50° for 10 hr. Yield: 0.70 g (87%). *Anal.* Calcd. for $C_{16}O_{16}Rh_6$: C, 18.04; Rh, 57.94. Found: C, 17.83; Rh, 57.69.

Properties

Hexadecacarbonylhexarhodium is a violet-brown solid* which is stable to air oxidation. It decomposes under nitrogen above 200°. It is very slightly soluble in organic solvents, the highest solubilities being in dichloromethane and chloroform.

The infrared spectrum of $Rh_6(CO)_{16}$ has sharp bands at 2075 (s), 2025 (m), and 1800 (s) cm^{-1} in Nujol due to ν(CO). The crystal-structure determination[7] has established that there are twelve terminal carbonyl groups and four bridging carbonyl groups, each of which bridges three metal atoms. These four bridging carbonyl groups are bonded to alternate faces of the octahedron of rhodium atoms and give rise to ν(CO) at 1800 cm^{-1}.

The compound $Rh_6(CO)_{16}$ and its substitution products with phosphines and phosphites appear to be potentially important catalysts and have been used, for example, for a variety of hydroformylation and hydrogenation reactions.[8–10]

References

1. W. Hieber and H. Lagally, *Z. Anorg. Chem.*, **251**, 96 (1943).
2. S. H. H. Chaston and F. G. A. Stone, *J. Chem. Soc. (A)*, **1969**, 500.
3. P. Chini and S. Martinengo, *Inorg. Chim. Acta*, **3**, 315 (1969).
4. B. R. James and G. L. Rempel, *Chem. Ind.*, **1971**, 1036.
5. P. Legzdins, R. W. Mitchell, G. L. Rempel, J. D. Ruddick, and G. Wilkinson, *J. Chem. Soc. (A)*, **1970**, 3322.
6. P. Legzdins, G. L. Rempel, H. Smith, and G. Wilkinson, *Inorganic Syntheses*, **13**, 90 (1972).
7. E. R. Corey, L. F. Dahl, and W. Beck, *J. Am. Chem. Soc.*, **85**, 1202 (1963).

*Large crystals of $Rh_6(CO)_{16}$ appear black with a violet nuance.

8. C. W. Bradford, *Platinum Mets. Rev.*, **16,** 50 (1972) and references therein.
9. P. Chini, S. Martinengo, and G. Garlaschelli, *Chem. Commun.*, **1972,** 709, and references therein.
10. J. P. Collman, L. S. Hegedus, M. P. Cooke, J. R. Norton, G. Dolcetti, and D. N. Marquardt, *J. Am. Chem. Soc.*, **94,** 1789 (1972).

16. DI-μ-CHLORO-BIS[TRICARBONYLCHLORO-RUTHENIUM(II)]

$$Ru_3(CO)_{12} \xrightarrow{CHCl_3} [Ru(CO)_3Cl_2]_2$$

Submitted by A. MANTOVANI* and S. CENINI†
Checked by R. M. HEINTZ‡ and D. E. MORRIS‡

There are several literature reports[1–6] on the synthesis of $[Ru(CO)_3Cl_2]_2$. Direct chlorination of $Ru_3(CO)_{12}$[1] gives a mixture of compounds which are difficult to separate. Also, $H_2Ru(CO)_4$ reacts with carbon tetrachloride to give impure $[Ru(CO)_3Cl_2]_2$.[2] Carbonylation of $RuCl_3 \cdot 3H_2O$ under pressure leads to the desired product;[3] however, some difficulties were experienced in repeating this preparation. The reaction of $RuCl_3 \cdot 3H_2O$ with formic acid in the presence of hydrochloric acid is a rather complicated method for the synthesis of $[Ru(CO)_3Cl_2]_2$.[4] The displacement of the diene ligand from $[(diene)RuCl_2]_n$ (diene = benzene, norbornadiene) by carbon monoxide to give $[Ru(CO)_3Cl_2]_2$ has also been reported,[5] but the yields of the product were not satisfactory. The best method of synthesis so far reported seems to be the reaction of $Ru_3(CO)_{12}$ with chloroform under a low pressure of nitrogen at an elevated temperature.[6] The details of this synthesis are given here.

Procedure

A suspension of $Ru_3(CO)_{12}$[§] (0.5 g) in chloroform (20 ml), stabilized with ethanol which favors the reaction,[6] is placed in a glass liner constructed to fit a 100-ml autoclave. A glass liner must be used to prevent the formation of decomposition products, presumably formed because of the presence of HCl.

*Cattedra di Chimica, Facoltà d'Ingegneria, Via Marzolo 9, 35100 Padova, Italy.
†Istituto di Chimica Generale, Centro C.N.R., Via G. Venezian 21, 20133 Milano, Italy.
‡Corporate Research Department, Monsanto Co., St. Louis, Mo. 63166.
§See Syntheses 13 and 14.

The autoclave is charged with nitrogen (5 atm) and then heated at 110° in an oil bath for 6 hr.*

After venting and cooling the autoclave, the glass liner and its contents are removed, and the suspension of the white $[Ru(CO)_3Cl_2]_2$ is collected on a filter. The solid is washed with a little chloroform and dried under vacuum. This first portion (0.32–0.37 g; 53–60% yield) gives an analytically pure product. *Anal.* Calcd. for $[Ru(CO)_3Cl_2]_2$: C, 14.1; Ru, 39.5; Cl, 27.7. Found: C, 14.6; Ru, 38.7; Cl, 28.1. The pale-yellow mother liquor, to which the chloroform washings from the previous filtration are added, is evaporated under vacuum to a small volume (ca. 5 ml) and cooled to ca. −10°. The pale-yellow precipitate is collected on a filter. This second portion of the product is recrystallized from a solution of hot 1,2-dichloroethane by adding *n*-hexane, and this yields an additional 0.05–0.1 g of product.

Properties

The white $[Ru(CO)_3Cl_2]_2$ turns to orange-brown at 215° and decomposes above 315°. It is slightly soluble in chloroform and 1,2-dichloroethane. It is readily soluble in methanol and tetrahydrofuran (THF), but this solvent also acts as a ligand to give $Ru(CO)_3(THF)Cl_2$.[3] In carbon tetrachloride its infrared spectrum shows $\nu(CO) = 2140$ (s), 2081 (s), and 2076 (s) cm^{-1}. An x-ray determination has shown that the molecular structure is of C_{2h} symmetry:[7]

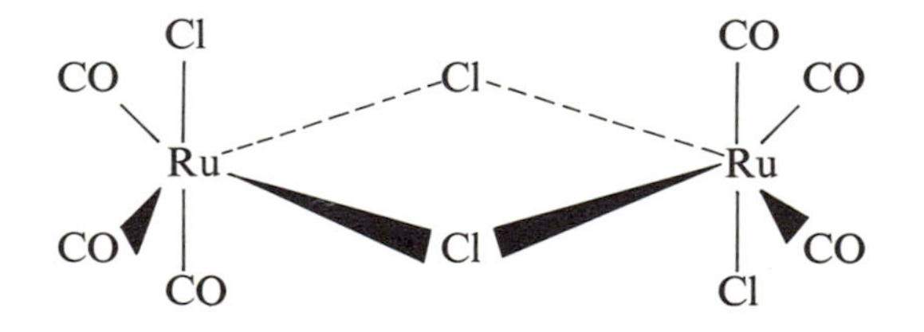

The chlorine bridges can be easily broken by THF or other ligands, e.g., pyridine, triphenylphosphine, and nitriles.[8]

The THF adduct is a good starting material for the synthesis of *cis*- or *trans*-$Ru(CO)_2L_2Cl_2$ complexes [L = PPh_3, $AsPh_3$, py; L_2 = bipy, $C_2H_4(PPh_3)_2$].[3]

References

1. B. F. G. Johnson, R. D. Johnston, and J. Lewis, *J. Chem. Soc. (A)*, **1969,** 792.
2. J. D. Cotton, M. I. Bruce, and F. G. A. Stone, *J. Chem. Soc. (A)*, **1968,** 2162.

*The checkers report that the pressure in the reactor increases from 5 to 10 atm on heating to 110°, and they find it desirable to carry out the reaction for 8 hr.

3. M. I. Bruce and F. G. A. Stone, *J. Chem. Soc. (A)*, **1967,** 1238.
4. R. Cotton and R. H. Farthing, *Aust. J. Chem.*, **24,** 903 (1971).
5. R. B. King and P. N. Kapoor, *Inorg. Chem.*, **11,** 336 (1972).
6. G. Braca, G. Sbrana, E. Benedetti, and P. Pino, *L. Chim. Ind. (Milan)*, **51,** 1245 (1969).
7. S. Merlino and G. Montagnoli, *Acta Cryst.*, **B24,** 424 (1965).
8. E. Benedetti, G. Braca, G. Sbrana, F. Salvetti, and B. Grassi, *J. Organomet. Chem.*, **37,** 361 (1972).

17. DICARBONYL(η-CYCLOPENTADIENYL)-(THIOCARBONYL)MANGANESE(I)

$$\eta\text{-}C_5H_5Mn(CO)_3 + C_8H_{14} \longrightarrow \eta\text{-}C_5H_5Mn(CO)_2(C_8H_{14}) + CO$$

$$\eta\text{-}C_5H_5Mn(CO)_2(C_8H_{14}) + (C_6H_5)_3P + CS_2 \longrightarrow \eta\text{-}C_5H_5Mn(CO)_2(CS) + (C_6H_5)_3PS + C_8H_{14}$$

Submitted by I. S. BUTLER,* N. J. COVILLE,* and A. E. FENSTER*
Checked by J. S. KRISTOFF†

Although thousands of transition metal carbonyl complexes have been synthesized over the past few years, only about 40 analogous transition metal thiocarbonyl complexes are known.[1] This is particularly surprising in view of the fact that following the discovery of the first metal-thiocarbonyl complex by Baird and Wilkinson in 1966,[2] a molecular-orbital calculation indicated that metal-thiocarbonyl complexes ought to be more stable than their carbonyl analogs.[3] The synthesis presented here describes the preparation in good yield of the first known thiocarbonyl complex of manganese by a method which future research may well prove to be a convenient general synthetic route to other transition metal thiocarbonyls. The starting materials are readily available, and a high-purity product is obtained.

■ **Caution.** *Metal carbonyls are highly toxic and must be handled in a hood and with care. The flammability of* CS_2 *should also be noted.*

Procedure

The intermediate *cis*-cyclooctene complex can be prepared by the method of Fischer and Herberhold.[4] Tricarbonyl(η-cyclopentadienyl)manganese(I)‡

*Chemistry Department, McGill University, P.O. Box 6070, Montreal, Quebec H3C 3G1, Canada.
†Department of Chemistry, Northwestern University, Evanston, Ill. 60201.
‡Strem Chemicals, Inc., Danvers, Mass. 01923.

(3.6 g, 0.017 mole) is dissolved in 400 ml of deaerated *n*-hexane, and the solution is placed in a nitrogen-purged ultraviolet irradiation vessel similar to that described by Strohmeier.[5,*] To this solution is added 25 ml (0.17 mole) of *cis*-cyclooctene. The reaction vessel is wrapped in aluminum foil and placed in an ice-water bath. The reaction mixture is then irradiated with ultraviolet light. (■ **Caution.** *Exposure of the eyes to the ultraviolet light must be avoided at all times.*) The progress of the reaction is monitored by following changes in the CO stretching region (2100–1800 cm^{-1}) of the infrared spectrum of the reaction mixture. After about 4 hr, the infrared spectrum indicates the presence of little, if any, of the tricarbonyl(η-cyclopentadienyl)manganese(I) starting material.† The resulting yellow solution is filtered under nitrogen to remove the brown decomposition product which has formed. The solvent is removed from the filtrate under reduced pressure (25°, 0.01 torr) to give a yellow-brown solid. This crude reaction product is placed in a sublimer, and any remaining tricarbonyl(η-cyclopentadienyl)manganese(I) is removed by sublimation (35°, 0.001 torr). The impure dicarbonyl(*cis*-cyclooctene)-(η-cyclopentadienyl)manganese(I) complex‡ (ca. 0.01 mole) and triphenylphosphine (3.2 g, 0.012 mole) are dissolved in 250 ml of spectrograde carbon disulfide§ contained in a 500-ml, two-necked, round-bottomed flask. The red solution obtained is refluxed under an atmosphere of nitrogen for 28 hr.¶ Removal of volatile material under reduced pressure (25°, 0.01 torr) yields a brown gum.** This gum is dissolved in 50 ml of deaerated spectrograde *n*-pentane, and the solution is filtered under nitrogen. The solid residue remaining on the filter (chiefly triphenylphosphine sulfide) is washed with 20-ml portions of *n*-pentane until the washings become colorless. Solvent removal under reduced pressure (25°, 0.01 torr) gives a yellow-brown solid. Extended high-vacuum sublimation (28°, 0.001 torr) of this crude product gives the desired complex as a bright-yellow solid [2.0 g, 55% yield based on η-$C_5H_5Mn(CO)_3$]. An analytical sample can be crystallized from *n*-pentane (m.p. 52–53°, uncorrected).

*The capacity of the vessel used by the authors was 500 ml, and a 450-W Hanovia ultraviolet lamp was placed in a *quartz* finger in the reaction vessel.

†In *n*-hexane solution, the CO stretching absorptions of the two complexes concerned are η-$C_5H_5Mn(CO)_3$, 2028 (s) and 1947 (vs); η-$C_5H_5Mn(CO)_2(C_8H_{14})$, 1964 (s) and 1905 (s).

‡Sublimation of this impure product (65°, 0.001 torr) gives the pure complex as a yellow solid (2.6 g, 60% yield).

§Freshly distilled under nitrogen from molecular sieves (Linde type 3A, 1/16 in.).

¶The progress of the reaction can again be conveniently monitored by following changes in the infrared spectrum of the reaction mixture in the CO stretching region.

**The checker found it more convenient to sublime the impure η-$C_5H_5Mn(CO)_2(C_8H_{14})$ after removal of unreacted η-$C_5H_5Mn(CO)_3$. In this way, a brown crystalline material, not a gum, is obtained after removal of CS_2. The product can then be directly sublimed at 30° onto a water-cooled probe to produce the pure thiocarbonyl.

Anal. Calcd. for $C_8H_5O_2SMn$: C, 43.6; H, 2.3; S, 14.6; mol. wt., 220.2. Found: C, 43.1; H, 2.6; S, 14.7; mol. wt., 220 (mass spectroscopy), 216 (osmometry).

Properties

Dicarbonyl(η-cyclopentadienyl)(thiocarbonyl)manganese(I), η-$C_5H_5Mn(CO)_2CS$, can be handled in air and light over a period of several days, but it eventually darkens. The complex is soluble in all common organic solvents. If the solvents are not deaerated, it decomposes rapidly. Its principal infrared absorptions are in CS_2 solution $\nu(CO) = 2006$ (s) and 1954 (s); $\nu(CS) = 1266$ (s) cm^{-1}; and in Nujol mull $\nu(CO) = 2010$ (s) and 1959 (s); $\nu(CS) = 1271$ (s) cm^{-1}. The ^1H nmr spectrum in CS_2 solution exhibits a single sharp η-cyclopentadienyl resonance at $\tau 5.18$. The major fragments in the mass spectrum at 70 eV are η-$C_5H_5Mn(CO)_2CS^+$ (*m/e* 220), η-$C_5H_5Mn(CS)^+$ (*m/e* 164), η-$C_5H_5Mn^+$ (*m/e* 120), and Mn^+ (*m/e* 55). The complex readily undergoes CO substitution with Group VA donor ligands, L, such as triphenylphosphine and trimethyl phosphite, to give η-$C_5H_5Mn(CO)(CS)L$ and η-$C_5H_5Mn(CS)L_2$.[6]

The *cis*-cyclooctene intermediate, η-$C_5H_5Mn(CO)_2(C_8H_{14})$, is a yellow, air-stable, crystalline compound, m.p., with decomposition, 108–110°(uncorrected). It is reasonably stable in solution and dissolves in all common organic solvents. In *n*-hexane solution, its $\nu(CO)$ absorptions are at 1964 (s) and 1905 (s) cm^{-1}. The principal fragments in its mass spectrum at 70 eV are η-$C_5H_5Mn(C_8H_{14})^+$ (*m/e* 230), η-$C_5H_5Mn^+$ (*m/e* 120), and Mn^+ (*m/e* 55). In the ^1H nmr spectrum in $CDCl_3$ solution, there are signals at τ 8.60 (br), 7.50 (br), 6.90 (br), and 6.05 (singlet, C_5H_5) with relative intensities of 10, 2, 2, and 5, respectively.[7] The complex undergoes S_N1 C_8H_{14} displacement with triphenylphosphine to form η-$C_5H_5Mn(CO)_2[(C_6H_5)_3P]$.[7]

References

1. I. S. Butler and A. E. Fenster, *Organomet. Chem. Rev.*, in press.
2. M. C. Baird and G. Wilkinson, *Chem. Comm.*, **1966**, 267.
3. W. G. Richards, *Trans. Faraday Soc.*, **63**, 257 (1967).
4. E. O. Fischer and M. Herberhold in "Essays in Coordination Chemistry," Birkhauser Verlag, Basel, 1965.
5. W. Strohmeier, *Angew. Chem., Int. Ed.*, **3**, 730 (1964).
6. I. S. Butler and N. J. Coville, *J. Organomet. Chem.*, in press.
7. R. J. Angelici and W. Loewen, *Inorg. Chem.*, **6**, 682 (1967).

18. TRANSITION METAL CARBONYL DERIVATIVES OF MAGNESIUM

Submitted by G. B. McVICKER*
Checked by T. J. MARKS† and A. M. SEYAM†

Transition metal carbonyl derivatives of magnesium can be prepared readily by the reductive cleavage of the metal-metal bond in numerous dimeric transition metal carbonyl complexes or by the removal of halogen from transition metal carbonyl halide complexes with an excess of 1% magnesium amalgam in the presence of a Lewis base. These preparations may be represented by the general equations

$$\text{M—M} + \text{Mg(Hg)} + x\text{B} \longrightarrow \text{B}_x\text{Mg[M]}_2$$

$$2\text{M—X} + \text{Mg(Hg)} + x\text{B} \longrightarrow \text{B}_x\text{Mg[M]}_2 + \text{MgX}_2$$

where M is a transition metal carbonyl residue such as $\text{—Fe(CO)}_2(\eta\text{-C}_5\text{H}_5)$, $\text{—Mo(CO)}_3(\eta\text{-C}_5\text{H}_5)$, —Co(CO)_4, —Mn(CO)_5, or a wide variety of phosphine-substituted derivatives; B is a Lewis base; and the subscript x is 2 or 4 when B is considered to be monobasic. The number of complexed Lewis bases was established by nmr measurements and chemical analyses and correlates well with the nucleophilicity of the transition-metal anion.[1] When M is a strongly nucleophilic anion, such as $\text{Fe(CO)}_2(\eta\text{-C}_5\text{H}_5)^-$, complexes containing two Lewis bases crystallize from solution, whereas less nucleophilic anions, such as Co(CO)_4^- or Mn(CO)_5^-, yield magnesium derivatives containing four complexed Lewis bases.

Their excellent hydrocarbon solubility makes these complexes useful as unique metallating agents, especially in the preparation of mixed metal-metal bonded complexes which, in the presence of polar solvents, undergo solvent-aided disproportionations. Although only three reactions are described in detail here, many additional magnesium–transition metal complexes can be prepared by following similar reduction procedures.

A. BIS(DICARBONYL-η-CYCLOPENTADIENYLIRON)BIS(TETRAHYDROFURAN)MAGNESIUM

$$[(\eta\text{-C}_5\text{H}_5)\text{Fe(CO)}_2]_2 + \text{Mg(Hg)} + 2\text{C}_4\text{H}_8\text{O} \longrightarrow (\text{C}_4\text{H}_8\text{O})_2\text{Mg[Fe(CO)}_2(\eta\text{-C}_5\text{H}_5)]_2$$

*Esso Research and Engineering Company, P.O. Box 45, Linden, N.J. 07036.
†Department of Chemistry, Northwestern University, Evanston, Ill. 60201.

Procedure

■ **Caution.** *Transition metal carbonyl derivatives of magnesium are pyrophoric and must be handled with extreme care (see Properties). The toxicity of the magnesium derivatives is not known but is probably of the same magnitude as the parent carbonyls. The magnesium–transition metal-carbonyl complexes exhibit low vapor pressures and therefore do not present a vapor-transport toxicity problem when an inert-atmosphere enclosure is employed.*

All reactions and sample preparations are carried out in an inert-atmosphere enclosure under dry nitrogen. Solvents and reagents are dried in the following manner. Benzene, tetrahydrofuran, and *n*-pentane are freshly distilled from lithium aluminum hydride; pyridine is distilled over barium oxide; and tetramethylethylenediamine is distilled over calcium hydride. Solvents used in preparing nmr and infrared samples are degassed by a freeze-thaw technique. Nmr spectra are obtained with torch-sealed nmr tubes. The commercial transition metal carbonyl complexes are recrystallized and vacuum-dried before use. Glassware is routinely flame-dried.

In a 250-ml one-necked flask equipped with a large magnetic stirring bar a 1 % magnesium amalgam is prepared by rapidly stirring 0.6 g of 200-mesh magnesium powder* into 60 g of mercury (a well-formed amalgam produces a mirror on the wall of the reaction flask). After the amalgam cools to room temperature, a solution containing 5.0 g of $[(\eta\text{-}C_5H_5)Fe(CO)_2]_2$† dissolved in 75 ml of THF (tetrahydrofuran) is added to the reaction flask. The flask is stoppered, and the deep-red solution is stirred vigorously for 18 hr at room temperature. The extent of the reaction is conveniently followed by noting the gradual formation of a yellow-green solution. The crude reaction mixture is suction-filtered through a medium-porosity glass frit. The suction-filtration apparatus consists of a 60-ml, sintered-glass filter funnel affixed to a heavy-walled filter flask with a rubber adapter. The yellow-green filtrate is concentrated to approximately 25 ml under reduced pressure (10^{-2} torr). The concentrated THF solution is flooded with *n*-pentane, which causes the immediate precipitation of a yellow-green solid. The yellow-green solid is collected by suction-filtering on a glass frit, is recrystallized twice from benzene, and is vacuum-dried (10^{-4} torr). The yield is typically 6.6–7.0 g (90–95 %). *Anal.* Calcd. for $C_{22}H_{26}Fe_2$-MgO_6: C, 50.6; H, 5.02; Fe, 21.4; Mg, 4.65; mol. wt., 522. Found: C, 49.7; H, 4.83; Fe, 22.5; Mg, 4.64; mol. wt. (cryoscopic in benzene), 528. An nmr spectrum of the complex in benzene solution consists of a sharp singlet at $\tau 5.30$ due to the $\eta\text{-}C_5H_5$ protons and two multiplets centered at $\tau 6.09$ and 8.56

*Research Organic/Inorganic Chemical Corporation, Sun Valley, Calif. 91352.
†Strem Chemicals, Inc., Danvers, Mass. 01923.

attributed to the complexed THF molecules. The intensity ratio of the THF signals to the C_5H_5 signal was calculated to be 1.60 and found to be 1.66. The infrared spectrum of the complex in THF solution is characterized by four strong bands in the carbonyl region at 1918, 1884, 1854, and 1713 cm^{-1}, whereas benzene solutions exhibit two strong carbonyl bands at 1921 and 1854 cm^{-1}. Such radical changes in the infrared spectra with solvent polarity are typically observed with transition metal carbonyl derivatives of magnesium and are indicative of complex solvation processes.

By a similar procedure, yellow-green solid $(C_4H_8O)_4Mg[Co(CO)_3P(C_4H_9)_3]_2$ is readily obtained by the reduction of $Co_2(CO)_6[P(C_4H_9)_3]_2$[2] with a 1% magnesium amalgam in THF solution. *Anal.* Calcd. for $C_{46}H_{86}Co_2MgO_{10}P_2$: C, 55.1; H, 8.64; Co, 11.8; Mg, 2.42; P, 6.17. Found: C, 54.6; H, 8.74; Co, 11.4; Mg, 2.45; P, 5.99. The ^1H nmr spectrum of the complex in benzene exhibited a multiplet centered at $\tau 8.53$ due to $P(C_4H_9)_3$ and two multiplets at $\tau 6.03$ and 8.45 due to the complexed tetrahydrofuran molecules. The ratio of the $P(C_4H_9)_3$ to THF signal intensities was calculated to be 1.69 and found to be 1.72. Benzene solutions of the complex exhibit carbonyl bands at 1952 (w), 1925 (sh), 1896 (vs), 1874 (vs), and 1863 (sh) cm^{-1}. For THF solutions carbonyl bands are found at 2040 (w), 1956 (w), 1936 (vs), 1866 (vs), 1833 (m), and 1706 (vs) cm^{-1}.

B. TETRAKIS(PYRIDINE)BIS(TETRACARBONYLCOBALT)-MAGNESIUM

$$Co_2(CO)_8 + Mg(Hg) + 4C_5H_5N \longrightarrow (C_5H_5N)_4Mg[Co(CO)_4]_2$$

Procedure

A 1% magnesium amalgam is prepared in a 250-ml one-necked flask by dissolving 0.5 g of 200-mesh magnesium powder in 50 g of mercury. To the cool amalgam is added 3.4 g of $Co_2(CO)_8$* dissolved in 100 ml of benzene. To this dark-red mixture is added a solution of 5.0 g of pyridine in 40 ml of benzene. The flask is stoppered, and the reaction is allowed to proceed at room temperature for 20 hr, after which the reaction mixture is dark yellow. The crude reaction mixture is suction-filtered through a medium-porosity glass frit. The filtrate is concentrated to approximately 30 ml under reduced pressure (10^{-2} torr). A yellow solid is precipitated by flooding the concentrated benzene solution with *n*-pentane. The solid is collected on a medium-porosity glass frit by

*Pressure Chemical Company, Pittsburgh, Pa. 15201.

suction filtering and is purified by recrystallizing from benzene and vacuum drying (10^{-4} torr). The yield of yellow solid is typically 4.8–5.8 g (70–85%). *Anal.* Calcd. for $C_{28}H_{20}Co_2MgN_4O_8$: C, 49.3; H, 2.95; Co, 17.3; Mg, 3.56; N, 8.21; mol. wt., 683. Found: C, 49.4; H, 3.26; Co, 17.1; Mg, 2.91; N, 7.79; mol. wt. (cryoscopic in benzene), 665. The infrared spectrum of the complex in benzene is characterized by two strong metal carbonyl bands at 1939 and 1764 cm^{-1}. A sharp, medium-intensity carbonyl band is also found at 2016 cm^{-1}.

A yellow, solid chelated magnesium-cobalt complex, $(TMEDA)_2Mg[Co(CO)_3(CH_3P(C_6H_5)_2)]_2$, where TMEDA represents *N,N,N',N,'*-tetramethylethylenediamine, is prepared by reducing $Co_2(CO)_6[P(CH_3)(C_6H_5)_2]_2$[2] with a 1% magnesium amalgam in benzene solution in the presence of a slight excess of TMEDA. *Anal.* Calcd. for $C_{44}H_{58}Co_2MgN_4O_6P_2$: C, 56.0; H, 6.20; Co, 12.5; Mg, 2.58. Found: C, 53.5; H, 6.14; Co, 12.9; Mg, 2.66. The ^1H nmr spectrum of the complex in *d*-acetonitrile exhibited multiplets centered at $\tau 2.81$ and 3.00 due to $P(C_6H_5)_2$ and a doublet centered at $\tau 8.45$ due to $P(CH_3)$. The complexed TMEDA molecule gave rise to signals at $\tau 7.81$ (methylene protons) and $\tau 8.06$ (N—CH_3 protons). The ratios of TMEDA/$P(C_6H_5)_2$ and TMEDA/$P(CH_3)$ signal intensities were calculated to be 1.60 and 5.33 and found to be 1.68 and 5.27, respectively.

C. BIS[DICARBONYL-η-CYCLOPENTADIENYL(TRIBUTYLPHOSPHINE)MOLYBDENUM]TETRAKIS(TETRAHYDROFURAN)MAGNESIUM

$$[(\eta\text{-}C_5H_5)Mo(CO)_3]_2 + I_2 \longrightarrow 2(\eta\text{-}C_5H_5)Mo(CO)_3I$$

$$(\eta\text{-}C_5H_5)Mo(CO)_3I + P(C_4H_9)_3 \longrightarrow (\eta\text{-}C_5H_5)Mo(CO)_2[P(C_4H_9)_3]I + CO$$

$$(\eta\text{-}C_5H_5)Mo(CO)_2[P(C_4H_9)_3]I + Mg(Hg) + 4C_4H_8O \longrightarrow (C_4H_8O)_4Mg[Mo(CO)_2\{P(C_4H_9)_3\}(C_5H_5)]_2$$

Procedure

$(\eta\text{-}C_5H_5)Mo(CO)_3I$ is prepared by cleaving the dimeric $[(\eta\text{-}C_5H_5)Mo(CO)_3]_2$ complex* with iodine in THF solution at room temperature.[3] The red phosphine-substituted complex, $(\eta\text{-}C_5H_5)Mo(CO)_2[P(C_4H_9)_3]I$, is obtained by allowing a slight excess of $P(C_4H_9)_3$ (freshly distilled) to react with $(\eta\text{-}C_5H_5)Mo(CO)_3I$ in refluxing benzene for 48 hr.[4] A dark-red solution containing 11.0 g of

*Pressure Chemical Company, Pittsburgh, Pa. 15201.

(η-C_5H_5)$Mo(CO)_2$[$P(C_4H_9)_3$]I dissolved in 200 ml of THF is added to a 500-ml one-necked flask containing 80 g of a 1% magnesium amalgam. The flask is stoppered, and the reaction mixture is vigorously stirred at room temperature. In about 30 min a light yellow-green slurry is formed, containing an off-white, suspended solid. The reaction is allowed to proceed for an additional 16 hr to ensure complete reaction. The crude reaction mixture is concentrated to approximately 40 ml under reduced pressure (10^{-2} torr) and suction-filtered through a medium-porosity glass frit to remove the suspended $MgI_2 \cdot x$THF produced during the reaction. The clear yellow-green filtrate is reduced to dryness under reduced pressure (10^{-2} torr), yielding a yellow-green residue. The residue is dissolved in a minimum quantity of benzene (40–50 ml), and the benzene extract is suction-filtered through a fine-porosity glass frit, removing the last trace of $MgI_2 \cdot x$THF. The filtrate is concentrated by slowly removing benzene under reduced pressure. As the solution becomes saturated, a yellow-green complex separates as well-defined crystals. The precipitated complex is collected on a glass frit by suction-filtering and recrystallized from benzene and vacuum-dried. The yield is typically 9.0–10.5 g (80–90%). *Anal.* Calcd. for $C_{54}H_{96}MgMo_2O_8P_2$: C, 56.3; H, 8.40; Mg, 2.11; Mo, 16.7; P, 5.38. Found: C, 55.8; H, 8.37; Mg, 2.37; Mo, 18.0; P, 5.51. The ^1H nmr spectrum of the complex in benzene exhibits a pair of multiplets centered at τ5.91 and 8.22 due to the complexed tetrahydrofuran molecules. A sharp singlet at τ4.64 is readily assigned to the C_5H_5 group. The $P(C_4H_9)_3$ ligand produces a multiplet centered near τ8.6. The $P(C_4H_9)_3$/tetrahydrofuran and $P(C_4H_9)_3/C_5H_5$ intensity ratios were calculated to be 1.69 and 5.4 and found to be 1.82 and 5.5, respectively. The infrared spectrum of the complex in THF solution is characterized by two strong carbonyl bands at 1810 and 1607 cm^{-1}.

Properties

Solid transition metal carbonyl complexes of magnesium can be isolated but must be handled in a good dry-box. It is best to store the compounds in sealed ampuls. The complexes are thermally stable, decomposing without melting at temperatures greater than 150°. The complexed Lewis bases are strongly held, as the compounds generally remain unchanged after prolonged periods of time under a vacuum of 10^{-4} torr at room temperature. The compounds are monomeric in hydrocarbon solvents. Conductivity measurements in THF indicate that the complexes are not dissociated to an appreciable extent in this solvent.[5] The magnesium–transition metal complexes are readily attacked by oxygen, producing magnesium oxide and the corresponding dimeric transition metal–carbonyl complex. The complexes react cleanly with X—Y-type molecules, e.g., alkyl halides and phenol, to produce alkyl– and hydrido–transition metal complexes.

References

1. R. E. Dessy, R. L. Pohl, and R. B. King, *J. Am. Chem. Soc.*, **88,** 5121 (1966).
2. L. H. Slaugh and R. D. Mullineaux, *J. Organomet. Chem.*, **13,** 469 (1968).
3. T. S. Piper and G. Wilkinson, *J. Inorg. Nucl. Chem.*, **3,** 104 (1956).
4. A. R. Manning, *J. Chem. Soc. (A)*, **1967,** 1984.
5. G. B. McVicker and R. S. Matyas, *Chem. Comm.*, **1972,** 972.

19. TRIS(PENTACARBONYLMANGANESE)-THALLIUM(III)

$$Mn_2(CO)_{10} \xrightarrow[THF]{Na/Hg} 2Na[Mn(CO)_5]$$

$$3Na[Mn(CO)_5] + 3TlNO_3 \xrightarrow{H_2O} Tl[Mn(CO)_5]_3 + 2Tl + 3NaNO_3$$

Submitted by A. T. T. HSIEH* and M. J. MAYS†
Checked by J. S. KRISTOFF‡ and D. F. SHRIVER‡

Tris(pentacarbonylmanganese)thallium(III) has been prepared in yields up to 45% by the reaction of sodium pentacarbonylmanganate with anhydrous thallium(III) chloride in tetrahydrofuran.[1] It has also been obtained from the reaction of trimethylthallium with pentacarbonylhydridomanganese.[2] The interaction of sodium pentacarbonylmanganate with thallium(I) salts in tetrahydrofuran[3] or water[4] provides a convenient synthesis of the complex in excellent yield. The procedure for the preparation of the complex in water is given below and is carried out anaerobically.[5] It may be readily scaled up or down with a corresponding decrease or increase in yield; yields up to 95% have been achieved with small-scale preparations.

Procedure

■ **Caution.** *Care should be exercised in handling thallium salts (harmful by*

*Department of Chemistry, Monash University, Clayton, Victoria 3168, Australia.
†University Chemical Laboratory, Lensfield Road, Cambridge CB2 1EW, England.
‡Department of Chemistry, Northwestern University, Evanston, Ill. 60201.

skin absorption) and decacarbonyldimanganese, which are highly toxic; solutions of the complex are potential skin irritants.

A 100-ml three-necked flask equipped with a stopcock at the bottom or a similarly modified round or pear-shaped separating funnel is used. After the reaction vessel has been flushed with nitrogen it is charged with 3.9 g (0.01 mole) of decacarbonyldimanganese,* 10–15 ml (excess) of sodium amalgam (1 %), and 40 ml of freshly distilled dry tetrahydrofuran. The mixture is stirred or shaken for about 1 hr, after which the amalgam is removed through the tap. The pale yellow-green solution is filtered through a medium-porosity sintered-glass filter into another flask equipped with a stopcock and evaporated to dryness *in vacuo*. Nitrogen is then carefully admitted, the white, solid sodium pentacarbonylmanganate is dissolved in 20 ml of degassed water, and the solution is treated at once with a solution of thallium(I) nitrate† (5.33 g, 0.02 mole) in 50 ml of degassed water. The voluminous black precipitate is collected by filtration‡ or, better, by centrifugation, washed well with 25 ml of degassed water, and briefly dried *in vacuo* over phosphorus pentoxide for 1–2 hr§ at room temperature. The black solid¶ is repeatedly extracted with small portions of dry degassed acetone (ca. 250 ml total) until the extract is no longer colored. The deep-red solution is filtered or centrifuged and reduced to a small volume (ca. 20–30 ml) *in vacuo* or under a stream of nitrogen. The crystals (ca. 5.0 g) are collected by filtration (fine-porosity sintered-glass filter), washed with 25 ml of petroleum ether (b.p. 30–40°) or pentane, dried *in vacuo* at room temperature, and stored under nitrogen. *Anal.* Calcd. for $C_{15}Mn_3O_{15}Tl$: C, 22.82; Tl, 25.9. Found: C, 22.7; Tl, 25.5. A further, less pure crop of the complex is obtained by evaporation of the mother liquor. The typical yield is 6.7 g (85 % based on decacarbonyldimanganese).

*Decacarbonyldimanganese is commercially available and can be prepared by literature methods.[6,7]

†Other water-soluble thallium(I) salts such as the formate, acetate, or sulfate can be used instead. An excess of the thallium(I) salt does not affect the yield of the product significantly.

‡The checkers recommend filtration of the black precipitate using a coarse frit (40–60 μm) on which a thin layer of filter aid has been placed.

§Prolonged drying (>3 hr) often results in the formation of pyrophoric materials.

¶If the reaction mixture has been briefly exposed to air during the earlier stages of this preparation, some yellow to brown solid will appear in this solid. Washing with petroleum ether (b.p. 30–40°) or pentane prior to acetone extraction is recommended but not absolutely necessary. The finely divided dark-gray metallic thallium left after extraction is in its active form, being pyrophoric when dried, and should therefore be stored under heptane or other suitable inert liquid.

Properties

The complex crystallizes as dark-red to almost black needlelike prisms, which are indefinitely stable at room temperature but decompose without melting at 180° to decacarbonyldimanganese and thallium metal. The crystalline form of the complex can be handled briefly in air without appreciable decomposition although it is extremely air-sensitive in solution. Its infrared spectrum in freshly prepared dichloromethane solution shows three intense ν(CO) absorptions at 2062 (vs), 2002 (vs), and 1993 (s, sh) cm^{-1}. The metal-metal bonds are readily cleaved by halogens, hydrogen halides, and organic halides.[1] It serves as a useful precursor to numerous pentacarbonylmanganese derivatives of main-group metals.[1,8]

References

1. A. T. T. Hsieh and M. J. Mays, *J. Organomet. Chem.*, **22**, 29 (1970).
2. A. G. Lee, personal communication (1969); see also ref. 1.
3. H.-J. Haupt and F. Neumann, *J. Organomet. Chem.*, **33**, C56 (1971).
4. A. T. T. Hsieh and M. J. Mays, *J. Organomet. Chem.*, **38**, 243 (1972).
5. S. Herzog, J. Dehnert, and K. Lühder, in "Technique of Inorganic Chemistry," H. B. Jonassen and A. Weissberger (eds.), Vol. 7, p. 119, Interscience Publishers, a division of John Wiley & Sons, Inc., New York, 1968; D. F. Shriver, "The Manipulation of Air-Sensitive Compounds," Chap. 7, McGraw-Hill Book Company, New York, 1969.
6. R. B. King, in "Organometallic Synthesis," J. J. Eisch and R. B. King (eds.), Vol. 1, p. 89, Academic Press, New York, 1965.
7. R. B. King, J. C. Stokes, and T. F. Korenowski, *J. Organomet. Chem.*, **11**, 641 (1968).
8. A. T. T. Hsieh, *Inorg. Chim. Acta*, **14**:87 (1975).

20. (ALKYLAMINO)DIFLUOROPHOSPHINE- AND HALODIFLUOROPHOSPHINETETRACARBONYLIRON COMPLEXES

Submitted by D. P. BAUER,* W. M. DOUGLAS,* and J. K. RUFF*
Checked by P. J. RUSSO,† H. SCHÄFER,† and A. G. MacDIARMID†

The study of fluorophosphinetetracarbonyliron compounds has long been hampered by the lack of a simple, high-yield preparation of appropriate starting materials. For example, trifluorophosphinetetracarbonyliron, $Fe(CO)_4PF_3$,

*Chemistry Department, University of Georgia, Athens, Ga. 30602.
†Chemistry Department, University of Pennsylvania, Philadelphia, Pa. 19174.

requires the use of gaseous PF_3 and iron pentacarbonyl at elevated temperatures, followed by vapor-phase chromatography of a complex reaction mixture in order to separate the desired product.[2,3] The procedure outlined below makes it possible to produce pure $Fe(CO)_4PF_3$ in large quantities and to provide phosphorus-substituted fluorophosphine complexes of general utility for the study of reactions at either phosphorus or iron. (Diethylamino)-difluorophosphinetetracarbonyliron is readily prepared by the cleavage of enneacarbonyldiiron, $Fe_2(CO)_9$. The phosphorus—nitrogen bond in this compound can be cleaved by anhydrous hydrogen chloride to produce $Fe(CO)_4PF_2Cl$, which with sodium fluoride in tetramethylene sulfone produces $Fe(CO)_4PF_3$.

Analogous reactions occur with bis(diethylamino)fluorophosphinetetracarbonyliron.[1] All the complexes described in these syntheses, although air- and moisture-sensitive, can be handled briefly in air, but it is recommended that their preparation and purification be carried out under an atmosphere of dry nitrogen.

■ **Caution.** *All the complexes described are potentially very toxic, and it is strongly recommended that all operations be carried out in a well-ventilated hood.*

A. (DIETHYLAMINO)DIFLUOROPHOSPHINE-TETRACARBONYLIRON

$$Fe_2(CO)_9 + F_2PN(C_2H_5)_2 \longrightarrow Fe(CO)_4F_2PN(C_2H_5)_2 + Fe(CO)_5$$

*Procedure**

Enneacarbonyldiiron[4] (36.4 g, 0.10 mole) is placed in a 500-ml, round-bottomed flask containing a magnetic Teflon-coated stirring bar and equipped with a nitrogen inlet tube. Then 50.0 ml of *n*-hexane is added to the solid, and 15.5 g (0.11 mole) of (diethylamino)difluorophosphine[5,6] is added to the flask. The mixture is heated to 40–50° for 10 to 12 hr with stirring. After cooling, the reaction mixture is filtered through a medium-porosity glass-frit-filter funnel to which some filter aid† had been added. An inverted filter funnel through which a stream of nitrogen passes is placed over the filter during this operation. Most of the hexane and pentacarbonyliron is removed *in vacuo* (at approximately 40°) using a rotary stripper connected to a water aspirator. The flask is filled with dry nitrogen and attached to the apparatus shown in Fig. 7. After

*The standard Pyrex vacuum line used in these syntheses was equipped with glass stopcocks and mercury manometers covered with Kel-F oil.

†Celite 545 is manufactured by Johns-Manville; packaged by Fisher Scientific Co.

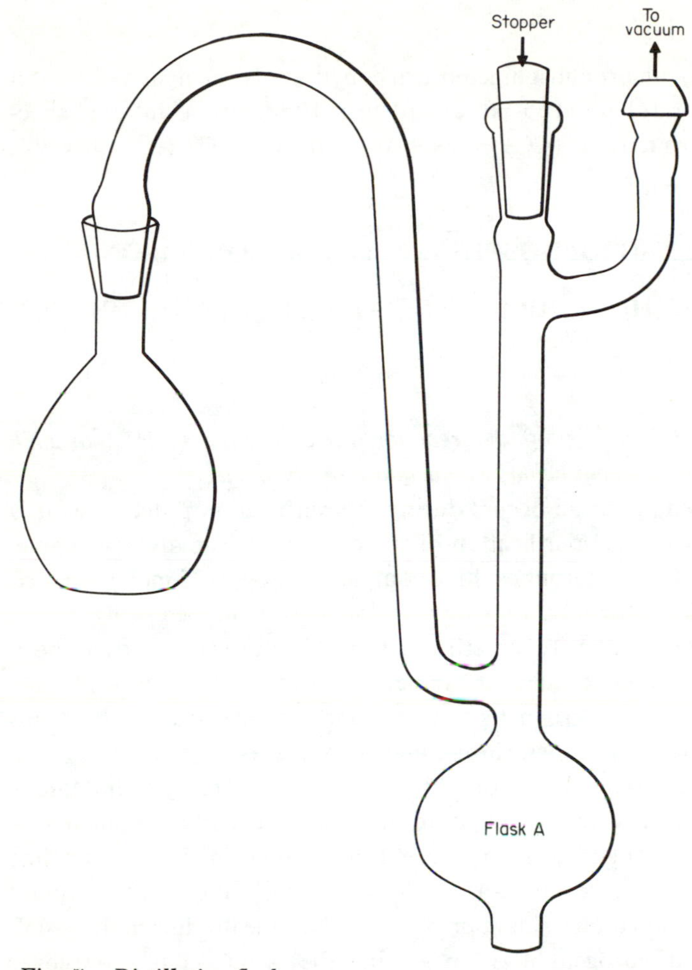

Fig. 7. Distillation flask.

the flask is evacuated, it is warmed to 50–60° with constant pumping (oil pump only). The liquid product collects at the bottom of flask *A*. After the apparatus is filled with dry nitrogen, the stopper is taken out and the product is removed with a long-needle hypodermic syringe. This distillation can be repeated if necessary to obtain a pure product. In a typical experiment, a 28.4-g (92%)* sample of the product was obtained. The compound may be conveniently identified by its ^{19}F nmr spectrum;[1] $\delta^{19}F = +28.0$ (CCl_3F external standard); $J_{PF} = 1119.6$ Hz. *Anal.* Calcd. for $Fe(CO)_4F_2PN(C_2H_5)_2$: C, 31.08; H, 3.24; N, 5.53; Fe, 18.12. Found: C, 30.8; H, 3.36; N, 4.67; Fe, 17.9.

*The checkers obtained a yield of 71%.

Properties

(Diethylamino)difluorophosphinetetracarbonyliron is a light-yellow oil with a musky odor. It boils at 55–60° at 0.01 torr. The solution* infrared spectrum shows absorptions at $\nu(CO)$ = 2069 (m), 2010 (s), 1978 (vs), 1964 (vs), and $\nu(PF)$ = 820 (m) cm^{-1}.

B. CHLORODIFLUOROPHOSPHINETETRACARBONYLIRON

$$Fe(CO)_4F_2PN(C_2H_5)_2 + 2HCl \longrightarrow Fe(CO)_4F_2PCl + [(C_2H_5)_2NH_2]Cl$$

Procedure

■ **Caution.** *All handling of the reaction vessel containing* HCl *at high pressure must be carried out behind an adequate safety shield.*

Since some decomposition occurs during the purification procedures, it is desirable to carry out the purification of the complex as rapidly as possible.

The (diethylamino)difluorophosphine complex (15.4 g, 0.05 mole) is added to a 100-ml, thick-walled glass pressure reaction vessel equipped with a Teflon Fischer-Porter valve and a Teflon stirring bar. After the complex has been degassed on a vacuum line, 7.3 g (0.2 mole) of anhydrous hydrogen chloride, measured in portions as a gas in a known volume, is condensed at −196° into the vessel. The stopcock is then closed, and the vessel is allowed to warm to room temperature. After stirring for 12 hr, the two-phase reaction mixture is cooled to −196°, at which temperature any liberated carbon monoxide is pumped off and the volatile materials are fractionated on the vacuum line through traps maintained at −45° (chlorobenzene) and −196° (liquid nitrogen). The fractionation yields approximately half the product in the −45° trap, and unreacted hydrogen chloride passes into the trap at −196°. To recover all the product, it is necessary to open the reactor in a nitrogen atmosphere and extract the amine hydrochloride formed with two 50.0-ml portions of pentane. After the solution has been filtered, the solvent is removed *in vacuo* until the volume is approximately 5 ml. This is then fractionated as above on the vacuum line. A long-needle hypodermic syringe is used to remove the product from the detachable trap on the vacuum line in which it has condensed.

The yield is 99%† based on the (diethylamino)difluorophosphine complex used. *Anal.* Calcd. for $Fe(CO)_4F_2PCl$: P, 11.38; Fe, 20.55; Cl, 13.03. Found: P, 11.1; Fe, 20.0; Cl, 13.1. Substitution of hydrogen chloride by hydrogen bro-

*All infrared spectra were obtained on cyclohexane solutions and were calibrated with polystyrene.

†The checkers obtained a yield of 76%.

mide in the above procedure produces bromodifluorophosphinetetracarbonyliron in good yield.

Properties

$Fe(CO)_4F_2PCl$ is an orange liquid with a musky odor. The solution infrared spectrum of $Fe(CO)_4PF_2Cl$ shows absorption bands at 2108 (m), 2098 (w), 2041 (s), 2034 (s), 2014 (s), 862 (sh), and 842 (s) cm^{-1}.

C. TRIFLUOROPHOSPHINETETRACARBONYLIRON

$$Fe(CO)_4PF_2Cl + NaF \longrightarrow Fe(CO)_4PF_3 + NaCl$$

Procedure

Approximately 150 ml of Sulfolane* is added to a 300-ml flask equipped with a Teflon stirring bar, reflux condenser, and nitrogen inlet tube. Then 5.27 g (0.019 mole) of $Fe(CO)_4PF_2Cl$ and 13.0 g (0.31 mole) of NaF are added.† The reaction mixture is heated for 3 hr at 50–55° with stirring, after which an oil pump is used to pump the product out of the reaction flask, while it is still warm, into a trap at −196°. This fraction is purified by fractionation through traps at −45° and −196°. A pure 3.4 g (70%) sample is retained in the trap at −45°. The ^{19}F nmr and infrared spectra correspond to those published in the literature.[2,7] The compound may be conveniently identified by means of its ^{19}F nmr spectrum:[1] $J_{PF} = 1329$ Hz.

Properties

The compound $Fe(CO)_4PF_3$ is a yellow air-sensitive liquid whose infrared spectrum [2101 (w), 2094 (w), 2021 (vs), 2004 (vs), 1996 (vs), 1970 (vw), 1960 (vw) cm^{-1}] indicates that it exists as a mixture of the axial and equatorial substituted isomers. All attempts to distinguish these isomers by nmr have failed, so that a rapid interconversion must be occurring.[2]

References

1. W. M. Douglas and J. K. Ruff, *J. Chem. Soc.* (A), **1971,** 3558.
2. R. J. Clark, *Inorg. Chem.,* **3,** 1395 (1964).

*Sulfolane (tetramethylenesulfone) is available from Special Products Division, Phillips Petroleum Co., Bartlesville, Okla. 74003. It is degassed on a vacuum line before use.

†Sodium fluoride is dried overnight at 130° before using.

3. T. Kruck, *Angew. Chem., Int. Ed.*, **6**, 53 (1967).
4. R. B. King, in "Organometallic Syntheses," J. J. Eisch and R. B. King (eds.), Vol. I, p. 93, Academic Press, New York, 1965.
5. R. Schmutzler, *Inorg. Chem.*, **3**, 415 (1964).
6. K. Issleib and W. Seidel, *Chem. Ber.*, **92**, 2681 (1959).
7. F. Ogilvie, R. J. Clark, and J. G. Verkade, *Inorg. Chem.*, **8**, 1904 (1969).

21. TRIS[BIS(2-METHOXYETHYL)ETHER]POTASSIUM AND TETRAPHENYLARSONIUM HEXACARBONYLMETALLATES(1−) OF NIOBIUM AND TANTALUM

$$MCl_5 + 6CO + 3(CH_3OCH_2CH_2)_2O + 6K \longrightarrow [K((CH_3OCH_2CH_2)_2O)_3][M(CO)_6] + 5KCl$$

Submitted by J. E. ELLIS* and A. DAVISON†
Checked by G. W. PARSHALL‡ and E. R. WONCHOBA‡

The syntheses of the hexacarbonylmetallate (1−) anions of niobium and tantalum were first reported in 1961 by Werner and Podall.[1] Their procedure involved the reductive carbonylation of $NbCl_5$ and $TaCl_5$ by sodium metal in bis-(2-methoxyethyl)ether(diglyme) under high pressures of carbon monoxide. The yields of $[Na(diglyme)_2][M(CO)_6]$ (M = Nb, Ta) obtained by this route were "low" and 32%, respectively. Later they reported that the rate of reductive carbonylation was greatly enhanced by the presence of small amounts of halides or carbonyls of Fe, Ru, or Os.[2] With this modified procedure they obtained $[Na(diglyme)_2][Nb(CO)_6]$ in 26% yield from a mixture of $NbCl_5$, Na metal, diglyme, $Fe(CO)_5$, and 4000 psi of carbon monoxide at 105° after 20 hr in a stirring autoclave.[2]

Although their method for preparing $[Na(diglyme)_2]$ $[V(CO_6)]$ is reliable,[3] numerous attempts to make the corresponding niobium and tantalum salts by this route have been unsuccessful. However, it has been found that 10–20% yields of these substances can be obtained consistently from the high-pressure

*Chemistry Department, University of Minnesota, Minneapolis, Minn. 55455.
†Chemistry Department, Massachusetts Institute of Technology, Cambridge, Mass. 02139.
‡Central Research Department, E. I. du Pont de Nemours & Co., Wilmington, Del. 19898.

reductive carbonylation of the freshly sublimed pentachlorides of these metals by sodium-potassium alloy in diglyme at room temperature.

The chemistry of these anions remains largely unexplored; however, recently reports on their interactions with main-group metal halides and Lewis bases, under photolytic conditions, have appeared in the literature.[4-6] The anions are easily oxidized under a variety of conditions. In acidic aqueous solution they form soluble, nonvolatile metal carbonyl–containing products which have not been characterized. In contrast, the less easily oxidized $[V(CO)_6]^-$ generates H_2 and water-insoluble $V(CO)_6$ under identical conditions.[7] Finally, the unsubstituted anions are presently the only precursors to substituted carbonyl anions of niobium and tantalum.[5] Although the reactions of these species with main-group metal halides closely parallel corresponding reactions with vanadium carbonyl anions, preliminary investigations indicate that their behavior toward oxidation or protonation is markedly different.[8]

A. TRIS[BIS(2-METHOXYETHYL)ETHER]POTASSIUM HEXACARBONYLNIOBATE (1—)

Procedure

Freshly distilled oxygen-free diglyme (600 ml, refluxed over CaH_2 for ca. 12 hr and distilled at atmospheric pressure, b.p. 162°) is introduced by syringe into a 1.5-l. stainless-steel autoclave (maximum rating: 11,000 psi at 38°, American Instrument Co.) previously purged with nitrogen. The autoclave is then charged with four 2.5-cm-diam. steel balls to help agitate the solution during the reaction and three 10-ml sealed pharmaceutical ampuls, secured to 2.5-cm-long steel bolts with copper wire, containing 47 g (0.174 mole) of sublimed $NbCl_5$.[9] Subsequently, 90 ml (ca. 1.8 mole K) of $NaK_{2.8}$ alloy* and then 2 ml of pentacarbonyliron are added by syringe. A protective blanket of nitrogen is maintained over the mouth of the autoclave during these operations.

■ **Caution.** *Sodium-potassium alloy readily inflames in a humid atmosphere and burns immediately on contact with water. Pentacarbonyliron is extremely toxic and should be handled in a well-ventilated hood.*

The sealed autoclave is carefully placed in the shaking assembly and initially pressurized to 5300 psi of CP grade carbon monoxide. (The autoclave is not

*The checker used a commercial 56% Na/K alloy obtained from Mine Safety Appliance, Evans City, Pa. 16033. The alloy is supplied in 60-ml ampuls which were added directly to the autoclave. This avoids the hazardous syringe transfer of the alloy.

initially flushed with carbon monoxide. The ampuls can usually be heard to break on pressurization of the autoclave.) Rocking action is then immediately initiated at room temperature. Shortly thereafter (2.5 hr) the autoclave is further pressurized to 6300 psi of carbon monoxide. After it has rocked for about 100 hr at room temperature, the autoclave is *cautiously* vented.

The autoclave contents are then filtered through a 2.5-cm layer of filter aid (prepared from a filter-aid–diglyme slurry), supported on a coarse-porosity sintered-glass disk, to give a deep-orange filtrate. (In the corresponding tantalum preparations the color of the filtrate is generally lighter.) Care should be taken to avoid sucking alloy into the filtrate in this latter step. The remaining solid in the autoclave is washed with additional diglyme (2 × 200 ml); the washings are filtered as above. Petroleum ether (ca. 150 ml) is then passed through the solids on the filter-aid layer until the eluent is colorless. Unreacted alloy present in the filtration apparatus is destroyed by cautious addition of a mixture of isopropyl alcohol and heptane (1:1, v/v). Heptane is added to moderate the reaction if necessary.

■ **Caution.** *Because of the hazardous nature of sodium-potassium alloy, this operation should be done on a dry tray so that if the mixture boils over, it will not roll into a sink or onto a damp bench.*

Direct introduction of 1200 ml of additional oxygen-free petroleum ether to the filtrate, followed by shaking, causes a voluminous precipitation of bright-yellow solid. After the solid settles, the light red-brown supernatant liquid is removed via a cannula. (The solid is often sufficiently fine to pass through and clog a medium-porosity sintered-glass disk; thus the solution is separated from the solid either by decantation or centifugation. The former procedure is more convenient with a preparation on this scale.) The resulting slurry of solid is washed with about 1 l. of petroleum ether in the above manner. After decantation of the supernatant, the slurry is washed thoroughly with 98% pentane (5 × 100 ml) and then dried *in vacuo*. A yield of 16.3 g (13.9%) of bright-yellow powder is obtained which gives acceptable analyses without further purification (m.p. 125–126°). *Anal.* Calcd. for $C_{24}H_{42}KNbO_{15}$: C, 41.03; H, 6.02; K, 5.56. Found: C, 40.9; H, 6.11; K, 5.69.

Properties

Attempts to recover $[K(diglyme)_3][Nb(CO)_6]$ from its solutions in solvents other than diglyme invariably lead to the formation of pyrophoric solids. Similarly, it has been reported that the only previously known compound containing this cation, $[K(diglyme)_3][Mo(CO)_5I]$, cannot be recrystallized from solvents other than diglyme, presumably because one or more of the coordinated diglyme molecules is lost.[11]

This moderately air-sensitive niobium salt is soluble in a wide variety of solvents, such as aromatic hydrocarbons, ethers, ketones, alcohols, and water, to give very air-sensitive solutions. The solid is also somewhat light-sensitive and should be stored in the dark. It is indefinitely stable at room temperature under nitrogen. The infrared spectrum of the compound in the carbonyl region is given in Table I.

TABLE I Infrared Solution and Mull Spectra of the Hexacarbonylmetallate(1−) Ions of Vanadium, Niobium, and Tantalum in the Carbonyl Region

Salt*	C—O stretching vibration,† cm^{-1}	
	Nujol mull	CH_2Cl_2
$[K(diglyme)_3][Nb(CO)_6]$	1850 (vs, b)	1855 (vs)
$[K(diglyme)_3][Ta(CO)_6]$	1850 (vs, b)	1855 (vs)
$[(C_6H_5)_4As][Nb(CO)_6]$	1854 (vs, b); 1831 (vs, b)	1855 (vs)
$[(C_6H_5)_4As][Ta(CO)_6]$	1852 (vs, b); 1828 (vs, b)	1855 (vs)
$[(C_6H_5)_4As][V(CO)_6]$	1855 (vs, b); 1830 (b)	1855 (vs)

*Diglyme = $(CH_3OCH_2CH_2)_2O$.
†vs = very strong; b = broad.

B. TRIS[BIS(2-METHOXYETHYL)ETHER]POTASSIUM HEXACARBONYLTANTALATE(1−)

Procedure

This salt is obtained in exactly the same manner as the corresponding niobium salt. In a representative preparation 64.0 g (0.179 mole) of sublimed $TaCl_5$, 90 ml of $NaK_{2.8}$ alloy, and 2 ml of pentacarbonyliron in 600 ml of freshly distilled diglyme under 6000 psi of carbon monoxide (initial pressure) yields, after 72 hr of rocking at room temperature, 20.2 g of $[K(diglyme)_3]$-$[Ta(CO)_6]$, a 14.0% yield based on the $TaCl_5$. The yellow powdery product gives an acceptable elemental analysis and a reasonably sharp melting point under nitrogen (125–127°; decomposes ca. 150°) without further purification. *Anal.* Calcd. for $C_{24}H_{42}KO_{15}Ta$: C, 36.48; H, 5.36; K, 4.95. Found: C, 36.8; H, 5.16; K, 5.15. The properties of this salt are very similar to those of the niobium compound. The infrared spectrum of the compound in the carbonyl region is given in Table I.

C. TETRAPHENYLARSONIUM HEXACARBONYLTANTALATE(1−)

$$[(C_6H_5)_4As]Cl + [K(diglyme)_3][Ta(CO)_6] \xrightarrow{H_2O} [(C_6H_5)_4As][Ta(CO)_6]\downarrow + KCl + 3\ diglyme$$

Procedure

A solution containing 1.97 g (2.5 mmoles) of $[K(diglyme)_3][Ta(CO)_6]$ in 100 ml of diglyme is filtered into a 1-l. flask containing 6 g (14.3 mmoles) of $[(C_6H_5)_4As]Cl$ in 15 ml of ethanol. Water (300 ml) is slowly added by syringe to the stirred solution, whereupon fine, glistening crystals of yellow product form. The crystals are collected on a filter, washed with water, and dried. A yield of 1.62 g (80%) of analytically pure solid is obtained (m.p. 193–194° with decomposition). *Anal.* Calcd. for $C_{30}H_{20}AsO_6Ta$: C, 49.23; H, 2.75. Found: C, 49.75; H, 2.90.

Properties

Crystalline $[(C_6H_5)_4As][Ta(CO)_6]$ is significantly more oxidatively stable than the corresponding tris(diglyme)potassium salt. The arsonium salt may be handled in air for at least $\frac{1}{2}$ hr at room temperature without deleterious effect. The salt dissolves in acetone, THF (tetrahydrofuran), methylene chloride, and organic solvents of high dielectric constant such as dimethylformamide to give extremely air-sensitive, orange-yellow solutions. It is insoluble in hydrocarbon solvents, diethyl ether, and water and slightly soluble in alcohols. The infrared spectrum of the compound in the carbonyl region is given in Table I.

D. TETRAPHENYLARSONIUM HEXACARBONYLNIOBATE (1−)

Procedure

Analytically pure $[(C_6H_4)_4As][Nb(CO)_6]$ is obtained by exactly the same procedure as that given for the corresponding tantalum salt. For example, from 0.65 g (0.93 mmole) of $[K(diglyme)_3][Nb(CO)_6]$ in 60 ml of diglyme and 2.0 g (4.8 mmoles) of $[(C_6H_5)_4As]Cl$ in 5 ml of ethanol, 0.45 g (76%) of deep-yellow crystalline product (darkens above 160°, no definite decomposition point) is obtained upon addition of 200 ml of water. *Anal.* Calcd. for $C_{30}H_{20}AsNbO_6$: C, 55.92; H, 3.13. Found: C, 55.7; H, 3.12.

The properties of this salt are very similar to those of the tantalum compound. The infrared spectrum of the compound in the carbonyl region is given in Table I.

References

1. R. P. M. Werner and H. E. Podall, *Chem. Ind. (Lond.),* **1961,** 144.
2. R. P. M. Werner, A. H. Filbey, and S. A. Manastyrskyj, *Inorg. Chem.,* **3,** 298 (1964).
3. R. B. King, in "Organometallic Syntheses," J. J. Eisch and R. B. King (eds.), Vol. 1, p. 82, Academic Press, New York, 1965.
4. A. Davison and J. E. Ellis, *J. Organomet. Chem.,* **23,** C1 (1970).
5. A. Davison and J. E. Ellis, *ibid.,* **31,** 239 (1971).
6. A. Davison and J. E. Ellis, *ibid.,* **36,** 113 (1972).
7. J. E. Ellis, Ph. D. thesis, Massachusetts Institute of Technology, Cambridge, Mass., 1970.
8. A. Davison and J. E. Ellis, *J. Organomet. Chem.,* **36,** 131 (1972).
9. G. Brauer, "Handbook of Preparative Inorganic Chemistry," Vol. 2, 2d ed., p. 1302, Academic Press, Inc., New York, 1965.
10. L. F. Fieser and M. Fieser, "Reagents for Organic Synthesis," p. 1102, John Wiley & Sons, Inc., New York, 1967.
11. E. W. Abel, M. A. Bennett, and G. Wilkinson, *Chem. Ind. (Lond.),* **1960,** 442.

Chapter Three

WERNER TYPE METAL COMPLEXES

22. (DINITROGEN OXIDE)PENTAAMMINERUTHENIUM(II) SALTS

$$[Ru(NH_3)_5(NO)]^{3+} + NH_2OH + OH^- \rightarrow [Ru(NH_3)_5(N_2O)]^{2+} + 2H_2O$$

Submitted by F. BOTTOMLEY*
Checked by J. N. ARMOR†

Dinitrogen oxide (nitrous oxide) is regarded as a fairly unreactive molecule. With the discovery of dinitrogen complexes of the transition metals[1,2] came the possibility of preparing analogous complexes of N_2O. Pratt and coworkers showed that N_2O could be reduced by a variety of transition metal complexes.[3] The reaction between $[Ru(NH_3)_5(H_2O)]^{2+}$ and N_2O was investigated by Diamantis[4,5] and by Taube[6–8] and their coworkers. From this reaction $[Ru(NH_3)_5(N_2O)][BF_4]_2$ was obtained,[5] and evidence for the formation of $[(Ru(NH_3)_5)_2(N_2O)]X_4$ (X = Br, BF_4) presented.[8] The direct reaction is not a good preparative route to $[Ru(NH_3)_5(N_2O)]^{2+}$ because of the necessity of manipulating reactive $[Ru(NH_3)_5(H_2O)]^{2+}$ under high pressures of N_2O[5] and because there is some uncertainty about the purity of the product.[8] During investigations of the behavior of $[Ru(NH_3)_5(NO)]^{3+}$ toward nucleophiles it was found that NH_2OH and $[Ru(NH_3)_5(NO)]X_3 \cdot H_2O$ react cleanly to give good yields of $[Ru(NH_3)_5(N_2O)]X_2$(X = Cl, Br, I).[9,10] The synthesis of $[Ru(NH_3)_5(N_2O)]^{2+}$

*Chemistry Department, University of New Brunswick, Fredericton E3B 5A3, New Brunswick, Canada.

†Chemistry Department, Boston University, Boston, Mass. 02215.

(which is still the only well-characterized complex containing N_2O) by this route is given in detail here.

Procedure

The cation $[Ru(NH_3)_5(NO)]^{3+}$ can be prepared by several methods.[11–13] Probably the most convenient[11] is to pass NO through an acid solution of $[Ru(NH_3)_6]Br_3$.* Precipitation of the product with excess HBr gives an essentially quantitative yield of $[Ru(NH_3)_5(NO)]Br_3 \cdot H_2O$. The I^- and Cl^- salts are prepared from this by metathesis using concentrated hydroiodic and hydrochloric acids, respectively.

Hydroxylammonium chloride, $[HONH_3]Cl$, is commercially available. The corresponding compounds $[HONH_3]Br$ and $[HONH_3]I$ are prepared by evaporating solutions of $[HONH_3]Cl$ in the appropriate HX acid until crystallization occurs. The product is recrystallized from the respective aqueous HX.

A mixture of $[HONH_3]Br$ (1.0 g) and $[Ru(NH_3)_5(NO)]Br_3 \cdot H_2O$ (0.094 g) is dissolved in the minimum quantity of ice-cold water (9 ml).† To the solution, kept ice-cold, are added four pellets (ca. 0.4 g) of NaOH. As the NaOH dissolves, pale-yellow $[Ru(NH_3)_5(N_2O)]Br_2$ precipitates. Occasional swirling of the solution assists dissolution of the NaOH. As soon as all the NaOH has dissolved, the precipitated $[Ru(NH_3)_5(N_2O)]Br_2$ is removed by filtration.‡ The solid is washed with ethanol and ether and dried *in vacuo* over P_2O_5. Yield: 0.057 g, 74%. *Anal.* Calcd. for $Br_2H_{15}N_7ORu$: H, 3.87; N, 25.13; Br, 40.97. Found: H, 3.9; N, 25.9; Br, 41.0.

The chloride and iodide salts are obtained similarly using $[Ru(NH_3)_5(NO)]Cl_3 \cdot H_2O$ (0.126 g) and $[HONH_3]Cl$ (1.0 g) dissolved in the minimum of water (4.0 ml) (yield 0.031 g, 28%) or $[Ru(NH_3)_5(NO)]I_3 \cdot H_2O$ (0.109 g) and $[HONH_3]I$ (0.26 g) in water (4.0 ml) (yield: 0.061 g, 71%). *Anal.* Calcd. for $Cl_2H_{15}N_7ORu$: H, 5.02; N, 32.55; Cl, 23.54. Found: H, 5.1; N, 31.9; Cl, 22.9. Calcd. for $H_{15}I_2N_7ORu$: H, 3.12; N, 20.26; I, 52.43. Found: H, 3.1; N, 20.8; I, 52.1.

*Salts of the complex $[Ru(NH_3)_6]^{3+}$ can be prepared[14,15] or purchased from Johnson Matthey, London.

†It is necessary that all the complex be dissolved, but the absolute minimum of solvent must be used. The checker found that about 30 ml of water is required to dissolve the mixture completely. The exact quantity of $[Ru(NH_3)_5(NO)]Br_3 \cdot H_2O$ is not critical, but the effect of scaling the reaction up by large factors, for example, 5 times, has not been investigated.

‡Since $[Ru(NH_3)_5(N_2O)]^{2+}$ rapidly decomposes in solution,[6] the mixture should not be allowed to stand before filtration.

Properties

The salts $[Ru(NH_3)_5(N_2O)]X_2$ (X = I, Br) are stable *in vacuo* for several days, but the very soluble chloride salt, which is difficult to obtain pure, often decomposes quite rapidly. Because of this instability, conventional microanalyses, unless performed immediately, give variable results. The decomposition product is probably $[Ru^{III}(NH_3)_5OH]^{2+}$, from the electronic spectra of the complexes.

The complexes evolve N_2O essentially quantitatively on oxidation by $Ce(SO_4)_2$. Because of their instability and concurrent oxidation of X^- to X_2, this reaction is best performed by adding $[Ru(NH_3)_5(N_2O)]I_2$ to a frozen, degassed solution of $Ce(SO_4)_2$ and allowing dissolution of $[Ru(NH_3)_5(N_2O)]I_2$ to take place on warming to room temperature. The I_2 liberated is easily trapped before analysis of the evolved gas.

The infrared spectra of the salts show, in addition to bands due to coordinated NH_3, a very intense band in the 1170-cm^{-1} region and a very weak band, the intensity of which varies with the anion, in the 2230-cm^{-1} region. These bands are assigned to ν_1 and ν_3 of coordinated N_2O, respectively.[10] A reasonable mechanism for the formation of the complexes and other chemical and physical properties are given in the literature.[10]

References

1. A. D. Allen and C. V. Senoff, *Chem. Commun.*, **1965**, 621.
2. A. D. Allen and F. Bottomley, *Acc. Chem. Res.*, **1**, 360 (1968).
3. R. G. S. Banks, R. J. Henderson, and J. M. Pratt, *J. Chem. Soc. (A)*, **1968**, 2886.
4. A. A. Diamantis and G. J. Sparrow, *Chem. Commun.*, **1969**, 469.
5. A. A. Diamantis and G. J. Sparrow, *ibid.*, **1970**, 819.
6. J. N. Armor and H. Taube, *J. Am. Chem. Soc.*, **91**, 6874 (1969).
7. J. N. Armor and H. Taube, *ibid.*, **92**, 2560 (1970).
8. J. N. Armor and H. Taube, *Chem. Commun.*, **1971**, 287.
9. F. Bottomley and J. R. Crawford, *Chem. Commun.*, **1971**, 200.
10. F. Bottomley and J. R. Crawford, *J. Am. Chem. Soc.*, **94**, 9092 (1972).
11. J. N. Armor, H. A. Scheidegger, and H. Taube, *J. Am. Chem. Soc.*, **90**, 5928 (1968).
12. K. Gleu and I. Budecker, *Z. Anorg. Allg. Chem.*, **269**, 202 (1952).
13. F. M. Lever and A. R. Powell, *J. Chem. Soc. (A)*, **1969**, 1477.
14. A. D. Allen, F. Bottomley, R. O. Harris, V. P. Reinsalu, and C. V. Senoff, *Inorganic Syntheses*, **12**, 7 (1970).
15. J. E. Fergusson and J. L. Love, *ibid.*, **13**, 211 (1972).

23. TETRAKIS(ISOTHIOCYANATO)BIS(2,2′-BIPYRIDINE)-NIOBIUM(IV)

$$NbCl_4 + 6KNCS \xrightarrow{CH_3CN} K_2[Nb(NCS)_6] + 4KCl$$

$$K_2[Nb(NCS)_6] + 2C_{10}H_8N_2 \xrightarrow{CH_3CN} Nb(NCS)_4(C_{10}H_8N_2)_2 + 2KNCS$$

Submitted by J. N. SMITH* and T. M. BROWN†
Checked by J. B. HAMILTON‡ and K. KIRKSEY‡

Numerous complexes of niobium(IV) and niobium(V) halides with various nitrogen-donor ligands have been reported in the literature. The products obtained from these reactions are critically dependent upon the reaction conditions.[1–5] Tetrakis(isothiocyanato)bis(2,2′-bipyridine)niobium(IV) can be prepared directly from the hexakis(isothiocyanato)niobate(IV) complex or by the reduction of the hexakis(isothiocyanato)niobate(V) ion according to the published method.[6] This method can also be extended to the preparation of other analogous complexes.

Since niobium pentachloride and tetrachloride decompose readily in the presence of oxygen or moisture, all solvents and vessels used in the reactions must be dried by accepted methods. The niobium chlorides should also be free of any oxy species. The potassium thiocyanate and 2,2′-bipyridine should be purified by recrystallization from water and ether, respectively, and then dried at a pressure of ca. 10^{-5} torr for 72 hrs.

Procedure

In an inert-atmosphere dry-box, 2.36 g (0.01 mole) of niobium tetrachloride is suspended in 25 ml of acetonitrile in a 100-ml round-bottomed flask equipped with a magnetic stirring bar. Niobium(IV) chloride is conveniently prepared by reduction of niobium(V) chloride with aluminum metal.[7–8] This mixture is stirred for ca. 1 hr and then 5.84 g (0.06 mole) of potassium thiocyanate dissolved in 25 ml of acetonitrile is added. Stirring is continued for 4 hr to allow for complete reaction. The red solution is filtered through a medium-porosity sintered-glass frit, separating the insoluble potassium chloride from the soluble $K_2[Nb(NCS)_6]$. The KCl precipitate is washed with three 5-ml portions of

*Motorola, Inc., Semiconductor Products Division, Phoenix, Ariz. 85005.
†Chemistry Department, Arizona State University, Tempe, Ariz. 85281.
‡Chemistry Department, Michigan State University, East Lansing, Mich. 48823.

acetonitrile, and the combined filtrates are transferred to a 100-ml round-bottomed flask. In a separate flask, 3.12 g (0.02 mole) of 2,2′-bipyridine is dissolved in ca. 15 ml of acetonitrile. This solution is added slowly to the $K_2[Nb(NCS)_6]$ solution. To ensure complete reaction, the solution is stirred for 4 hr. The brown insoluble $Nb(NCS)_4(C_{10}H_8N_2)_2$ is then removed by filtration using a medium-porosity sintered-glass frit. The product is washed with four 20-ml portions of acetonitrile and then transferred to a drying tube equipped with a ball-joint connector and stopcock. The complex is dried at a pressure of ca. 10^{-5} torr for 24 hr. The yield is ca. 95% based on niobium tetrachloride. *Anal.* Calcd. for $Nb(NCS)_4(C_{10}H_8N_2)_2$: C, 45.21; H, 2.51; N, 17.48; S, 19.94. Found: C, 45.1; H, 2.5; N, 17.7; S, 19.9.

Properties

Tetrakis(isothiocyanato)bis(2,2′-bipyridine)niobium(IV) is a medium-brown crystalline solid. It is only very slightly soluble in 1,2-dichloroethane, dichloromethane, and acetonitrile. These extremely dilute solutions of the complex readily decompose in the presence of oxygen or moisture; however, in the solid state the complex appears to be air-stable. It does not melt or decompose below ca. 275°. The infrared spectrum, ultraviolet spectrum, magnetic moment, and x-ray powder diffraction pattern of the product obtained by this procedure are essentially identical to those obtained for the product prepared by the reduction of the hexakis(isothiocyanate)niobate(V) complex.[6]

The infrared spectrum contains bands characteristic of isothiocyanato complexes and also the bands expected for coordinated 2,2′-bipyridine. The ultraviolet spectrum (in $C_2H_4Cl_2$) displays three charge-transfer bands at 245, 300, and 404 nm. The transitions near 300 and 250 nm can be ascribed to intraligand $\pi \rightarrow \pi^*$ transitions within the aromatic ligand, whereas the low-energy transition is the isothiocyanate-niobium charge transfer. The complex has a magnetic moment of 1.72 B.M. at 298K.

Discussion

The method described above has been found suitable for the synthesis of $Nb(NCSe)_4(C_{10}H_8N_2)_2$, $Nb(NCS)_4(dmbipy)_2$, and $Nb(NCSe)_4(dmbipy)_2$, where dmbipy is 4,4′-dimethyl-2,2′-bipyridine. These niobium (IV) compounds can also be prepared from the hexakis(isothiocyanate)niobate(V) complex.[6] The sensitivity of the materials to air and moisture makes the method somewhat more time-consuming, but the use of niobium tetrachloride as a starting material is avoided. The tantalum(V) complexes, $Ta(NCS)_5(C_{10}H_8N_2)$

and $Ta(NCSe)_5(C_{10}H_8N_2)$, can also be prepared by a modification of this method. These complexes are obtained by reacting a stoichiometric amount of either $K[Ta(NCS)_6]$ or $K[Ta(NCSe)_6]$ with 2,2′-bipyridine in 1,2-dichloroethane. In this case the product is extracted away from the insoluble KCNS with 1,2-dichloroethane. Reduction of tantalum(V) was not observed in these reactions.

References

1. C. Djordjevic and V. Katovic, *J. Chem. Soc. (A)*, **1970,** 3382.
2. R. E. McCarley, B. G. Hughes, J. C. Boatman, and B. A. Torp, *Adv. Chem. Ser.*, **37,** 243 (1963).
3. M. Allbutt, K. Feenan, and G. W. A. Fowles, *J. Less-Common Met.*, **6,** 299 (1964).
4. R. E. McCarley and J. C. Boatman, *Inorg. Chem.*, **2,** 547 (1963).
5. R. E. McCarley and B. A. Torp, *Inorg. Chem.*, **2,** 540 (1963).
6. J. N. Smith and T. M. Brown, *Inorg. Chem.*, **11,** 2697 (1972).
7. H. Schafer, C. Goser, and L. Bayer, *Z. Anorg. Allg. Chem.*, **265,** 258 (1951).
8. F. Fairbrother, "The Chemistry of Niobium and Tantalum," p. 122, American Elsevier Publishing Company, Inc., New York, 1967.

24. MALONATO COMPLEXES OF CHROMIUM(III)

Submitted by J. C. CHANG*
Checked by R. GODDARD† and J. H. WORRELL†

The malonato complexes of chromium(III) are analogous to the oxalate complexes of chromium(III). Since malonic acid is a weaker acid than oxalic acid, the malonato complexes are expected to be more labile than the oxalato complexes. The dicarboxylate complexes of chromium(III) form a group of anionic complexes which are suitable for the study of octahedral complex reactivity.

A. POTASSIUM TRIS(MALONATO)CHROMATE(III) TRIHYDRATE

$$2Cr(OH)_3 + 6H_2C_3H_2O_4 + 3K_2CO_3 \longrightarrow 2K_3[Cr(C_3H_2O_4)_3] \cdot 3H_2O + 3CO_2 + 3H_2O$$

*Chemistry Department, University of Northern Iowa, Cedar Falls, Iowa 50613.
†Chemistry Department, University of South Florida, Tampa, Fla. 33620.

Procedure

Potassium tris(malonato)chromate(III) trihydrate can be prepared by the method of Lapraik,[1] reacting freshly precipitated chromium(III) hydroxide with malonic acid and potassium carbonate. Lapraik obtained the crystals by drying the aqueous solution over sulfuric acid, but the crystals are more easily obtained by precipitation in ethanol.

A solution of 30 ml of concentrated (15*M*) ammonium hydroxide in 170 ml of H_2O is added, with stirring, to a solution of commercial chromium(III) chloride hexahydrate (13.3 g, 0.05 mole) in 200 ml of water. The precipitated chromium(III) hydroxide is centrifuged.*

The freshly prepared $Cr(OH)_3$ is placed in a 250-ml graduated beaker; then a hot (ca. 80°) solution of malonic acid (15.6 g, 0.15 mole) in 100 ml of water is added to the solid chromium(III) hydroxide in 20-ml portions to dissolve the chromium(III) hydroxide. The solution is evaporated† at ca. 80° until the volume is reduced to ca. 25 ml, and potassium carbonate (10.4 g, 0.075 mole) is added to the solution in small portions. The solution is filtered and poured slowly into 300 ml of absolute ethanol in a 400-ml beaker. Upon stirring the mixture, an oil is obtained. The solvent is decanted, and 200 ml more of absolute ethanol is added with stirring until the oil begins to solidify. It may be necessary to triturate the solid with absolute ethanol several times to obtain the complex as a fine green powder. The solid is collected by filtration, washed with absolute ethanol and ether, and air-dried. The yield is 22.2 g (84.5%).‡ The product is slightly hygroscopic. *Anal.* Calcd. for $K_3[Cr(C_3H_2O_4)_3] \cdot 3H_2O$: Cr, 9.82; malonate, 57.82. Found: Cr, 9.80; malonate, 57.5.

B. POTASSIUM *cis*- AND *trans*-DIAQUABIS(MALONATO)-CHROMATE(III) TRIHYDRATE

$$K_2Cr_2O_7 + 5H_2C_3H_2O_4 \longrightarrow 2K[Cr(C_3H_2O_4)_2(H_2O)_2] \cdot 3H_2O + 2CO_2 + HCOOH + H_2O$$

*The $Cr(OH)_3$ suspension is poured into two 230-ml centrifuge bottles and centrifuged for 5 min at 1900 rpm. The supernate is decanted and replaced with distilled water; then the $Cr(OH)_3$ is stirred and recentrifuged for 5 min at 1900 rpm. This process is repeated 10 times to remove all traces of residual chloride ion, which is tested for by the addition of silver nitrate solution to the decanted supernate. Preparation of the $Cr(OH)_3$ requires 2–3 hr.

†Evaporation requires about 3 hr. One should not use an air stream over the solution to hasten evaporation.

‡Yield is a direct function of how efficiently the initial $Cr(OH)_3$ can be washed and collected.

Procedure

A solution of potassium dichromate (11.8 g, 0.04 mole) in boiling water (25 ml) and a solution of malonic acid (20.8 g, 0.20 mole) also in boiling water (25 ml) are mixed and allowed to react until no more carbon dioxide evolves. Then water is added to the mixture to make a volume of ca. 200 ml.

To prepare the cis isomer, the solution is evaporated at 80° to ca. 50 ml. After cooling, the concentrated solution is added very slowly, with stirring, to 300 ml of 95% ethanol in a 400-ml beaker to precipitate the product.* It is collected by filtration, washed with 95% ethanol and ether, and air-dried. Yield: ca. 20 g (66%).

To prepare the trans isomer, the solution is evaporated to only one-half the original volume and allowed to evaporate slowly at room temperature. Within 2 days a purple-violet precipitate will form.† Decant the liquid, wash the precipitate with 5 ml of cold water (5°), and quickly collect by filtration; wash with 95% ethanol and ether and air-dry. Yield: 1.0–1.5 g (ca. 4%).

Anal. Calcd. for $K[Cr(C_3H_2O_4)_2(H_2O)_2]\cdot 3H_2O$: Cr, 13.50; malonate, 52.97. Found: for the cis isomer, Cr, 13.8; malonate, 52.2; for the trans isomer: Cr, 13.5; malonate, 52.6.

Properties

Potassium tris(malonato)chromate(III) trihydrate is a blue-green compound which is very soluble in water. The cis isomer of potassium diaquabis(malonato)chromate(III) trihydrate is purple and also very soluble in water. The trans isomer is violet and fairly soluble in water, but it is more soluble than the analogous oxalate compound. All three compounds exhibit two absorption maxima in their visible absorption spectra, listed in Table I.

TABLE I

Complex	λ_{max}, nm	Molar absorptivity	λ_{max}, nm	Molar absorptivity	Ref.
$[Cr(C_3H_2O_4)_3]^{3-}$	570	68.7	420	55.6	2
cis-$[Cr(C_3H_2O_4)_2(H_2O)_2]^-$	566	49.9	417	41.4	2
	565	49.4	415	41.0	3
	565	50.8	418	42.1	4
trans-$[Cr(C_3H_2O_4)_2(H_2O)_2]^-$	560	19.6	404	21.0	2
	555	17.6	401	20.3	4
	560	21	405	22	5

*If an oil forms, decant the alcohol and triturate with an additional 200 ml of 95% ethanol or until a solid forms.

†If allowed to stand longer, coprecipitation of cis and trans compounds results, as evidenced by electronic spectra.

References

1. W. Lapraik, *Chem. News*, **67**, 219 (1893).
2. J. C. Chang, *J. Inorg. Nucl. Chem.*, **30**, 945 (1968).
3. K. R. Ashley and K. Lane, *Inorg. Chem.*, **9**, 1795 (1970).
4. M. J. Frank and D. H. Huchital, *Inorg. Chem.*, **11**, 776 (1972).
5. M. Casula, G. Illuminati, and G. Ortaggi, *Inorg. Chem.*, **11**, 1062 (1972).

25. SELENOUREAMETAL COMPLEXES

Submitted by G. B. AITKEN* and G. P. McQUILLAN†
Checked by G. L. SEEBACH‡ and J. R. WASSON‡

Selenourea is less stable than urea or thiourea, as it is sensitive to air and light, particularly in solution, and tends to dimerize in certain systems.[1] In general, the donor properties of selenourea are similar to those of thiourea,[2,3] and complexes with cobalt(II),[4,5] palladium(II), platinum(II),[6] zinc, cadmium, and mercury[5] acceptors can be obtained by appropriate modification of the methods used for the preparation of the analogous thiourea derivatives.

Acetone, 1-butanol, ethanol, methanol, and water are useful solvents for selenourea; the solutions in acetone or 1-butanol seem to be the least air-sensitive. Aqueous solutions can be stabilized by the addition of acid.[6] Commercial samples of selenourea are sometimes contaminated with elemental selenium, which can be removed by filtration of the ligand solution.[§] Solid selenourea may be exposed to the atmosphere for short periods without appreciable decomposition, but as a general rule it is advisable to handle selenourea and selenourea complexes in a nitrogen atmosphere and to avoid prolonged exposure to light.

■ **Caution.** *Selenourea is toxic and should be handled with care. Experiments with selenourea should be conducted in an efficient fume hood or in a glove box vented to a fume-extraction system. The explosive hazard of perchlorate salts of metal complexes should also be noted.*

*Chemistry Department, Universiti Sains Malaysia, Penang, Malaysia.
†Chemistry Department, University of Aberdeen, Aberdeen AB9 2UE, Scotland.
‡Chemistry Department, University of Kentucky, Lexington, Ky. 40506.

§The checkers recommend filtration of the ligand solution before all preparations. The infrared spectrum can be used as a check on ligand purity. Principal infrared bands in (solid) selenourea are 3350–3140 (s, br), 1610 (s), 1480 (m), 1400 (s), 1085 (w), 1040 (w), 730 (m), 640 (m, sh), 585 (s, br), 525 (s, br), 390 (s), 340 (vw) cm^{-1}.[5,7]

A. TETRAKIS(SELENOUREA)COBALT(II) PERCHLORATE

$$Co(ClO_4)_2 \cdot 6H_2O + 4(NH_2)_2CSe \longrightarrow [Co\{(NH_2)_2CSe\}_4](ClO_4)_2 + 6H_2O$$

Procedure

A quantity of 1-butanol (ca. 100 ml) is deoxygenated by boiling and cooling in a nitrogen atmosphere. Of the deoxygenated solvent, 40 ml is transferred to a 100-ml two-necked flask fitted with a nitrogen inlet arranged to reach almost to the bottom of the flask* and a reflux condenser. A stream of oxygen-free nitrogen is maintained through the system throughout the procedure described below. Selenourea† (0.49 g, 0.004 mole) is added to the flask, and the mixture is heated to the reflux temperature. If traces of insoluble material are evident in the refluxing solution, it is allowed to cool and filtered under nitrogen in a glove box. A solution of cobalt(II) perchlorate hexahydrate (0.365 g, 0.001 mole) in oxygen-free 1-butanol (20 ml) is added to the refluxing selenourea solution. The reaction mixture immediately becomes deep green; it is left to reflux for 30 min and is then reduced in volume by distilling off about 30 ml of solvent. The concentrated solution is cooled to room temperature, and an approximately equal volume of petroleum ether is added to precipitate the complex. The product is separated by filtration in the glove box, washed with petroleum ether, and dried over phosphorus pentoxide *in vacuo*. Yield: 0.5 g, 70% approx. *Anal.* Calcd. for $[Co\{(NH_2)_2CSe\}_4](ClO_4)_2$: Co, 7.9; C, 6.4; H, 2.1; N, 14.9. Found: Co, 7.8; C, 6.7; H, 2.1; N, 15.0.

Properties

The complex is obtained as deep olive-green platelike crystals, m.p. 142° (decomposes). The solid compound decomposes slowly on exposure to the atmosphere; the process is greatly accelerated if the material is finely divided. The preparation of samples for physical measurements must be carried out in a nitrogen atmosphere. The complex forms very air-sensitive solutions in 1-butanol or nitromethane. The infrared spectrum exhibits characteristic "ionic perchlorate" absorptions at 1100 and 625 cm^{-1}, and there are strong ligand bands at 1625, 1400, 575, 500, and 380 cm^{-1}.

*The N_2 inlet must be *below* the surface of the solvent.

†Obtained from Fluka AG, Buchs, Switzerland.

B. SULFATOTRIS(SELENOUREA)COBALT(II)

$$CoSO_4 \cdot 7H_2O + 3(NH_2)_2CSe \longrightarrow [Co\{(NH_2)_2CSe\}_3(SO_4)] + 7H_2O$$

Procedure

A solution of selenourea (0.74 g, 0.006 mole) in acetone (30 ml) is prepared in a 100-ml two-necked flask equipped with a nitrogen supply and a reflux condenser, as described in Sec. A. The solution is filtered, if necessary, to remove traces of elemental selenium. A solution of cobalt(II) sulfate heptahydrate (0.42 g, 0.0015 mole) in 10 ml of deoxygenated methanol is diluted with an equal volume of deoxygenated acetone and added to the refluxing selenourea solution. An olive-green precipitate separates immediately. The reaction mixture is cooled to room temperature, and the product is separated by filtration in a nitrogen atmosphere in a glove box. The compound is washed with acetone and dried at 50° *in vacuo* to remove traces of coordinated methanol. Yield: 0.73 g, 70%. *Anal.* Calcd. for $[Co\{(NH_2)_2CSe\}_3(SO_4)]$: Co, 11.25; C, 6.9; H, 2.3; N, 16.1. Found: Co, 10.9; C, 6.7; H, 2.5; N, 16.3.

Properties

The complex is obtained as relatively air-stable olive-green crystals, m.p. 125°C (decomposes). It may be exposed to the atmosphere for short periods but should be stored under nitrogen. The compound is insoluble in acetone, 1-butanol, methanol, and nitromethane. The monodentate sulfato ligand gives rise to infrared absorptions at 1140 (s), 1050 (sh), 970 (w), 640 (m), and 610 (m) cm^{-1}, and there are characteristic selenourea bands at 1640 (s,br), 1510 (m), 1405 (m), 540 (m,br), and 380 (m) cm^{-1}.

Mercury(II) Halide Complexes

Complexes of thiourea with mercury(II) halides have been known for almost a century,[8] and compounds containing one, two, three, or four molecules of coordinated thiourea can be isolated from aqueous solution.[9] With selenourea in acetone, it is possible to prepare 1:2 complexes $Hg[(NH_2)_2CSe]_2X_2$ (X = C, Br, I) and a dimeric 1:1 complex $[Hg\{(NH_2)_2CSe\}Cl_2]_2$ in excellent yields.

C. DICHLOROBIS(SELENOUREA)MERCURY(II)

$$HgCl_2 + 2(NH_2)_2CSe \longrightarrow Hg[(NH_2)_2CSe]_2Cl_2$$

Procedure

An oxygen-free solution of selenourea (0.49 g, 0.004 mole) in acetone (25 ml) is prepared in a two-necked flask equipped with a nitrogen supply and a reflux condenser. A solution of mercury(II) chloride (0.55 g, 0.002 mole) in deoxygenated acetone (15 ml) is added to the refluxing selenourea solution. A white precipitate separates immediately. The reaction mixture is cooled to room temperature and filtered under nitrogen in a glove box. The complex is washed with acetone and dried over P_4O_{10} *in vacuo*. Yield: 1.0 g, 96%. *Anal.* Calcd. for $Hg[(NH_2)_2CSe]_2Cl_2$: N, 10.8; Cl, 13.7. Found: N, 10.9; Cl, 13.6.

D. DIBROMOBIS(SELENOUREA)MERCURY(II)

$$HgBr_2 + 2(NH_2)_2CSe \longrightarrow Hg[(NH_2)_2CSe]_2Br_2$$

Procedure

This is prepared in exactly the same way as the chloride complex, using 0.49 g (0.004 mole) of selenourea and 0.72 g (0.002 mole) of mercury(II) bromide. Yield: 1.08 g, 90%. *Anal.* Calcd. for $Hg[(NH_2)_2CSe]_2Br_2$: N, 9.2; Br, 26.3. Found: N, 9.1; Br, 26.5.

E. DI-μ-CHLORO-DICHLOROBIS(SELENOUREA)DIMERCURY(II)

$$2HgCl_2 + 2(NH_2)_2CSe \longrightarrow [Hg\{(NH_2)_2CSe\}Cl_2]_2$$

Procedure

In this reaction it is necessary to maintain an excess of mercury(II) chloride, to avoid contamination of the product with the 1:2 complex $Hg[(NH_2)_2CSe]_2$-Cl_2. The selenourea solution must therefore be added to the mercury(II) chloride solution, and not the reverse. A solution of mercury(II) chloride (0.95 g, 0.0035 mole) in oxygen-free acetone (30 ml) is prepared in a 100-ml two-necked flask equipped with a nitrogen supply and a reflux condenser and is heated to the reflux temperature. The solution is stirred vigorously, using either a rapid stream of nitrogen or a magnetic stirrer, and a solution of selenourea (0.37 g, 0.003 mole) in oxygen-free acetone (15 ml) is added. A white solid precipitates immediately. The reaction mixture is kept at the reflux temperature for 30 min and then cooled. The product is removed by filtration under nitrogen in a glove box, washed with acetone, and dried over P_4O_{10}

in vacuo. Yield: 1.05 g, 89%. *Anal.* Calcd. for $[Hg\{(NH_2)_2CSe\}Cl_2]_2$: N, 7.1; Cl, 18.0. Found: N, 6.9; Cl, 17.8.

Properties

The mercury complexes are obtained as white microcrystalline solids which begin to darken and decompose at 180–200°. The compounds are more stable than other selenourea derivatives and may be exposed to the atmosphere for several hours without noticeable decomposition. They are insoluble, or only very slightly soluble, in water, ethanol, acetone, chloroform, and nitromethane. Unlike the corresponding thiourea derivatives, the selenourea complexes do not dissolve in water or acetone containing an excess of ligand. A dimeric chlorine-bridged structure is proposed for the 1:1 complex because its far-infrared spectrum closely resembles those of other 1:1 mercury(II) chloride complexes known to possess such structures.[5] The mercury-terminal chlorine stretching vibration in the 1:1 complex gives rise to a strong infrared band at 276 cm^{-1}. This band has no direct counterpart in the spectrum of the 1:2 complex (for which the highest mercury-chlorine or mercury-ligand band occurs at 204 cm^{-1}),[5] and it can therefore be used to distinguish between the two compounds.

References

1. A. Chiesi, G. Grossoni, M. Nardelli, and M. E. Vidoni, *Chem. Commun.*, **1969,** 404.
2. A. Yamaguchi, R. B. Penland, S. Mizushima, T. J. Lane, C. Curran, and J. V. Quagliano, *J. Am. Chem. Soc.*, **80,** 527 (1958).
3. F. A. Cotton, O. D. Faut, and J. T. Mague, *Inorg. Chem.*, **3,** 17 (1964).
4. O. Piovesana and C. Furlani, *J. Inorg. Nucl. Chem.*, **30,** 1249 (1968).
5. G. B. Aitken, J. L. Duncan, and G. P. McQuillan, *J. Chem. Soc. (Dalton)*, **1972,** 2103.
6. P. J. Hendra and Z. Jovic, *Spectrochim. Acta*, **24A,** 1713 (1968).
7. G. B. Aitken, J. L. Duncan, and G. P. McQuillan, *J. Chem. Soc.* (*A*), **1971,** 2695.
8. R. Maly, *Ber.*, **9,** 172 (1876); A. Claus, *ibid.*, 226.
9. I. Aucken, *Inorganic Syntheses*, **6,** 26 (1960).

26. NICKEL TETRAFLUOROOXOVANADATE(IV) HEPTAHYDRATE

Submitted by A. KLEIN,* P. CARROLL,† and G. MITRA*
Checked by J. SELBIN,‡ J. H. NIBERT,‡ and H. J. SHERRILL‡

$$Ni(NO_3)_2 + (NH_4)_2[VOF_4] \xrightarrow[HF+NH_4F]{excess} Ni[VOF_4] + 2NH_4NO_3$$

*Chemistry Department, King's College, Wilkes-Barre, Pa. 18711.
†Department of Chemistry, Temple University, Philadelphia, Pa. 19122.
‡Department of Chemistry, Louisiana State University, Baton Rouge, La. 70803.

Complex compounds of transition metals with simple d^1 configuration possessing O_h symmetry have been studied extensively.[1-3] Virtually all the known compounds of Ti(III) and V(IV) have been investigated. The preparation given below is for a previously unreported tetravalent vanadium compound. During the course of this investigation $Ni[VOF_4]\cdot 7H_2O$ was isolated in good yield using a relatively simple technique. This is possibly the only well-defined transition metal salt of tetrafluorooxovanadate(IV), $[VOF_4]^{2-}$, isolated up to now. The corresponding compounds with iron(II) and cobalt(II) could not be isolated. The nickel salt of the ion was formed by the addition of nickel dinitrate hexahydrate to a solution containing $[VOF_4]^{2-}$ and other ions.

■ **Caution.** *Hydrofluoric acid causes severe burns, and care should be taken that it does not come in contact with the skin. Because of the hazardous nature of* HF *and* SO_2, *the syntheses should be carried out in a well-ventilated hood.*

Procedure

In a platinum evaporating dish, 23 g (0.2 mole) of ammonium metavanadate (mol. wt. = 117) is dissolved in about 50 ml of 35% HF (prepared by diluting 48% hydrofluoric acid). The solution is heated over a hot plate, and the pentavalent vanadium is reduced to the tetravalent state by bubbling SO_2 gas through the heated solution for 20 min. Sulfur dioxide is generated by the action of concentrated sulfuric acid on sodium hydrogen sulfite, or the SO_2 is obtained from a gas cylinder. The intermediate compound $(NH_4)_2[VOF_4]\cdot 2H_2O$ is then obtained by adding a large excess (ca. 4: 1*M*) of ammonium hydrogen difluoride (mol. wt. = 57) either as the solid or in a minimum of hot water to the hot solution, crystallizing on an ice bath, and washing with a 50% methanol solution until the washings are colorless. About 25 g of this crude product is obtained. This impure $(NH_4)_2[VOF_4]\cdot 2H_2O$ is then dissolved in about 50 ml of 35% hydrofluoric acid in a platinum evaporating dish. To this solution is added 45 g of $Ni(NO_3)_2\cdot 6H_2O$ (0.16 mole) dissolved in a minimum of warm water (ca. 12 ml). The solution is then quickly cooled by placing it on an ice bath, and the blue-green crystals of $Ni[VOF_4]\cdot 7H_2O$ are obtained. The compound should be washed with ice-cold 50% methanol until the washings are colorless or faintly blue. The compound should be dried in an atmosphere of nitrogen. Approximately 22 g of the compound is isolated. *Anal.* Calcd. for $Ni[VOF_4]\cdot 7H_2O$: Ni, 17.91; VO, 20.42; F, 23.19. Found: Ni, 17.8; VO, 20.4; F, 23.8.

Properties

The stability of crystalline $Ni[VOF_4]\cdot 7H_2O$ in air is questionable, but it can be kept under kerosene in a bottle for months.* The compound is partly soluble in water, and the dissociation constants for the $[VOF_4]^{2-}$ ion can be measured using a conductimetric technique. It is found that the ion $[VOF_4]^{2-}$ is fairly stable. The same conclusion is reached by measuring the free fluoride ion in a solution of potassium tetrafluorooxovanadate(IV) with the fluoride-specific ion electrode. The infrared spectrum of $Ni[VOF_4]\cdot 7H_2O$ shows the characteristic V=O stretching absorption at 950 cm^{-1}. The electronic absorption spectrum has shown three peaks with maxima at 752, 642, and 390 nm. Differential thermal analysis gives evidence that the seven water molecules come off in two steps, one at 150° and the other at 230°. The quantitative determination for water performed at 150° gives a value of 32.74%, indicating that six water molecules are driven off at this temperature while the seventh leaves at 230°. The monohydrated form of this compound cannot be established definitely since the crystal structure apparently breaks up with the escape of water molecules.

References

1. J. Selbin and L. H. Holmes, *J. Inorg. Nucl. Chem.*, **24,** 1111 (1962).
2. C. J. Ballhausen and H. B. Gray, *Inorg. Chem.*, **1,** 111 (1962).
3. Clark, R. J. H., "The Chemistry of Titanium and Vanadium," American Elsevier Publishing Company, Inc., New York, 1968.

27. POLYMERIC CHROMIUM(III)-BIS(PHOSPHINATES)

Submitted by H. D. GILLMAN,† P. NANNELLI,† and B. P. BLOCK†
Checked by E. E. FLAGG‡

Among the most interesting poly(metal phosphinates) are those based on chromium(III). In an earlier volume a procedure for the preparation of poly-[aqua-bis(μ-diphenylphosphinato)-hydroxochromium(III)] from chromium-(II) chloride was presented.[1] It has since been found that an identical product can be prepared directly from chromium(III) starting materials,[2] a route which avoids the necessity for working with an air-sensitive chromium(II) intermediate. At the same time it was reported that the aquahydroxobis-

*The compound can also be preserved in a vacuum or under an atmosphere of nitrogen.
†Technological Center, Pennwalt Corporation, King of Prussia, Pa. 19406.
‡Dow Chemical Company, Midland, Mich. 48640.

(phosphinates) of chromium(III) have a more complicated structure than implied by the formula $[Cr(H_2O)(OH)(OPRR'O)_2]_x$ and that under appropriate conditions they can be dehydrated to hydroxobis(phosphinates) of chromium(III), a family of chromium(III)-bis(phosphinate) polymers with somewhat more certain composition. The general procedures that follow illustrate the synthesis of polymers of both types and avoid the use of chromium-(II) intermediates. For convenience the water-containing species will be considered to correspond to the simple formula, although such a formulation represents only part of the picture.

A. POLY[AQUAHYDROXO-BIS(μ-R,R′-PHOSPHINATO)-CHROMIUM(III)]

$$2RR'P(O)OH + K_2CO_3 \longrightarrow 2KOPRR'O + H_2O + CO_2$$

$$2CrCl_3 \cdot 6H_2O + 4KOPRR'O + K_2CO_3 \longrightarrow 2/x[Cr(H_2O)(OH)(OPRR'O)_2]_x + 6KCl + CO_2 + 9H_2O$$

Procedure

Although all operations should be performed in a well-ventilated hood because of the use of tetrahydrofuran, only simple laboratory ware and a magnetic stirrer with hot plate (or equivalent) are required. A solution of 0.0200 mole of the appropriate phosphinic acid and 2.07 g (0.0150 mole) of potassium carbonate in 50 ml of 1:1 v/v water-tetrahydrofuran is added with stirring to a solution of 0.0100 mole of a soluble simple chromium(III) salt in 50 ml of 1:1 v/v water-tetrahydrofuran. The mixture is brought to boiling, and the tetrahydrofuran is allowed to evaporate. An oily product separates upon evaporation of the tetrahydrofuran. After most of the tetrahydrofuran has evaporated, an additional 50 ml of water is added, and the suspension is kept boiling until the precipitate can be ground easily with a spatula.* The solid is then collected on a filter, thoroughly washed with water, and allowed to dry as indicated. The yield is essentially quantitative. Table I summarizes further details for the preparation of three polymers of the aquahydroxo type. No difficulties have been encountered in this general procedure with quantities up to 10 times the amounts specified.

*If the phosphinate groups contain long alkyl chains, the resulting polymer tends to remain oily; upon cooling to room temperature it solidifies to a waxy solid that can be ground, albeit with difficulty.

TABLE I Preparation of $[Cr(H_2O)(OH)(OPRR'O)_2]_x$

R	R′	Cr salt	Drying conditions	Yield	Analysis* C	H	Cr	P
C_6H_5	C_6H_5†	$CrCl_3 \cdot 6H_2O$	In air, 3 days	5.2 g (100%)	55.01 (55.29)	4.53 (4.45)	9.74 (9.97)	11.67 (11.88)
CH_3	C_6H_5‡	$CrCl_3 \cdot 6H_2O$	In air, 3 days	3.85 g (97%)	42.38 (42.33)	4.81 (4.82)	12.90 (13.09)	15.36 (15.59)
C_8H_{17}	C_8H_{17}§,¶	$Cr(NO_3)_3 \cdot 9H_2O$	In vacuum, 1 day	6.39 g (96%)	57.78 (57.72)	10.57 (10.75)	7.94 (7.81)	9.52 (9.30)

*Calculated values in parentheses.

†Available from several suppliers of reagents or see *Inorganic Syntheses,* **8**, 71 (1966).

‡The authors have had the most success with the Friedel-Crafts reaction described by Biddle et al.,[3] among the various procedures in the literature for the preparation of this acid.

§Prepared by the peroxide-catalyzed addition of hypophosphorous acid to 1-octene, as described by Peppard et al.[4]

¶The checker substituted diheptylphosphinic acid for dioctylphosphinic acid and obtained a 95% yield of the corresponding polymer. *Anal.* Calcd.: C, 55.2; H, 10.4. Found: C, 55.5; H, 10.6.

Properties

The properties of the diphenylphosphinate chromium(III) polymers are given with the previously reported procedure.[1] Although the methylphenylphosphinate and dioctylphosphinate chromium(III) polymers are generally similar to the diphenylphosphinate chromium(III) polymer, there are significant differences. Freshly prepared samples are soluble in chloroform, benzene, and tetrahydrofuran but not as readily nor to the same extent. Samples of the methylphenyl and dioctyl polymers become less soluble with age, but solubility is restored by prolonged agitation after addition of a few drops of water or alcohol to the solvent. Initial intrinsic viscosities in chloroform are 0.4 dl/g for the methylphenyl polymer and 0.3 dl/g for the dioctyl polymer, and there is no appreciable change on standing. The methylphenyl polymer, like the diphenyl polymer, apparently undergoes only dehydration at temperatures below 300° in air, whereas extensive decomposition of the dioctyl polymer starts at about 200° in air. None of the three melts before decomposition.

B. POLY[HYDROXO-BIS(μ-R,R′-PHOSPHINATO)-CHROMIUM(III)

$$[Cr(H_2O)(OH)(OPRR'O)_2]_x \longrightarrow [Cr(OH)(OPRR'O)_2]_x + xH_2O$$

Procedure

Heating the water-containing polymers to constant weight under vacuum

TABLE II Preparation of $[Cr(OH)(OPRR'O)_2]_x$

R	R′	Dehydration conditions	Analysis*			
			C	H	Cr	P
C_6H_5	C_6H_5	Vacuum, 200°, 10 hr	57.25 (57.27)	4.32 (4.21)	10.20 (10.33)	12.50 (12.31)
CH_3	C_6H_5	Vacuum, 200°, 10 hr	44.13 (44.34)	4.77 (4.52)	13.30 (13.71)	16.56 (16.34)
C_8H_{17}	C_8H_{17}†	Vacuum, 160°, 1 day	58.98 (59.33)	10.68 (10.74)	7.78 (8.03)	9.62 (9.56)

*Calculated values in parentheses.

†The checker dehydrated the aquated diheptylphosphinate analog under vacuum at 160° for 1 day. *Anal.* Calcd.: C, 56.8; H, 10.4. Found: C, 56.3; H, 10.0.

converts them quantitatively to hydroxobis(phosphinate) polymers. Table II summarizes the details for the dehydration of the three polymers given in Sec. A.*

Properties

The polymer $[Cr(OH)\{OP(C_6H_5)_2O\}_2]_x$ is similar to its parent polymer in that it is also a green solid soluble in chloroform, benzene, and tetrahydrofuran. Its freshly prepared solutions in chloroform have intrinsic viscosities of 0.08 to 0.28 dl/g, values somewhat higher than found for similar solutions of the parent polymer. There is an increase in intrinsic viscosity when the solutions are aged. In contrast, the methylphenylphosphinate and dioctylphosphinate chromium(III) polymers of this type are virtually insoluble. This suggests that in their formation by removal of the water there is some degree of cross-linking, whereas the diphenyl polymer appears to remain substantially linear. These polymers undergo extensive decomposition at the same temperatures as their parent polymers but do not, of course, lose any weight at lower temperatures through loss of water.

References

1. K. D. Maguire, *Inorganic Syntheses,* **12,** 258 (1970).
2. P. Nannelli, H. D. Gillman, and B. P. Block, *J. Polym. Sci. (A-1)*, **9,** 3027 (1971).
3. P. Biddle, J. Kennedy, and J. L. Williams, *Chem. Ind. (Lond.)*, **1957,** 1481.
4. D. F. Peppard, G. W. Mason, and S. Lewey, *J. Inorg. Nucl. Chem.*, **27,** 2065 (1965).

*The temperature used for dehydration must necessarily be lower than the decomposition temperature of the polymer being prepared. The checker found that there was noticeable decomposition of the diheptylaquahydroxo polymer at 160°. A preliminary observation of thermal stability is desirable in order to select a temperature for the dehydration of other aquahydroxo polymers with alkyl side groups.

28. RESOLUTION OF THE *cis*-BROMOAMMINEBIS(ETHYLENEDIAMINE)COBALT(III) ION

Submitted by G. B. KAUFFMAN* and E. V. LINDLEY, JR.*
Checked by M. P. GRANCHI† and B. E. DOUGLAS†

The first successful resolution of coordination compounds was reported by Werner[1] in 1911 and involved two series of complexes, the *cis*-chloroammine-bis(ethylenediamine)cobalt(III) series (with V. L. King)[2] and the *cis*-bromo-amminebis(ethylenediamine)cobalt(III) series (with E. Scholze). Of the two, the bromoammine series[3] is easier to resolve because of the greater difference in solubility between the diastereoisomers that are formed with the resolving agent, silver (+)-α-bromocamphor-π-sulfonate. For both series, the (+)-bromocamphorsulfonate of the (+)-antipode is less soluble than that of the (−)-antipode. The resolutions provided unequivocal proof of the octahedral configuration first postulated by Werner[4] for cobalt(III) in 1893.

The method given here employs the commercially available ammonium (+)-α-bromocamphor-π-sulfonate instead of the more expensive silver salt used by Werner, and it gives higher yields of enantiomers.[5,6] The chloride, bromide, and nitrate of the (+) series can be prepared from the precipitated (+)(+)-bromocamphorsulfonate diastereoisomer by metathesis with the appropriate concentrated acid, whereas the chloride, bromide, and nitrate of the (−)-series can be prepared by treatment with the appropriate concentrated acid of the (−)-dithionate precipitated from the diastereoisomer filtrate.

A. (+)-*cis*-BROMOAMMINEBIS(ETHYLENEDIAMINE)COBALT(III) (+)-α-BROMOCAMPHOR-π-SULFONATE[(+)(+)-DIASTEREOISOMER]

$$2\ cis\text{-}[Coen_2NH_3Br]Br_2 \cdot H_2O + 4NH_4(+)\text{-}[O_3SOC_{10}H_{14}Br] \longrightarrow (+)\text{-}cis\text{-}[Coen_2NH_3Br](+)\text{-}[O_3SOC_{10}H_{14}Br]_2 \downarrow + (-)\text{-}cis\text{-}[Coen_2NH_3Br](+)\text{-}[O_3SOC_{10}H_{14}Br]_2 + 4NH_4Br + 2H_2O$$

*Department of Chemistry, California State University Fresno, Fresno, Calif. 93740. The authors wish to acknowledge the assistance of the National Science Foundation Undergraduate Research Participation Program (Grant GY-9916) and the California State University, Fresno Research Committee and Stenographic Service Center.

†Department of Chemistry, University of Pittsburgh, Pittsburgh, Pa. 15260.

Procedure

Eighteen grams (0.0533 mole) of ammonium (+)-α-bromocamphor-π-sulfonate* is dissolved in 60 ml of water which has been heated to 35°. A solution of 12.44 g (0.0274 mole) of *cis*-bromoamminebis(ethylenediamine)cobalt(III) bromide 1-hydrate[8] in 110 ml of water at 35° is similarly prepared.† The solutions are mixed, filtered if necessary, and allowed to stand in a refrigerator (ca. 5°) until crystallization of the pasty mass of red-violet (+)(+)-diastereoisomer appears complete (ca. 1 day). The crude diastereoisomer (ca. 12 g) is recrystallized from 290 ml of water at 70° by placing the impure product in a sintered-glass funnel connected to a suction-filtration flask. Suction is applied, and the hot water is poured over the finely ground product with stirring. The filter flask is immersed in an ice bath to cool the filtrate quickly. After all the crude diastereoisomer has dissolved, the solution is refrigerated overnight. The purified product is collected by suction filtration and air-dried. The yield is 4.91 g. The filtrate is rotary-evaporated at ca. 35° to a volume of ca. 180 ml and is refrigerated to yield a second crop of 2.11 g. The total yield is 7.02 g (57.2%). For a 0.800% aqueous solution in a 2-dm tube, $\alpha_C^{23} = +1.0468°$, whence $[\alpha]_C^{23} = +65.42°$‡(Werner[1] reported $[\alpha]_C^{22} = +65.7°$; the checkers report $[\alpha]_C^{25} = +62.6°$). For a 0.150% aqueous solution in a 1-dm tube, the checkers report $[\alpha]_D^{25} = +141.7°$.§

B. (−)-*cis*-BROMOAMMINEBIS(ETHYLENEDIAMINE)COBALT(III) DITHIONATE

$$[(-)\text{-}cis\text{-}[Coen_2NH_3Br](+)\text{-}[O_3SOC_{10}H_{14}Br]_2 + Na_2S_2O_6 \cdot 2H_2O \rightarrow$$
$$(-)\text{-}cis\text{-}[Coen_2NH_3Br]S_2O_6 \downarrow + 2Na(+)\text{-}[O_3SOC_{10}H_{14}Br] + 2H_2O$$

Procedure

To the mother liquor from which the (+)(+)-diastereoisomer was crystallized in Sec. A, a solution of 2.00 g of sodium dithionate 2-hydrate in 8 ml of water is added, and the mixture is allowed to stand for ½ hr in an ice bath. The

*Obtainable from Aldrich Chemical Co., Inc., Milwaukee, Wis. 53233 (cat. no. B6000–2), or from Pfaltz and Bauer, Inc., Flushing, N.Y. 11368. If desired, the salt can be prepared from (+)-3-camphor (Eastman Organic Chemicals, Rochester, N.Y. 14650).[7]

†Inasmuch as the salt dissolves slowly, use of a mechanical stirrer is recommended.

‡A high-intensity white light coupled with a Klett no. 66 filter (640–700 nm) was used by the authors for rotations measured for the Fraunhofer C line. The polarimeter half-shade angle was maintained at 15° for all these polarimetric measurements.

§Rotations for the sodium D line (half-shade angle, 3.5°) were measured by the checkers.

resulting crystalline racemic dithionate (ca. 0.4 g) is removed by suction filtration and discarded. An additional 3.2 g of sodium dithionate 2-hydrate dissolved in 12 ml of water is added to the filtrate, and the mixture is allowed to stand for 3 hr in an ice bath. The resulting fine violet crystals of (−)-*cis*-bromoamminebis(ethylenediamine)cobalt(III) dithionate are collected on a 4-cm Büchner funnel and air-dried. An additional 4.0 g of sodium dithionate 2-hydrate in 16 ml of water is added to the filtrate, and the mixture is allowed to stand overnight in an ice bath. The resulting crystalline (−)-dithionate is collected on a 4-cm Büchner funnel and air-dried. The combined yield of (−)-dithionate is 4.50 g (75.3%).* The compound is too insoluble for an accurate determination of its optical activity.

C. (+)-*cis*-BROMOAMMINEBIS(ETHYLENEDIAMINE)COBALT(III) CHLORIDE

$$(+)\text{-}cis\text{-}[Coen_2NH_3Br](+)\text{-}[O_3SOC_{10}H_{14}Br]_2 + 2HCl \longrightarrow$$
$$(+)\text{-}cis\text{-}[Coen_2NH_3Br]Cl_2 + 2H(+)\text{-}[O_3SOC_{10}H_{14}Br]$$

Procedure

Two and four-tenths grams (0.00268 mole) of (+)-*cis*-bromoamminebis(ethylenediamine)cobalt(III) (+)-α-bromocamphor-π-sulfonate from Sec. A is dissolved in 4 ml of concentrated hydrochloric acid. Almost immediately, fine red-violet crystals of the chloride begin to precipitate from the red-violet solution. The resulting mixture is allowed to stand in an ice bath until precipitation appears complete (ca. 5 min), and the crystals are collected by suction filtration. Additional product is obtained from the filtrate by addition of 16 ml of absolute ethanol. The combined precipitate is washed with 8 ml of absolute ethanol and 8 ml of diethyl ether and is then air-dried. The crude chloride is purified by dissolving it in 2 ml of water, filtering the resulting solution if necessary, and adding to it 2 ml of concentrated hydrochloric acid. Four milliliters of absolute ethanol is then added, and the solution is allowed to stand in an ice bath until crystallization appears complete (ca. 15 min). The shiny, dark red-violet, needlelike crystals of the (+)-chloride are collected by suction-filtration, washed with absolute ethanol and ether, and air-dried. The yield of (+)-chloride is 0.807 g (86.6%). For a 0.400% aqueous solution in a 1-dm tube, $\alpha_C^{25} = +0.1900°$, whence $[\alpha]_C^{25} = +47.5°$ (Werner[1] reported $[\alpha]_C^{22} = +50.6°$; the checkers report $[\alpha]_C^{25} = +48.1°$). For a 0.120% aqueous solution in a 1-dm tube, the checkers report $[\alpha]_D^{25} = +155.7°$.

*The checkers report 4.17 g (69.8%).

D. (−)-*cis*-BROMOAMMINEBIS(ETHYLENEDIAMINE)COBALT(III) CHLORIDE

$$(-)\text{-}cis\text{-}[Coen_2NH_3Br]S_2O_6 + 2HCl \rightarrow (-)\text{-}cis\text{-}[Coen_2NH_3Br]Cl_2 + H_2S_2O_6$$

Procedure

One and two-tenths grams (0.00275 mole) of (−)-*cis*-bromoamminebis-(ethylenediamine)cobalt(III) dithionate from Sec. B is ground thoroughly with 4 ml of concentrated hydrochloric acid. The solution is allowed to stand in an ice bath for ca. 30 min, and the resulting deep red-violet crystals are collected by suction filtration. Additional product is obtained from the filtrate by addition of 12 ml of absolute ethanol. The combined crystals are dissolved in a minimum volume of water (ca. 4 ml), 2 ml of concentrated hydrochloric acid is added, and the solution is refrigerated overnight. The resulting crystals are collected by suction filtration, washed with absolute ethanol and ether, and then air-dried. The yield of (−)-chloride is 0.538 g (56.4%). For a 0.200% aqueous solution in a 2-dm tube, $[\alpha]_C^{29} = -0.1886°$, whence $[\alpha]_C^{29} = -47.15°$ (Werner[1] did not prepare the (−)-chloride, but he reported $[\alpha]_C^{22} = +50.6°$ for the (+)-chloride; the checkers report $[\alpha]_C^{25} = -50.6°$). The checkers report $[\alpha]_D^{26} = -155.2°$.

Properties

The active *cis*-bromoamminebis(ethylenediamine)cobalt(III) salts, like the racemic compounds, are similar in properties to the corresponding *cis*-chloroamminebis(ethylenediamine)cobalt(III) salts, but they have been less extensively investigated. They are quite stable, and their solutions do not racemize on standing at room temperature or even on heating to incipient boiling.[1] They can be converted from one salt to another without loss of optical activity.[1] Treatment of the active bromide or (+)-α-bromocamphor-π-sulfonate with liquid ammonia yields the corresponding optically active *cis*-diammine-bis(ethylenediamine)cobalt(III) salts.[9] A partial resolution of the racemic bromide into optical isomers on (+)- and (−)-quartz has been reported.[10] The specific magnetic susceptibility has been found to be 3.49×10^{-7}.[11]

Measurements of light absorption, optical rotatory dispersion, the Cotton effect, and circular dichroism have been made.[12–15] The mechanism, kinetics, and stereochemistry of the aquation and basic hydrolysis of $[Coen_2NH_3Br]^{2+}$ complexes have been investigated.[16] Change in optical activity with aquation has also been determined.[17]

References

1. A. Werner, *Ber.*, **44,** 1887 (1911). For a discussion and English translation of this paper, see G. B. Kauffman, "Classics in Coordination Chemistry, Pt. I. The Selected Papers of Alfred Werner," pp. 155–173, Dover Publications, Inc., New York, 1968.
2. V. L. King, "Über Spaltungsmethoden und ihre Anwendung auf komplexe Metall-Ammoniakverbindungen," dissertation, Universität Zürich, Buchdruckerei J. J. Meier, Zürich, 1912.
3. A. Werner, *Ann.*, **386,** 176, 178, 179 (1912).
4. A. Werner, *Z. Anorg. Chem.*, **3,** 267 (1893). See Kauffman, reference 1, p. 5.
5. E. V. Lindley, Jr., Chemistry 190 Independent Study Report, California State University, Fresno, May 1968.
6. Martin L. Tobe, personal communication to George B. Kauffman, July 12, 1967.
7. G. B. Kauffman, *J. Prakt. Chem.*, [4] **33,** 295 (1966).
8. M. L. Tobe and D. F. Martin. *Inorganic Syntheses*, **8,** 198 (1966).
9. A. Werner and Y. Shibata, *Ber.*, **45,** 3288, 3290 (1912).
10. H. Kuroya, M. Aimi, and R. Tsuchida, *J. Chem. Soc. Jap.*, **64,** 995 (1943).
11. E. Rosenbohm, *Z. Phys. Chem.*, **93,** 695 (1919); L. C. Jackson, *Phil. Mag.*, [7] **2,** 87 (1926); [7] **4,** 1074 (1927).
12. Y. Shibata, *J. Coll. Sci. Imp., Univ. Tokyo,* **37,** Art. 2, 26 (1915).
13. J.-P. Mathieu, *Compt. Rend.*, **199,** 278 (1934); *Bull. Soc. Chim. Fr.*, [5] **3,** 463, 476 (1936).
14. J. G. Brushmiller, E. L. Amma, and B. E. Douglas, *J. Am. Chem. Soc.*, **84,** 111, 3227 (1962).
15. T. Bürer, *Helv. Chim. Acta,* **46,** 242 (1963).
16. R. S. Nyholm and M. L. Tobe, *J. Chem. Soc.*, **1956,** 1707.
17. J.-P. Mathieu, *Bull. Soc. Chim. Fr.*, [5] **4,** 697 (1937).

29. BIS(ALKYLPHOSPHINE)TRICHLOROTITANIUM(III) COMPLEXES

Submitted by C. H. KOLICH* and C. D. SCHMULBACH*
Checked by J. TOWARNICKY† and E. P. SCHRAM†

A series of bis(alkylphosphine)trichlorotitanium(III) complexes can be prepared by direct combination of the components in toluene at elevated temperature. The bis(phosphine) complexes are soluble in aromatic hydrocarbon solvents such as benzene and toluene. Molecular weight measurements and epr spectroscopy indicate that the complexes are five-coordinate monomers in benzene.[1] The weakly basic character of the hydrocarbon solvent diminishes

*Department of Chemistry and Biochemistry, Southern Illinois University, Carbondale, Ill. 62901.
†Chemistry Department, The Ohio State University, Columbus, Ohio 43210.

the extent of solvent competition with substrate and reactants for vacant coordination sites on the metal complex. Since the titanium in the solution phase is only five-coordinate, the complexes possess at least one vacant coordination site, a condition deemed necessary for catalytic activity.[2] The only other reported phosphine-titanium(III) complex is $TiCl_3[(C_2H_5)_2P]_2C_2H_4$.[3]

The methyl phosphines used in the following reactions are conveniently prepared from phosphine and methyl iodide as described by Jolly.[4,*] Phosphine can be readily generated by the thermal decomposition of phosphoros acid.[5] The procedure outlined below is for the preparation of small quantities (0.2–0.5 g) of the bis(phosphine) complexes.

■ **Caution.** *Phosphine and the alkyl phosphines used in these experiments are toxic, malodorous compounds which must be handled with great care using vacuum-line manipulation. The pyrophoric nature of the methyl phosphines makes it advisable to prepare and use them on as small a scale as possible.*

A. BIS(METHYLPHOSPHINE)TRICHLOROTITANIUM(III)

$$TiCl_3 + 2P(CH_3)H_2 \xrightarrow[85°]{\text{toluene}} TiCl_3[PH_2(CH_3)]_2$$

Procedure

An all-glass reaction vessel, as illustrated in Fig. 8, is used. The vessel is evacuated and dried by gentle heating with a hand torch on the vacuum line. The apparatus is then charged with 0.34 g (2.2 mmoles) of titanium(III) chloride (Alfa Inorganic) in a nitrogen-filled glove bag.† The vessel is again placed under vacuum, and 7.0 ml of toluene‡ is condensed onto the solid. A 4.40-mmole sample of gaseous methylphosphine (v.p. = 72 torr at −63°) is manometrically measured out in a calibrated volume of the vacuum system and condensed into the reaction vessel. While the reaction mixture is held at −196°, the vessel is sealed at constriction *A* (Fig. 8). After warming to room temperature, the sealed vessel is placed in an oven held at 85°. The solvent becomes yellow-

*The phosphines $(CH_3)_3P$ and $(C_2H_5)_3P$ are available at Orgmet, East Hampstead, N.H. 03828. The synthesis of $(CH_3)_3P$ is given in Synthesis 41 (Chap. 6).

†Fuming of the titanium(III) chloride during transfer must be avoided. Use of a fresh bottle of titanium(III) is recommended. Heat the titanium(III) chloride at 200° for 2 hr or more under a dynamic vacuum of 10^{-5} torr to remove traces of adsorbed titanium tetrachloride. The nitrogen used in the glove bag should be dried by passage through a 2–3-ft glass column filled with an efficient desiccant, e.g., anhydrous magnesium perchlorate or tetraphosphorus decoxide.

‡Toluene should be dried over tetraphosphorus decoxide overnight and then degassed immediately before use by condensation into a U trap on the vacuum line.

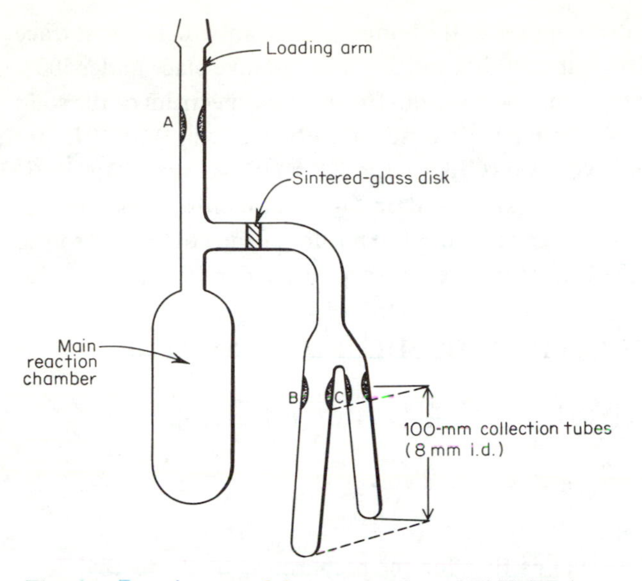

Fig. 8. Reaction vessel for preparation of $TiCl_3(PR_3)_2$ *complexes.*

brown within 2–4 days. The apparatus is removed from the oven after 10 days, and the hot liquid is filtered through the coarse-porosity fritted disk into the side arm. The deep yellow-brown filtrate should be equally divided between the two collection tubes.* The solvent is then condensed back into the main reaction chamber, leaving the solid product in the collection tubes. Additional product can be isolated from the remaining reaction solid by heating the entire vessel at 85° for 10 min. The extracted product can then be transferred to the collection tubes by repeating the filtration and evaporation processes. This extraction procedure applies only to the slightly soluble methylphosphine adduct. Final drying of the product is accomplished by holding the main reaction chamber at −196° while gently warming the side arm.† The collection tubes are then sealed off at constrictions *B* and *C*. The yield of bis(methylphosphine)-trichlorotitanium(III) is 0.40 g (72%). *Anal.* Calcd. for $TiCl_3[P(CH_3)H_2]_2$: C, 9.60; H, 4.03; Cl, 42.49. Found: C, 9.25; H, 4.15; Cl, 42.2.

Properties

Bis(methylphosphine)trichlorotitanium(III) is a dark-red crystalline solid.

*The liquid level should be well below the constrictions if the collection tubes are the prescribed size.

†The checkers report that gentle heating when distilling off the solvent assures greater crystallinity of the product.

The compound slowly decomposes as the temperature is raised. The first trace of melting occurs at 420°, but complete melting does not take place under 500°. The adduct is sensitive to water and oxygen. The infrared spectrum of the solid (mineral-oil mull) shows characteristic absorption bands (cm^{-1}) at 1075 (m) (H—P—C deformation); 965 (vs) (CH_3 rock); 720 (s) (P—C stretch); 410 (s) (Ti—P stretch); 400 (sh,s); 375 (s) (Ti—P or Ti—Cl stretch); 335 (s) (Ti—Cl stretch); and 298 (s) (Ti—Cl stretch). Brief exposure of the compound to the atmosphere results in the loss of infrared absorption under 500 cm^{-1}.

B. BIS(DIMETHYLPHOSPHINE)TRICHLOROTITANIUM(III)

$$TiCl_3 + 2P(CH_3)_2H \xrightarrow[85°]{\text{toluene}} TiCl_3[PH(CH_3)_2]_2$$

Procedure

The procedure is the same as that for the preparation of the methylphosphine complex. A mixture of 0.11 g (0.70 mmole) of titanium(III) chloride, 1.40 mmole of dimethylphosphine (v.p. = 34 torr at −45°), and 7.0 ml of toluene is heated overnight at 85°. The hot mixture is then filtered into the side arm of the reaction vessel; upon cooling, a dark red-brown solid separates from the blue-green filtrate. Solvent and excess phosphine are removed as in Sec. A to give 0.14 g (72%) of bis(dimethylphosphine)trichlorotitanium(III), m.p. 177–189° with decomposition. *Anal.* Calcd. for $TiCl_3[P(CH_3)_2H]_2$; C, 17.26; H, 5.07; Cl, 38.21; P, 22.25; Ti, 17.21. Found: C, 17.1; H, 5.28; Cl, 37.9; P, 22.0; Ti, 17.4.

Properties

Bis(dimethylphosphine)trichlorotitanium(III) is a red-brown crystalline solid. The complex is sensitive to water and oxygen. The infrared spectrum of the solid (mineral-oil mull) shows characteristic absorption bands (cm^{-1}) at 985 (vs) (H—P—C deformation); 955 (vs) (CH_3 rock); 840 (m); 745 (m); 730 (m) (P—C stretch); 375 (vs) (Ti—P stretch); 350 (s) (Ti—P or Ti—Cl stretch); and 312 (s) (Ti—Cl stretch).

C. BIS(TRIMETHYLPHOSPHINE)TRICHLOROTITANIUM(III)

$$TiCl_3 + 2P(CH_3)_3 \xrightarrow[85°]{\text{toluene}} TiCl_3[P(CH_3)_3]_2$$

Procedure

The procedure described in Sec. A is used. A 0.11-g (0.70-mmole) sample of titanium(III) chloride is combined with 1.75 mmoles of trimethylphosphine (v.p. = 49 torr at $-45°$) in 7.0 ml of toluene. The mixture is held at 85° overnight and then filtered into the side arm of the reaction vessel. Removal of solvent and excess phosphine from the yellow-brown filtrate yields 0.16 g of bis(trimethylphosphine)trichlorotitanium(III), m.p. 230–255° (decomposes) (76% yield). *Anal.* Calcd. for $TiCl_3[P(CH_3)_3]_2$: C, 23.52; H, 5.92; Cl, 34.71; P, 20.22; Ti, 15.63. Found: C, 23.3; H, 6.02; Cl, 34.9; P, 20.0; Ti, 15.5.

Properties

Bis(trimethylphosphine)trichlorotitanium(III) is a red-brown crystalline solid that is highly air-sensitive. Characteristic absorption bands (cm^{-1}) in the infrared spectrum of the solid (mineral-oil mull) appear at 955 (vs) (CH_3 rock); 735 (s) (P—C stretch); 370 (vs) (Ti—P stretch); 340 (s) (Ti—P or Ti—Cl stretch); and 295 (m) (Ti—Cl stretch).

D. BIS(TRIETHYLPHOSPHINE)TRICHLOROTITANIUM(III)

$$TiCl_3 + 2P(C_2H_5)_3 \xrightarrow[85°]{\text{toluene}} TiCl_3[P(C_2H_5)_3]_2$$

Procedure

In the manner described in Sec. A, 0.18 g (1.2 mmoles) of titanium(III) chloride is combined with 0.35 ml (2.4 mmoles) of triethylphosphine in 7.0 ml of toluene. The triethylphosphine (Pfaltz and Bauer) is stored under vacuum in a calibrated tube and condensed directly into the reaction vessel without further purification. The mixture is heated at 85° for 15 min and then filtered while still hot into the side arm of the reaction vessel. Evaporation of the gray-green filtrate to dryness gives 0.40 g (87% yield) of bis(triethylphosphine)trichlorotitanium(III), m.p. 122.5–127.5°. *Anal.* Calcd. for $TiCl_3[P(C_2H_5)_3]_2$: C, 36.90; H, 7.74; Cl, 27.23; P, 15.86; Ti, 12.26; mol. wt., 390. Found: C, 36.6; H, 7.81; Cl, 27.1; P, 15.5; Ti, 12.5; mol. wt. (cryoscopically determined in benzene), 402 ± 14.

Properties

Bis(triethylphosphine)trichlorotitanium(III) is a gray-green solid that is extremely air-sensitive. Characteristic absorption bands (cm^{-1}) in the infrared spectrum of the solid (mineral-oil mull) occur at 1050 (vs) (CH_3 rock); 965 (m) (CH_3 rock); 870 (s,br); 765 (vs,br) (P—C stretch); 320 (vs) (Ti—P stretch); and 305 (vs) (Ti—P or Ti—Cl stretch).

References

1. C. D. Schmulbach, C. H. Kolich, and C. C. Hinckley, *Inorg. Chem.*, **11,** 2841 (1972).
2. J. P. Collman, *Acc. Chem. Res.*, **1,** 136 (1968).
3. J. Chatt and R. G. Hayter, *J. Chem. Soc.*, **1963,** 1343.
4. W. L. Jolly, *Inorganic Syntheses*, **11,** 124 (1968).
5. S. D. Gokhale and W. L. Jolly, *ibid.*, **9,** 56 (1967).

Chapter Four

OTHER TRANSITION METAL COMPLEXES

30. REAGENTS FOR THE SYNTHESIS OF η-DIENE COMPLEXES OF TRICARBONYLIRON AND TRICARBONYLRUTHENIUM

Submitted by A. J. P. DOMINGOS,* J. A. S. HOWELL,* B. F. G. JOHNSON,* and J. LEWIS
Checked by N. GRICE† and R. PETTIT†

Previous syntheses of tricarbonyl(η-diene)iron complexes have relied mainly on the reaction of $Fe(CO)_5$, $Fe_3(CO)_{12}$, or $Fe_2(CO)_9$ with the free diene. The use of the first two carbonyls suffers from the prolonged reflux times and/or ultraviolet irradiation necessary to obtain reaction and the consequent low yields and mixtures of complexes obtained with heat- and ultraviolet-sensitive dienes.[1] The latter reagent, although utilized at lower temperatures, may react with polyenes ($n \geqslant 3$) to give mixtures containing, in addition to the expected product, binuclear derivatives containing a metal—metal bond.[2]

The only readily available ruthenium carbonyl, $Ru_3(CO)_{12}$, reacts in a complex manner with many polyolefins to give, in addition to mononuclear derivatives, complexes retaining the Ru_3 cluster and products resulting from hydrogen abstraction.[3]

Described here is the preparation of (benzylideneacetone‡)tricarbonyliron(0) and (1,5-cyclooctadiene)tricarbonyl ruthenium(0). Both compounds func-

*University Chemical Laboratory, Lensfield Road, Cambridge CB2 1EW, England.
†Chemistry Department, University of Texas at Austin, Austin, Tex. 78712.
‡4-Phenyl-3-buten-2-one.

tion as convenient sources of the metal tricarbonyl moiety by displacement of the organic ligand under mild conditions.

A. (BENZYLIDENEACETONE)TRICARBONYLIRON(0)

$$C_6H_5CH{=}CHCOCH_3 + Fe_2(CO)_9 \xrightarrow[60^\circ]{\text{toluene}} (C_6H_5CH{=}CHCOCH_3)Fe(CO)_3 + Fe(CO)_5 + CO$$

While a great number of tricarbonyl(η-diene)iron complexes have been reported and their reactivity investigated, much less is known of the corresponding heterodiene complexes. In recent years, synthesis of several tricarbonyl(heterodiene)iron systems involving[3] η coordination of the heterodiene unit has been achieved.[4–6] Among the tetracarbonyl(η-olefin)iron complexes prepared by Weiss[7] was tetracarbonyl(cinnamaldehyde)iron, which converts on heating to the η-bonded tricarbonyl(cinnamaldehyde)iron.[8] The preparation and synthetic utility of (benzylideneacetone)tricarbonyl iron, an analogous complex of an α,β-unsaturated ketone, are reported here.

Procedure

■ **Caution.** *The reaction should be carried out in a well-ventilated hood, since volatile, toxic iron pentacarbonyl is obtained as a by-product.*

In a 250-ml flask, benzylideneacetone* (10.4 g, 0.07 mole) and $Fe_2(CO)_9$[9] (26 g, 0.07 mole) are heated in toluene† (100 ml) under nitrogen with magnetic stirring for 4–5 hr at 60°. After removal of the solvent and iron pentacarbonyl under vacuum, the residue is dissolved in 20 ml of 10% ethylacetate-toluene, filtered through kieselguhr, and chromatographed on silica gel (2.5 by 40 cm). Elution with the same solvent develops a red band which, on removal of the solvent, gives the product as orange-red crystals (6.07 g, 32%, m.p. 88–89°). The complex may be recrystallized from hexane. *Anal.* Calcd. for $C_{13}H_{10}O_4Fe$: C, 54.5; H, 3.5. Found: C, 54.4; H, 3.6.

Properties

(Benzylideneacetone)tricarbonyliron is stable indefinitely when stored under nitrogen and is soluble in most common organic solvents. The infrared spectrum exhibits three metal carbonyl frequencies at 2065, 2005, and 1985 cm^{-1} (cyclohexane). The nmr spectrum (Hα, τ3.98, doublet; Hβ, τ6.90, doublet;

*Koch-Light Laboratories, Colnbrook, Bucks., England.

†The checkers recommend the use of benzene in place of toluene.

$J = 6.0$ Hz; CH_3, $\tau 7.50$, singlet; Ph, $\tau 2.73$, multiplet) is consistent with its η-bonded structure.

The complex functions as a convenient source of the tricarbonyliron moiety by displacement of the α,β-unsaturated ketone. For example, reaction with 1,3-cycloheptadiene results in a 78% yield of (η-1,3-cycloheptadiene)tricarbonyliron. More importantly, it may be used in syntheses of tricarbonyl(diene)iron complexes where the iron carbonyls are not satisfactory. Several complexes of sensitive heptafulvenes have been prepared in this way, and the reagent has been used in the synthesis of tricarbonyliron complexes of several steroids.

These reactions may be carried out under mild conditions (60°, benzene, 4–8 hr) and are clean, in that no colloidal iron or other iron carbonyls are produced. The work-up is thus much easier, and the reactions can be monitored by infrared spectroscopy. In addition, the well-defined stoichiometry of the displacement is in marked contrast to the uncertain stoichiometry of the reactions of the iron carbonyls, which are commonly used in large excess.

B. (1,5-CYCLOOCTADIENE)TRICARBONYLRUTHENIUM(0)

$$1,5\text{-}C_8H_{12} + Ru_3(CO)_{12} \xrightarrow[\text{reflux}]{\text{benzene}} (1,5\text{-}C_8H_{12})Ru(CO)_3$$

In contrast to the very large number of tricarbonyl(η-diene)iron complexes described in the literature,[10–12] the corresponding ruthenium compounds have received very little attention. This may reflect the well-documented tendency of ruthenium to form metal—metal bonds as opposed to iron.[13] In particular, while the metal—metal bonds in $Fe_3(CO)_{12}$ are easily broken, $Ru_3(CO)_{12}$ undergoes a variety of reactions in which the Ru_3 cluster is retained.

Described here is the preparation of (1,5-cyclooctadiene)tricarbonylruthenium and its use in the synthesis of tricarbonyl(η-diene)ruthenium complexes.

Procedure

In a 100-ml flask, $Ru_3(CO)_{12}$* (960 mg, 1.5 mmoles) is reacted for 8 hr† with 1,5-cyclooctadiene‡ (15 ml, 0.12 mole) in refluxing benzene (75 ml) under nitrogen and the excess of diene distilled (ca. 30°, 7 torr). Unreacted $Ru_3(CO)_{12}$ (307 mg, 32%) is separated by treating the residue with pentane. The

*See Syntheses 13 and 14.

†Refluxing for longer periods results in significant isomerization to (1,2,3,6-η-C_8H_{12})Ru-$(CO)_3$.

‡Koch-Light Laboratories, Colnbrook, Bucks., England.

pentane solution is concentrated and chromatographed* with pentane on a silica gel column (1.5 by 30 cm) to separate traces of (1,2,3,6-η-C_8H_{12})$(RuCO)_3$ and small amounts of $Ru_3(CO)_{12}$. The major band, on concentration, gives a yellowish oil which is sublimed at 0° and 0.005 torr onto a cold finger at −30°. A second chromatography and sublimation yield (1,5-cyclooctadiene)tricarbonylruthenium (684 mg, 58%) as white crystals melting at room temperature. *Anal.* Calcd. for $C_{11}H_{12}O_3Ru$; C, 44.5; H, 4.1. Found: C, 44.5; H, 4.3.

Properties

(1,5-Cyclooctadiene)tricarbonylruthenium is unstable even at −20° under nitrogen but may be kept in a frozen benzene solution for a month without decomposition. The infrared spectrum exhibits metal carbonyl absorptions at 2043, 1982 (sh), and 1967 (br) cm^{-1} (heptane). The nmr spectrum shows multiplets at τ6.28 and τ7.6–8.2 due to the olefinic and methylenic protons, respectively.

The complex functions as the most suitable source of the tricarbonylruthenium unit in syntheses of tricarbonyl(η-diene)ruthenium complexes. Derivatives of 1,3-cyclohexadiene, 1,3-cycloheptadiene, cycloheptatriene, cyclooctatetraene, 2,4,6-cycloheptatrien-1-one, bicyclo[3.2.1]octa-2,6-diene, bicyclo-[3.2.1]octa-2,4-diene, and butadiene have been prepared by displacement of 1,5-cyclooctadiene.

Although alternative procedures exist for the preparation of some of these compounds, the ligand-displacement method possesses two distinct advantages: (1) The absence of side reactions leads to high yields and facilitates the purifications. (2) The mild conditions required for the reaction to go to completion (refluxing benzene, 30 min) avoid extensive decomposition of the complex formed; this makes the method invaluable where the complexes are very unstable.

References

1. For example, W. McFarlane, L. Pratt, and G. Wilkinson, *J. Chem. Soc.*, **1963**, 2162.
2. G. F. Emerson, R. Pettit, J. E. Mahler, and R. Collins, *J. Am. Chem. Soc.*, **86**, 3591 (1964).
3. For example, A. J. P. Domingos, B. F. G. Johnson, and J. Lewis, *J. Organomet. Chem.*, **36**, C43 (1972); M. I. Bruce, M. A. Cairns, and M. Green, *J. Chem. Soc., Dalton Trans.*, **1972**, 1293.
4. S. Otsuka, T. Yoshida, and A. Nakamura, *Inorg. Chem.*, **6**, 20 (1967).
5. H. T. Dieck and H. Beck, *Chem. Comm.*, **1968**, 678.
6. M. Dekker and G. R. Knox, *Chem. Comm.*, **1967**, 1243.

*The chromatography is continuously monitored by passage of the eluant through an infrared solution cell.

7. E. Weiss, K. Stark, J. E. Lancaster, and H. D. Murdoch, *Helv. Chim. Acta,* **46,** 288 (1963).
8. K. Stark, J. E. Lancaster, H. D. Murdoch, and E. Weiss, *Z. Naturforsch.,* **19,** 284 (1964).
9. R. Pettit, G. F. Emerson, and J. Mahler, *J. Chem. Educ.,* **40,** 175 (1963).
10. R. B. King, in "Organometallic Syntheses," J. J. Eisch and R. B. King (eds.), Vol. I, p. 93, Academic Press, 1965.
11. R. Pettit and G. F. Emerson, *Adv. Organomet. Chem.,* **1,** 1 (1964).
12. H. W. Quinn and J. H. Tsai, *Adv. Inorg. Radiochem.,* **12,** 217 (1969).
13. M. I. Bruce and F. G. A. Stone, *Angew. Chem., Int. Ed.,* **7,** 427 (1968).

31. BIS(η-CYCLOPENTADIENYL)NIOBIUM COMPLEXES

Submitted by C. R. LUCAS*
Checked by J. A. LABINGER† and J. SCHWARTZ†

The observation of catalytic behavior by lower-valent compounds of Group VA elements[1,2] has made it desirable to expand our knowledge of the chemistry of these and neighboring elements. The organic chemistry of the early transition elements is relatively little explored in comparison with that of later members of the series. This situation has arisen for a number of reasons, some at least of which are related to the lack of good preparative methods for starting materials. Preparations of dichlorobis(η-cyclopentadienyl)niobium[3,4] and bis(η-cyclopentadienyl)hydrido(triethylphosphine)niobium[1] are described in the literature, but both products are obtained in low yields and involve prolonged or tedious procedures or the use of equipment not readily available. In contrast, the preparations reported below provide relatively rapid routes to good yields of suitable starting materials for the investigation of organoniobium chemistry. For example, approximately 40 g of dichlorobis(η-cyclopentadienyl)niobium can be prepared in 1 day by this new method.

A. DICHLOROBIS (η-CYCLOPENTADIENYL)NIOBIUM(IV)

$$NbCl_5 + 6NaC_5H_5 \longrightarrow 5NaCl + (C_5H_5)_4Nb + \text{organic products}$$

$$2(C_5H_5)_4Nb + 4HCl + [O] \xrightarrow[O_2]{Br_2} [\{(\eta\text{-}C_5H_5)_2NbCl\}_2O]Cl_2{}^{\dagger} + 2C_5H_6$$

*Department of Chemistry, Memorial University of Newfoundland, St. John's, Newfoundland AIC 5S7, Canada.

†Department of Chemistry, Princeton University, Princeton, N.J. 08540.

†The ionic substances containing niobium(V) which exist in these red solutions have not all been identified. In addition to the ion indicated, species of the form $[(\eta\text{-}C_5H_5)_2Nb(X)Y]^+$ (X, Y = Cl, Br, OH, or O—) are probably present in an unknown equilibrium with each other.

$$5Cl^- + [\{(\eta\text{-}C_5H_5)_2NbCl\}_2O]^{2+} + SnCl_3^- + 2H^+ \longrightarrow 2(\eta\text{-}C_5H_5)_2NbCl_2 + SnCl_6^{2-} + H_2O$$

Procedure

The reaction is carried out in a 3-l. three-necked flask. Through the central neck, a mechanical stirrer is fitted, and through another is added the niobium pentachloride.* To the third neck is attached a T piece, one side of which is connected to an inert-gas supply and the other to a bubbler.

Under an atmosphere of dry nitrogen, dry cyclopentadienylsodium† (100 g, 1.14 moles) is prepared and suspended at room temperature in dry benzene‡ (400 ml). With vigorous mechanical stirring, powdered niobium pentachloride (57 g, 0.21 mole) is added in small portions, care being taken to prevent the reaction temperature from exceeding ca. 70° after each addition. After 1 hr the suspension has cooled to room temperature and is a uniform purple-brown color. It is stirred for another $\frac{1}{2}$ hr and then poured in air onto concentrated hydrochloric acid (1l.). The resulting mixture is heated gently (in the hood) with occasional swirling until all the benzene has been evaporated. To the boiling brown suspension, small quantities of bromine are then carefully added (ca. 60 ml in all) until the solid phase no longer acts like a sticky lump (some scraping of the vessel walls may be necessary) but as a suspended powder. Excess bromine is then boiled away. The hot red§ liquid is decanted in air through a filter; extraction with hot concentrated hydrochloric acid and bro-

*Niobium pentachloride is available commercially from Alfa Inorganics, Ltd. The checkers found that sublimation of the commercial material *in vacuo* was necessary before use in the preparation.

†White cyclopentadienylsodium can be prepared by cracking dicyclopentadiene (for details, see "Organometallic Syntheses," J. J. Eisch and R. B. King (eds.), Vol. I, pp. 64 ff., Academic Press, New York, 1965) under an atmosphere of dry nitrogen directly into a suspension of sodium sand in dry tetrahydrofuran from which dissolved oxygen has been removed. The apparatus for this operation should be assembled so that the distillate of cyclopentadiene monomer is added directly under nitrogen to the sodium. The distillation should proceed at a rate of ca. 3 ml/min until almost all the sodium has dissolved. The bulk of volatile substances may then be removed *in vacuo* on a hot-water bath to give a sticky white residue. Final removal of the last traces of tetrahydrofuran from this product is essential. This can be accomplished by heating the residue to 100° *in vacuo* until no more liquid distils into a liquid-nitrogen trap. Breaking up large lumps assists solvent removal. Large batches (300 g) may require heating for up to 72 hr.

‡Freshly distilled from calcium hydride and free from dissolved oxygen.

§If the cyclopentadienylsodium is not dry and white, the hydrochloric acid extracts will be pink or yellow instead of red, and the major product of this preparation will be a white precipitate of Nb_2O_5.

mine is repeated until the extracts are nearly colorless. After about 10 extractions, the collected filtrates are brought to the boiling point to redissolve any precipitated matter, and they are placed under an inert atmosphere.* To the hot solution is added under an inert atmosphere a hot solution of tin dichloride dihydrate (47.5 g, 0.21 mole) in concentrated (12*M*) hydrochloric acid (200 ml), and the mixture is left standing overnight. The brown tabular crystals of dichlorobis(η-cyclopentadienyl)niobium are separated by filtration, and the product is washed several times with water and once with acetone (50 ml) before being dried *in vacuo*. Yield is 45 g (75% based on $NbCl_5$). *Anal.* Calcd. for $C_{10}H_{10}Cl_2Nb$: C, 40.8; H, 3.4; Cl, 24.1; Nb, 31.6. Found: C, 40.8; H, 3.5; Cl, 24.2; Nb, 31.3.

Properties

Crystals of the product obtained by this method may be weighed and handled in air although they should be stored and used in further reactions under an inert atmosphere. The compound is only very sparingly soluble in common organic solvents. It is paramagnetic, and its infrared spectrum (Nujol) is 3090 (m), 1011 (m), 822 (s), 725 (m), 308 (m), 290 (s), and 268 (s) cm^{-1}.

B. BIS(η-CYCLOPENTADIENYL)(TETRAHYDROBORATO)-NIOBIUM(III)

$$2(\eta\text{-}C_5H_5)_2NbCl_2 + 4NaBH_4 \longrightarrow 4NaCl + 2(\eta\text{-}C_5H_5)_2NbBH_4 + H_2 + B_2H_6$$

Procedure

Standard Schlenk-tube techniques[5] for the handling of air-sensitive compounds should be employed in this preparation.†

A mixture of dichlorobis(η-cyclopentadienyl)niobium (12.0 g, 40.8 mmoles) and sodium tetrahydroborate (4.60 g, 121 mmoles) is suspended at room temperature under an atmosphere of dry nitrogen† in 1,2-dimethoxyethane‡ (250 ml). Vigorous effervescence occurs. After magnetic stirring for 1½ hr, the

*The checkers find that concentration of the extracts improves the yield.

†Due to extreme air-sensitivity of the compounds, rigorous exclusion of air from solutions must be practiced throughout this preparation or difficulties will be encountered during attempted crystallization of the product.

‡Freshly distilled from calcium hydride and free from dissolved oxygen.

solution is dark-green, and all the brown crystals have disappeared. The mixture is filtered and the residue washed with dry 1,2-dimethoxyethane (60 ml). (■ **Caution.** *The residue is pyrophoric in air.*) The solvent is removed *in vacuo* on a hot-water bath from the combined filtrates. The resulting residue is extracted with toluene (freshly distilled from calcium hydride) at room temperature until the extracts are nearly colorless. Dilution with an equal volume of *n*-heptane and cooling to $-78°$ gives green tabular crystals. Yield is 8.25 g (85% based on $(\eta\text{-}C_5H_5)_2NbCl_2$). *Anal.* Calcd. for $C_{10}H_{14}BNb$: C, 50.5; H, 5.9; B, 4.5; Nb, 39.0. Found: C, 50.4; H, 5.7; B, 4.9; Nb, 39.2.

Properties

The compound sublimes *in vacuo* at 80° with some simultaneous decomposition to an involatile residue. It is soluble in ethers and aromatic hydrocarbons. The compound is pyrophoric in air particularly when it has been purified by sublimation, but it is stable indefinitely under a dry nitrogen atmosphere at room temperature. The infrared spectrum (Nujol) is 3100 (w), 2470 (sh), 2420 (s), 2310 (m), 1720 (m), 1440 (s), 1280 (m), 1265 (sh), 1160 (s), 1010 (m), 885 (m), 870 (m), 850 (m), 838 (m), 800 (s), 648 (m), 380 (sh), 370 (m), and 360 (sh) cm^{-1}. The nmr spectrum 60 MHz in benzene-d_6 (TMS internal standard) shows a single peak at $\tau 5.00$. No resonance due to borohydrido-hydrogens was detectable, but this is not unusual.[6,7] The compound's mass spectrum has a parent ion peak at $m/e = 238$.

C. BIS(η-CYCLOPENTADIENYL)HYDRIDO(DIMETHYLPHENYLPHOSPHINE)NIOBIUM(III)

$$(\eta\text{-}C_5H_5)_2NbBH_4 + 2R_3P \longrightarrow (\eta\text{-}C_5H_5)_2Nb(H)(PR_3) + H_3BPR_3$$

$$(\eta\text{-}C_5H_5)_2Nb(H)(PR_3) + HCl \longrightarrow [(\eta\text{-}C_5H_5)_2Nb(H)_2(PR_3)]Cl$$

$$[(\eta\text{-}C_5H_5)_2Nb(H)_2(PR_3)]^+ + OH^- \longrightarrow (\eta\text{-}C_5H_5)_2Nb(H)(PR_3) + H_2O$$

Procedure

Standard Schlenk-tube techniques[5] for the handling of air-sensitive compounds should be used in this preparation.

Dimethylphenylphosphine (1.28 ml, 9.08 mmoles) is added at room temperature under nitrogen to a green solution of bis(η-cyclopentadienyl) (tetrahydroborato)niobium (1.07 g, 4.54 mmoles) in benzene (50 ml). The red solution which results is stirred magnetically for 15 min, and the solvent is then removed *in vacuo* at room temperature.* The residue is dissolved in ether (50 ml), and the solution chilled to $-50°$. Hydrogen chloride gas is passed through the solution until the red color is discharged. The white precipitate is separated, washed twice with ether (25 ml), and then stirred vigorously at room temperature with 3*N* aqueous sodium hydroxide solution (50 ml) and ether (50 ml). The red ether layer is separated, and the solvent is removed *in vacuo* at room temperature. The red residue is recrystallized from petroleum ether at $-78°$. Yield is 1.15 g, or 70%. *Anal.* Calcd. for $C_{18}H_{22}PNb$: C, 59.7; H, 6.1; P, 8.5. Found: C, 58.8; H, 6.1; P, 8.2.

Properties

In the solid state, thermal decomposition begins at 94°, but in solution the compound is unstable above 40°. No obvious color change accompanies thermal decomposition in solution, but upon removal of the solvent, a red oil is obtained which has no infrared absorptions in the terminal metal-hydrogen stretching region. Even large crystals of the compound are very sensitive to oxidation by air. The compound is soluble in hydrocarbons, ethers, alcohols, and acetone. It reacts with halogenated solvents. Its infrared spectrum (Nujol) is 3100 (w), 3070 (w), 1630 (s), 1440 (s), 1180 (m), 1120 (sh), 1108 (s), 1010 (s), 990 (s), 940 (s), 918 (sh), 905 (s), 872 (m), 838 (m), 805 (m), 780 (sh), 748 (s), 718 (m), 700 (s), 675 (m), 508 (s), 421 (m), and 335 (m) cm^{-1}. The nmr spectrum of the compound in benzene-d_6 or acetone-d_6 (TMS internal standard) shows a complex band centered at $\tau 2.65$, a doublet with splitting $J_{PH} = 2.0$ Hz centered at $\tau 5.34$, a doublet with splitting $J_{PH} = 6.8$ Hz centered at $\tau 8.53$, and a doublet with splitting $J_{PH} = 28.6$ Hz centered at $\tau 17.53$.

D. BIS(η-CYCLOPENTADIENYL)DIHYDRIDO-(DIMETHYLPHENYLPHOSPHINE)NIOBIUM (V) HEXAFLUOROPHOSPHATE OR TETRAFLUOROBORATE

$$(\eta\text{-}C_5H_5)_2Nb(H)PR_3 + HA \longrightarrow [(\eta\text{-}C_5H_5)_2Nb(H)_2PR_3]A$$
$$(A = BF_4 \text{ or } PF_6)$$

Procedure

Standard Schlenk-tube techniques[5] for the handling of air-sensitive compounds should be employed in this preparation.

*The compound is thermally unstable in solution above ca. 40°. Overheating results in isolation of a red oil, which contains no terminal metal-hydrogen bonds, from solutions which are similar in appearance to those of the desired product.

To a red solution of (η-$C_5H_5)_2Nb(H)[P(CH_3)_2(C_6H_5)]$ (0.52 g, 1.4 mmoles) in ether (30 ml) at $-78°$ is added under an atmosphere of dry nitrogen 5 drops of 60% aqueous tetrafluoroboric or hexafluorophosphoric acid. After the mixture has stood for 1 hr, the colorless supernatant liquid is decanted from the white precipitate which is washed twice with small portions of ether. The precipitate is dried *in vacuo* at $-20°$. Yield is 100%. *Anal.* Calcd. for $C_{18}H_{23}PNb$: C, 48.0; H, 5.2; P, 8.5. Found: C, 47.6; H, 4.9; P, 8.7.

Properties

In the solid state at room temperature the compound is stable for only about 1 hr, but it can be stored indefinitely under nitrogen at $-20°$. The compound should always be handled under an inert atmosphere as it is air-sensitive. Its infrared spectrum consists of strong bands at 1010 and 815 cm^{-1} arising from vibrations of the η-cyclopentadienyl ligand and bands at 1100, 750, 700, and 490 cm^{-1} due to vibrations of coordinated dimethylphenylphosphine. Bands at 1050 and 530 cm^{-1} in spectra of the tetrafluoroborate salt or at 830 cm^{-1} in those of the hexafluorophosphate salt arise from vibrations of the anion; a band at 1740 cm^{-1} is due to a niobium-hydrogen stretching vibration. The nmr spectrum (TMS internal standard) at $-36°$ of an acetone-d_6 solution of the tetrafluoroborate salt is composed of a complex band centered at $\tau 2.4$, a doublet with $J_{PH} = 2.0$ Hz centered at $\tau 4.42$, a doublet with $J_{PH} = 7.1$ Hz centered at $\tau 8.52$, and a doublet with $J_{PH} = 31.5$ Hz centered at $\tau 13.96$.

E. BROMOBIS(η-CYCLOPENTADIENYL)DIHYDRIDO-(DIMETHYLPHENYLPHOSPHINE)NIOBIUM(III)

$$(\eta\text{-}C_5H_5)_2Nb(H)[P(CH_3)_2(C_6H_5)] + C_4H_9Br \longrightarrow (\eta\text{-}C_5H_5)_2Nb(Br)[P(CH_3)_2(C_6H_5)] + C_4H_{10}$$

Procedure

Standard Schlenk-tube techniques[5] for the handling of air-sensitive compounds should be employed in this preparation.

To a red solution at room temperature of (η-$C_5H_5)_2Nb(H)[P(CH_3)_2(C_6H_5)]$ (0.25 g, 0.68 mmole) in petroleum ether (125 ml) from which dissolved oxygen has been removed, 1-bromobutane (0.20 ml, 1.1 mmoles) is added under an atmosphere of nitrogen, and the solution is stirred briefly. After the mixture has stood overnight, the colorless supernatant liquid is decanted from the green precipitate. The compound is recrystallized by dissolving it in dry tetrahydrofuran and adding dry ether at room temperature. Yield is 0.27 g (90% based on metal hydride). *Anal.* Calcd. for $C_{18}H_{21}BrPNb$: C, 49.0; H, 4.8; Br, 15.0; P, 5.8. Found: C, 49.9; H, 4.9; Br, 14.4; P, 5.9.

Properties

The compound is air-sensitive and should be handled under an inert atmosphere. It is readily oxidized to $(\eta\text{-}C_5H_5)_2NbX_2$ (X = halogen or pseudohalogen) by substances containing X. The choice of solvent and conditions for the preparation are therefore critical since the product must be precipitated as it forms, or it will react further to give $(\eta\text{-}C_5H_5)_2NbX_2$. The compound is soluble in tetrahydrofuran, aromatic hydrocarbons, and acetone. Its infrared spectrum is very similar to that of the metal hydride from which it is derived except for the absence of the band at 1630 cm^{-1}, which is due to the metal-hydrogen stretching vibration. The nmr spectrum of a benzene-d_6 solution of the compound (TMS internal standard) consists of a complex band centered at τ2.7, a doublet with splitting J_{PH} = 2.1 Hz centered at τ5.21, and a doublet with splitting J_{PH} = 7.5 Hz centered at τ8.50.

References

1. F. N. Tebbe and G. W. Parshall, *J. Am. Chem. Soc.*, **93,** 3793 (1971).
2. E. K. Barefield, G. W. Parshall, and F. N. Tebbe, *J. Am. Chem. Soc.*, **92,** 5234 (1970).
3. F. W. Siegert and H. J. de Liefde Meijer, *J. Organomet. Chem.*, **23,** 177 (1970).
4. W. E. Douglas and M. L. H. Green, *J. Chem. Soc., Dalton Trans.*, **1972,** 1796.
5. R. B. King, in "Organometallic Syntheses," J. J. Eisch and R. B. King (eds.), Vol. I, p. 4, Academic Press, New York, 1965.
6. M. L. H. Green, H. Munakata, and T. Saito, *J. Chem. Soc. (A)*, **1971,** 469.
7. M. Grace, H. Beall, and C. H. Bushweller, *Chem. Commun.*, **1970,** 701.

32. DICHLORO(1,3-PROPANEDIYL)PLATINUM AND *trans*-DICHLOROBIS(PYRIDINE)(1,3-PROPANEDIYL)-PLATINUM DERIVATIVES

Submitted by G. W. LITTLECOTT,* F. J. McQUILLIN,* and K. G. POWELL*
Checked by S. D. ITTEL†

*Department of Organic Chemistry, The University, Newcastle upon Tyne NE1 7RU, England.
†Department of Chemistry, Northwestern University, Evanston, Ill. 60201.

Dichloro(1,3-propanediyl)platinum and derivatives carrying substituents in the 1,3-propanediyl residue have been prepared by displacement of ethylene from bis(ethylene)tetrachlorodiplatinum by cyclopropane or substituted cyclopropanes.[1,2] This reasonably general procedure,[2] which is also appreciably more convenient than that originally used by Tipper,[3] has also been applied in the particular case of cyclopropane by Brown.[4] However, cyclopropanes bearing strongly electron-withdrawing substituents, for example, —CN, —CO_2Me, —COMe, —COC_6H_5, fail to react,[2] and certain *cis*-1,2-disubstituted cyclopropanes are isomerized to give olefinic products.[2]

A. DICHLORO(1,3-PROPANEDIYL)PLATINUM

$$n[(C_2H_4)PtCl_2]_2 + 2nR\text{—}\triangleleft \longrightarrow 2[(RC_3H_5)PtCl_2]_n + 2nC_2H_4$$

■ **Caution.** *The Carius tube (or glass pressure vessel) must be used with care and proper shielding because of the possibility of an explosion at excess pressure.*

Procedure

The preparation[5] of tetrachlorobis(ethylene)diplatinum is accelerated if carried out with ethylene under pressure, either in a Carius tube or (because of the danger with a Carius tube), preferably, in an 80-ml glass pressure vessel.

When a Carius tube is used, potassium tetrachloroplatinate (2 g) in water (9 ml) and concentrated hydrochloric acid (1 ml) in a 150-ml Carius tube is cooled in liquid nitrogen, and ca. 1.5 ml of liquid ethylene is condensed into the tube, which is sealed. The mixture is allowed to warm to room temperature, and the tube is shaken mechanically for 36 hr. During this time, the original deep-red solution changes color to yellow.

Alternatively,* the same quantities of potassium tetrachloroplatinate, water, and concentrated hydrochloric acid are stirred magnetically in an 80-ml pressure glass vessel under an ethylene pressure of not more than 80 psi for 24 hr.

The contents of the Carius tube (opened only after cooling, e.g., in liquid nitrogen) or of the glass pressure vessel are then evaporated *in vacuo* in a rotary evaporator at room temperature. The dry residue is then crushed and

*The checker recommends this alternative procedure be performed in a Lab-Crest Aerosol Reaction vessel from Fischer and Porter, Warminster, Pa. 18974.

extracted into ethanol (20 ml) containing concentrated hydrochloric acid (0.8 ml). After filtration (Büchner funnel), the filtrate is evaporated at 40° in a rotary evaporator. As the solution becomes viscous, the rotating flask is warmed to 60°. Further evaporation results in the sudden crystallization of the orange crystalline product. The crystals are then dried in a vacuum desiccator and recrystallized from toluene to give tetrachlorobis(ethylene)diplatinum as a yellow solid (yield 1.3 g, 92%).

Dichloro(1,3-propanediyl)platinum is obtained by passing a slow stream of cyclopropane from a cylinder through tetrachlorobis(ethylene)diplatinum (0.2 g) stirred in dichloromethane (15 ml) for 20 min. The pale-yellow precipitate is collected on a filter, washed with a little chloroform and then with ether, and dried in vacuo. Yield: 0.15 g, 75%. This material is a highly insoluble tetramer which has been characterized by elemental analysis and infrared spectrum.[3,6]

Dichloroplatinum derivatives from substituted cyclopropanes are obtained by the following general procedure.[1,2] Tetrachlorobis(ethylene)diplatinum (0.1 g) under dry ether (5 ml) with an excess of a substituted cyclopropane, e.g., phenylcyclopropane (0.2 g), is warmed under gentle reflux for 2 hr, and the pale-yellow product is collected on a filter. These products are insoluble chlorine-bridged polymers which melt with decomposition (see Table I), and which have been characterized by elemental analysis[2] and infrared spectra.[2] *Anal.* Calcd. for dichloro(2-phenyl-1,3-propanediyl)platinum: C, 28.1; H, 2.6; Cl, 18.5; Pt, 50.8. Found: C, 28.2; H, 2.7; Cl, 18.3, Pt, 51.1.

B. *trans*-DICHLOROBIS(PYRIDINE)(2-PHENYL-1,3-PROPANEDIYL)PLATINUM

$$[(RC_3H_5)PtCl_2]_n + 2n(C_5H_5N) \longrightarrow n(RC_3H_5)Pt(C_5H_5N)_2Cl_2$$

The bis(pyridine) derivatives are prepared by stirring dichloro(1,3-propanediyl)- or dichloro-substituted(1,3-propanediyl)platinum (0.2 g) with pure pyridine (0.2 g) in dry chloroform (2 ml) for a few minutes. The solution is passed through a short column of silica gel (6 g) and washed through with chloroform (10 ml). Evaporation of the eluate *in vacuo* and recrystallization of the residue gives the derivatives listed in Table I. *Anal.* Calcd. for *trans*-dichlorobis(pyridine)(2-phenyl-1,3-propanediyl)platinum: C, 42.1; H, 3.7; Cl, 13.9; N, 5.2; Pt, 36.0. Found: C, 41.7; H, 3.7; Cl, 13.2; N, 5.2; Pt, 36.6.

Properties

Properties are summarized in Table I.

TABLE I

R	m.p., °C	Yield, %	m.p., °C	Yield, %	nmr,*τ
	$(RC_3H_5)PtCl_2$		$(RC_3H_5)Pt(py)_2Cl_2$		
H	148†	70	145‡	80	7.34
2-Phenyl	135†	90	130‡	65	7.03, 5.90
2-*o*-Nitrophenyl	148†	80	220‡	80	7.02, 5.44
2-Benzyl	123†	90	114§	55	7.30, 6.79
1-*p*-Tolyl	125†	85	107§	85	7.05, 5.07
2-*n*-Hexyl	120†	80	125§	80	7.58, 7.25
R′, R″ =	$(R'R''C_3H_4)PtCl_2$		$(R'R''C_3H_4)Pt(py)_2Cl_2$		
trans-1,2-Diphenyl	163†	80	116‡	65	4.88, 5.25 6.79
trans-1-*n*-butyl-2-methyl	119†	60	¶		7.07, 7.6

*1,3-propanediyl ^1H nmr signals in $CDCl_3$ relative to TMS.
†Decomposed.
‡From ethanol.
§From benzene-hexane.
¶Unstable in solution.

Dichloro(1,3-propanediyl)platinum and its bis(pyridine) derivative have been studied by a number of authors.[6] Dichloro(1,3-propanediyl)platinum, and the corresponding substituted 1,3-propanediyl platinum compounds release the parent cyclopropane on treatment with potassium cyanide, potassium iodide, a tertiary phosphine, carbon monoxide, and other ligands.[2,6] Reduction by means of hydrogen or lithium aluminum hydride yields chiefly isomeric substituted propanes.[7] Dichlorobis(pyridine)(1,3-propanediyl)platinum in refluxing benzene yields a pyridinium ylid complex,[6,8] $(CH_3CH_2CHNC_5H_5)$-$PtpyCl_2$.

References

1. W. J. Irwin and F. J. McQuillin, *Tetrahedron Lett.*, **1968,** 1937.
2. F. J. McQuillin and K. G. Powell, *J. Chem. Soc., Dalton Trans.*, **1972,** 2123.
3. C. F. H. Tipper, *J. Chem. Soc.*, **1955,** 2045.
4. D. B. Brown, *J. Organomet. Chem.*, **24,** 787 (1970).
5. J. Chatt and M. L. Searle, *Inorganic Syntheses*, **5,** 210 (1957).
6. D. M. Adams, J. Chatt, R. G. Guy, and N. Sheppard, *J. Chem. Soc.*, **1961,** 738; S. E. Binns, R. H. Cragg, R. D. Gillard, B. T. Heaton, and M. F. Pilbrow, *J. Chem. Soc.* (*A*), **1969,** 1227; R. D. Gillard, M. Keeton, R. Mason, M. F. Pilbrow, and D. R. Russell, *J. Organomet. Chem.*, **33,** 247 (1971).
7. W. J. Irwin and F. J. McQuillin, *Tetrahedron Lett.*, **1968,** 2195.
8. M. Keeton, R. Mason, and D. R. Russell, *J. Organomet. Chem.*, **33,** 259 (1971).

33. OLEFIN(β-DIKETONATO)SILVER(I) COMPOUNDS

Submitted by W. PARTENHEIMER* and E. H. JOHNSON†
Checked by H. A. TAYIM‡

Silver olefin compounds have been studied extensively for over 30 years. The majority of the compounds are stable only in solution and many have been characterized via stability constant studies. Solid relatively unstable salts have been prepared having the stoichiometry $Ag_m(\text{olefin})_nX_m$, where X is usually the nitrate, tetrafluoroborate, or perchlorate anion, $m = 1, 2$, and $n = 1, 2, 3$.[1] Recently, relatively stable, nonionic, monomeric complexes of the type β-diketonato(olefin)silver(I) have been characterized.[2]

A. (1,1,1,5,5,5-HEXAFLUORO-2,4-PENTANEDIONATO)(1,5-CYCLOOCTADIENE)-SILVER(I)

$$AgNO_3 + C_8H_{12} \xrightarrow{H_2O} Ag(C_8H_{12})NO_3$$

$$C_5H_2F_6O_2 + NaOH \xrightarrow{H_2O} NaC_5HF_6O_2 + H_2O$$

$$Ag(C_8H_{12})NO_3 + NaC_5HF_6O_2 \xrightarrow{H_2O} Ag(C_8H_{12})(C_5HF_6O_2)\downarrow + NaNO_3$$

Procedure

Commercially available 1,5-cyclooctadiene§ and 1,1,1,5,5,5-hexafluoro-2,4-pentanedione¶ are used as received. The preparations are performed in air. The operations should be performed in a hood since both reactants are very malodorous.

A solution of (1,5-cyclooctadiene)silver(I) nitrate is prepared as follows: 5.94 g (34.9 mmoles) of silver nitrate, 40 ml of distilled water, and a magnetic stirring bar are placed in a 125-ml Erlenmeyer flask. The solution is stirred magnetically until the silver nitrate is dissolved. To this solution is added 4 ml of 1,5-cyclooctadiene, and the solution is stirred vigorously for 10 min. A white precipitate of (1,5-cyclooctadiene)silver(I) nitrate forms.

A solution of sodium hexafluoroacetylacetonate is prepared as follows:

*Amoco Research Center, Naperville, Ill. 60540.
†Dow Research Laboratories, Midland, Mich. 48640.
‡Chemistry Department, American University of Beirut, Beirut, Lebanon.
§Aldrich Chemical Company, Cedar Knolls, N.J. 07927.
¶Peninsular Chemresearch, Gainesville, Fla. 32601.

1.10 g (27.6 mmoles) of sodium hydroxide, 20 ml of water, and a magnetic stirring bar are placed in a 50-ml beaker. The sodium hydroxide is dissolved by magnetic stirring and then stirred continuously while 5.87 g (28.2 mmoles) (ca. 4 ml) of hexafluoroacetylacetone is slowly added via a 5-ml syringe. The tip of the syringe is immersed in the solution during the addition of the hexafluoroacetylacetone.

The (1,5-cyclooctadiene)silver(I) nitrate solution is stirred vigorously and the sodium hexafluoroacetylacetonate solution is added, yielding a flocculent white precipitate.* The white product is collected on a sintered-glass filter, washed with two 40-ml aliquots of distilled water, and then air-dried. The product can be purified by dissolving it in a minimum amount of cyclohexane at room temperature and allowing the solution to evaporate under a stream of air. Yield: 5.44 g (47%). *Anal.* Calcd. for $AgC_{13}H_{13}O_2F_6$: C, 36.90; H, 3.10; Ag, 25.50; mol. wt., 423. Found: C, 36.8; H, 3.22; Ag, 25.5; mol. wt., 435 (bromoform).†

Properties

1,5-Cyclooctadiene(hexafluoroacetylacetonato)silver(I) loses olefin slowly when exposed to air over long periods of time. It should be stored in a tightly closed container and kept in a refrigerator. It is soluble in halogenated solvents. The nmr spectrum of the compound has chemical shifts of $\tau 3.70$, 3.93, and 7.38 with relative intensities of 4, 1, and 8, respectively. Its infrared spectrum has characteristic absorbances at 1667 and 1634 cm^{-1}.

B. OLEFIN (β-DIKETONATO)SILVER(I)

$$AgNO_3 + \text{olefin} \xrightarrow{H_2O} Ag(\text{olefin})NO_3$$

$$Ag(\text{olefin})NO_3 + Na(\beta\text{-diketonate}) \xrightarrow{H_2O} Ag(\text{olefin})(\beta\text{-diketonate}) + NaNO_3$$

For β-diketone = 1,1,1-trifluoro-2,4-pentanedione, olefin = 1,5-cyclooctadiene or 1,3,5,7-cyclooctatetraene. For β-diketone = 1,1,1,5,5,5-hexafluoro-

*Occasionally a dark oil forms, which may be due to the presence of a slight excess of sodium hydroxide. This oil can be extracted with two 10-ml aliquots of benzene; the benzene extract is then allowed to evaporate at room temperature in a beaker to yield the desired white product.

†The checker reports the product to be analytically pure before the recrystallization, as well as a significantly higher yield (ca. 76%).

2,4,-pentanedione, olefin = 1,3,5,7-cyclooctatetraene, *cis*-cyclooctene, cyclohexene, or cycloheptene.

Procedure

The method of preparation of the above silver(I) compounds is identical to that used in Sec. A. The compounds can be purified in an identical manner except that a chloroform-cyclohexane mixture is used. The compounds are obtained in approximately 40% yields. *Anal.* Calcd. for $AgC_{13}H_{15}O_2F_6$(cyclooctene complex): C, 36.72; H, 3.56; mol. wt., 425. Found: C, 36.1; H, 3.47; mol. wt., 425 (bromoform). Satisfactory carbon and hydrogen analyses were not obtainable for the remainder of the compounds, since they lose olefin readily.*

Properties

All the compounds lose olefin when exposed to air. They should be stored in tightly closed containers and kept in a refrigerator. The compounds are soluble in halogenated solvents. Nmr data are given in Table I. Infrared spectra have two intense characteristic absorptions in the region 1610–1680 cm^{-1} (one may be a shoulder).

TABLE I ^1H nmr Data for Olefin(β-diketonato)silver(I) Compounds

Hexafluoroacetylacetonato derivatives			Trifluoroacetylacetonato derivatives		
Compound	τ value, ppm	Relative intensity	Compound	τ value, ppm	Relative intensity
Cyclooctatetraene	3.58	8	Cyclooctadiene	3.93	4
	3.90	1			
Cyclooctene	3.90	2		4.46	1
	4.09	1		7.92	3
	7.55	4			
	8.41	8			
Cyclohexene	3.64	2		7.48	8
	4.02	1	Cyclooctatetraene	3.83	8
	7.71	4		4.43	1
	8.27	4		7.91	3

References

1. C. D. M. Beverwijk, G. J. M. van der Kerk, A. J. Leusink, and J. G. Noltes, *Organomet. Chem. Rev.*, **5**, 215 (1970).
2. W. Partenheimer and E. H. Johnson, *Inorg. Chem.*, **11**, 2840 (1972).

*The checker reports that analytically pure products (determined via silver analyses) can be obtained in high yields before the recrystallization step. He also prepared the norbornadiene derivatives.

34. TRICHLOROMETHYLTITANIUM AND TRIBROMOMETHYLTITANIUM

$$TiX_4 + \tfrac{1}{2}(CH_3)_2Zn \longrightarrow CH_3TiX_3 + \tfrac{1}{2}ZnX_2 \qquad (X = Cl, Br)$$

Submitted by R. J. H. CLARK* and M. A. COLES*
Checked by P. W. SMITH†

As a result of their role in Ziegler-Natta catalysis,[1] much interest has been developed in organotitanium compounds. Although the alkylcyclopentadienyl derivatives (Cp_2TiR_2) are reasonably stable, the alkyltitanium halides ($RTiX_3$) show thermal, oxidative, and hydrolytic instability and have been little studied. Traditional preparative routes using aluminum alkyls or Grignard or lithium reagents are complicated or accompanied by side reactions. The most stable parent alkyl derivatives, the methyltitanium trihalides, are best prepared using dimethylzinc as alkylating agent. The syntheses described here are adaptations of the procedures described by de Vries[2] and Thiele.[3] The hazards associated with the use of dimethylzinc have been reduced by using vacuum transfer techniques. The preparation of reagents requires a few hours, but the actual reactions can be performed in 2.5 hr (CH_3TiCl_3) and 5 hr (CH_3TiBr_3), and the resultant products may be stored and used over a period of months. The scale described here produces 4 ampuls containing *in toto* ca. 24 mmoles of product. The scale can be increased by up to a factor of 2, but beyond this the increased size of the apparatus and the increased volumes of solvent to be transferred present difficulties.

General Procedure

The hydrolytic and oxidative instability of the methyltitanium trihalides necessitates the use of carefully purified reagents and care in transfer of materials. A combined vacuum-nitrogen line,[4] such that vacuum or purified nitrogen can be admitted to the apparatus via a single length of heavy-walled rubber tubing, is desirable. The purified reagents are stored under vacuum in ampuls fitted with poly(tetrafluoroethylene) (PTFE) Rotaflo stopcocks (Fig. 9*a*) and vacuum-distilled before use, masses being noted by difference.‡

*Christopher Ingold Laboratories, University College London, 20 Gordon St., London WC1H OAJ, England.

†Chemistry Department, University of Tasmania, Hobart, Tasmania 7001, Australia.

‡With titanium tetrachloride, care must be taken to ensure complete distillation from the barrel of the ampul *after* the PTFE plunger is closed. This is necessary to prevent fouling of the seal by subsequent hydrolysis of residual titanium tetrachloride.

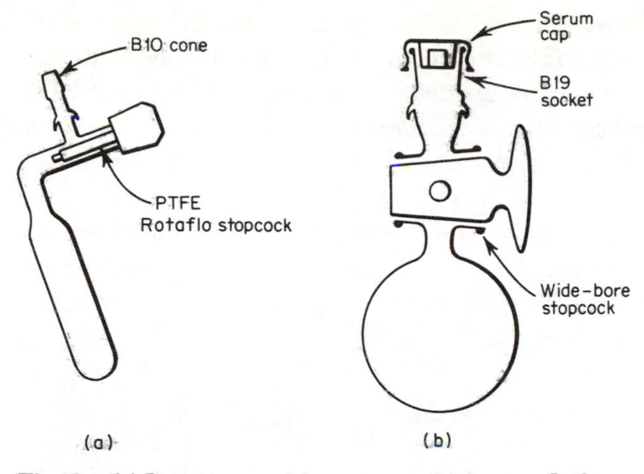

Fig. 9. (a) Storage ampul for reagents. (b) Storage flask.

Solvents, after drying and degassing, are stored under nitrogen over a potassium mirror in a storage flask capped by wide-bore stopcocks and serum caps (Fig. 9*b*). Anaerobic liquid transfers are performed by using 20-ml syringes, the source and receiver being capped by Subaseal serum caps. After baking at ca. 150°, the syringes are allowed to cool before use with nitrogen purging of the needle and barrel. A positive pressure of nitrogen (from a mercury bubbler escape in the nitrogen line) is maintained in the reaction apparatus during syringe additions and reaction.

Preparation of Reagents

Titanium tetrachloride (British Drug Houses) is stirred over copper powder under nitrogen at ca. 80° for 3 hr before vacuum distillation into a storage ampul[5] (Fig. 9*a*). Immediately before use, a $TiCl_4$-hexane solution (ca. 1 mmole of $TiCl_4$ per milliliter of hexane) is prepared by vacuum distillation of a weighed quantity of titanium tetrachloride from the storage ampul into a 100-ml flask by short-path vacuum distillation. The flask is filled with nitrogen, removed, and quickly stoppered with a serum cap. The required volume of hexane is syringed into the flask.

Titanium tetrabromide (Alfa Inorganics) is stored in sealed glass ampuls under vacuum. A pentane solution (ca. 1 mmole of $TiBr_4$ per milliliter of pentane) is prepared prior to use as above, a suitable ampul being cracked open and quickly transferred to the short-path vacuum-distillation apparatus. Gentle flaming aids the distillation. The mass of titanium tetrabromide transferred is obtained by a difference weighing of the glass ampul.

Dimethylzinc, prepared and purified by standard methods,* is stored in a storage ampul (Fig. 9*a*), and the required quantity is vacuum-distilled into a similar ampul before reaction (■ **Caution.** *Dimethylzinc is very volatile* (bp. 46°) *and extremely pyrophoric. The transfer of dimethylzinc, even as a dilute hydrocarbon solution, by syringe techniques invariably results in some decomposition.*)

n-Hexane and *n*-pentane† are dried by being refluxed over lithium tetrahydridoaluminate under nitrogen for 4 hr, degassed by alternate freezing and melting *in vacuo*, and finally distilled under nitrogen onto a freshly prepared potassium mirror‡ in a storage flask (Fig. 9*b*).

A. TRICHLOROMETHYLTITANIUM

Procedure

The reaction apparatus (Fig. 10), each 150-ml flask containing a PTFE-coated magnetic stirring bar and with the dimethylzinc ampul attached, is flamed under vacuum (taps 2 and 5 open). When cool, it is filled with nitrogen. The titanium tetrachloride solution (7.6 g, 40 mmoles of $TiCl_4$) is syringed into flask *A* through the serum cap and tap 2. Hexane (20 ml) is added in the same manner so that the total volume of solution is ca. 60 ml. The magnetically stirred solution is cooled in a solid carbon dioxide–acetone bath and evacuated via taps 3 and 4 (taps 2 and 5 closed). Dimethylzinc (1.9 g, 20 mmoles) is admitted to the evacuated apparatus (taps 3 and 4 closed) over $\frac{1}{2}$ hr by occasionally opening ampul tap 1. A darkening of the stirred solution is apparent after about 5 min, and a dark-brown slurry is produced on complete addition of dimethylzinc. Nitrogen is then admitted to the apparatus, the solid CO_2–acetone bath is removed, and the mixture is stirred for 1 hr further, by

*Small quantities (ca. 2 g) of dimethylzinc are conveniently prepared by the sealed-tube method of N. K. Hota and C. J. Willis, *J. Organomet. Chem.*, **9,** 169 (1967). Larger quantities can be prepared by the method of R. R. Renshaw and C. E. Greenlow, *J. Am. Chem. Soc.*, **42,** 1472 (1920), the Zn/Cu couple being prepared from zinc powder and copper citrate [R. C. Kung and P. J. C. Tang, *J. Am. Chem. Soc.*, **76,** 2262 (1954)]. In a large-scale preparation the careful distillation required to separate the dimethylzinc product from unreacted iodomethane (methyl iodide) can be avoided by evacuation (with some loss of dimethylzinc) of the initially formed iodomethylzinc before pyrolysis to dimethylzinc.

†Fisons (S.L.R. grade), pentane b.p. 35–37°, hexane b.p. 67–69°.

‡A small pellet of oil-free potassium is introduced into the storage flask, which is then evacuated. The potassium pellet is locally heated with a small gas flame until it boils and coats the flask walls with a metallic film.

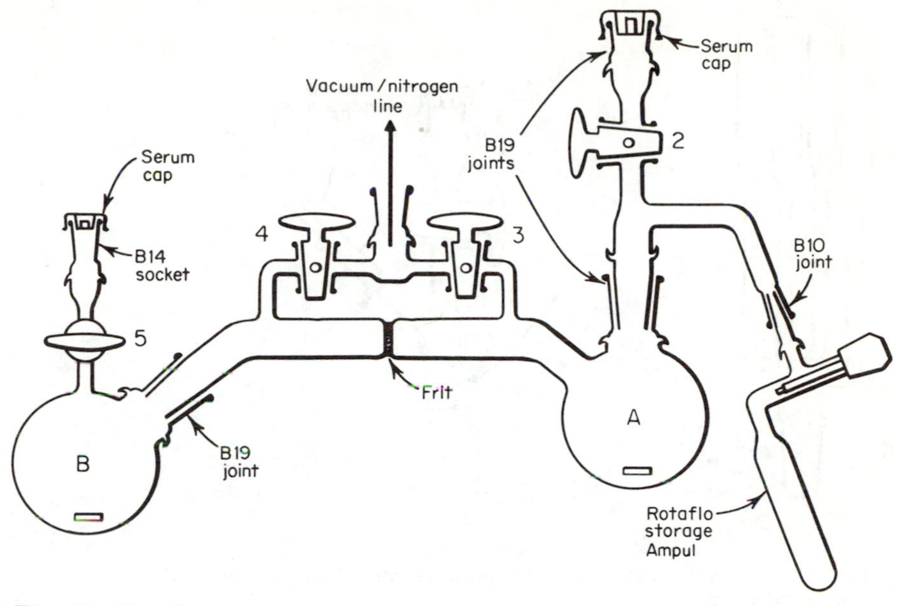

Fig. 10. Reaction apparatus.

which time it has warmed to room temperature and changed to an orange-yellow solution containing suspended zinc dichloride.

The precipitate of zinc dichloride is filtered by tilting the apparatus and applying a slight vacuum via tap 4 (tap 3 closed); the orange-yellow filtrate collects in flask *B*, which is cooled to $-78°$. After a few minutes the cooled filtrate deposits dark-violet crystals of trichloromethyltitanium, which quickly coat the flask walls. Crystallization can be initiated by stirring the filtrate. When crystallization is complete, as much as possible of the mother liquor is decanted into flask *A* via the frit by tilting the apparatus and applying a slight vacuum at tap 3 (tap 4 closed). Care must be taken not to carry any crystals onto the frit, as they readily redissolve on slight warming. The product is dissolved in fresh hexane (10 ml) syringed into flask *B* via the serum cap and tap 5; the cooling, crystallization, and decantation steps are then carried out as before. The serum cap on flask *B* is replaced by a tubing connection to the vacuum line and the flask is cooled in a liquid-nitrogen bath. The latter is then removed and with a fast flow of nitrogen via tap 4, flask *B* is disconnected from the apparatus and quickly attached to the nitrogen-filled product manifold (Fig. 11*a*). Flask *B* is plunged into liquid nitrogen and evacuated via tap 5. With tap 5 closed, flask *B* is then allowed to warm toward room temperature until the solid trichloromethyltitanium has dissolved in the occluded hexane to

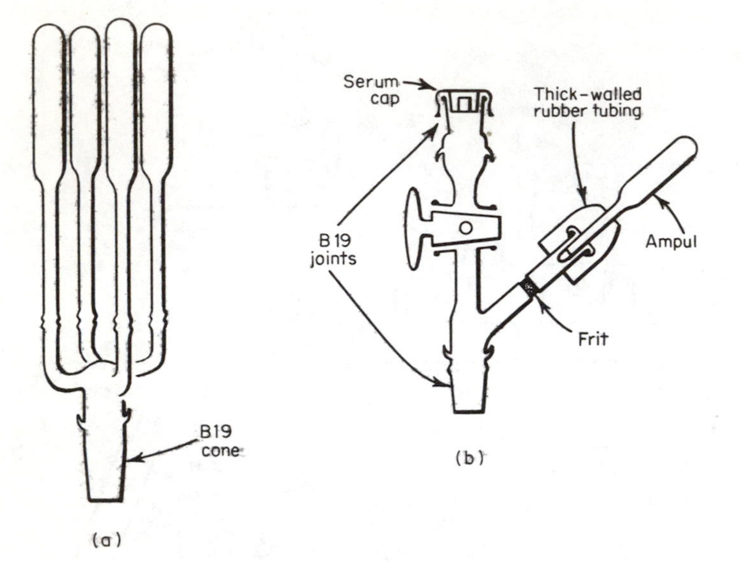

Fig. 11. (a) Product manifold. (b) Apparatus for the transfer of product to a reaction vessel.

produce a dark-orange solution. The manifold is then quickly inverted and cooled in liquid nitrogen, whereupon the concentrated trichloromethyltitanium solution partly pours and partly distills into the ampul arms. When transference to the ampuls is complete, tap 5 is opened to vacuum and the ampuls are sealed at the constrictions. The ampuls are stored in the dark at −78°.

B. TRIBROMOMETHYLTITANIUM

Procedure

The apparatus and reaction procedure are identical to those for the preparation of trichloromethyltitanium, although the work-up is modified. Dimethylzinc (1.9 g, 20 mmoles) is distilled *in vacuo* into the cold (−78°) stirred suspension of titanium tetrabromide (13.9 g, 38 mmoles) in pentane (80 ml) over $\frac{3}{4}$ hr. A similar sequence of color changes is apparent, and an orange solution and suspension are obtained after warming the mixture to room temperature over 1 hr under nitrogen. The orange solution is filtered free of zinc dibromide and reduced in volume to ca. 20 ml by vacuum evaporation of pentane from the stirred filtrate cooled to −78° in flask *B*. A liquid-nitrogen trap interposed between tap 5 and the vacuum line condenses the evaporated pentane. The removal of solvent can be hastened by allowing flask *B* to warm above

$-78°$, but some tribromomethyltitanium is lost by codistillation with the pentane. When the bulk of the solvent has been removed and solid is apparent, the apparatus is filled with nitrogen and the liquid-nitrogen trap is removed. Further stirring at $-78°$ maximizes the yield of tribromomethyltitanium in flask *B*. The mother liquor is carefully decanted into flask *A* via the frit by the application of a slight vacuum at tap 3 (tap 4 closed). The resultant solid is transferred to ampuls as before.

Tribromomethyltitanium is far more reluctant to crystallize than the chloro analog, and concentrated solutions are required before any solid is obtained. Rapid cooling of these concentrated solutions often produces a pale-yellow solid, but on standing or slight warming above $-78°$ this changes to dark-violet crystals. Greater volumes of solvent are required in this preparation, as the solubility of titanium tetrabromide at $-78°$ is low. If hexane is used as the solvent, considerable losses of the product occur at the higher temperature required for reasonable evaporation of this solvent. The slight excess of dimethylzinc over the 1:1 ratio is used to ensure complete consumption of the tetrabromide, which is difficult to remove from the product.

The yields of ampuled trichloromethyltitanium and tribromomethyltitanium are 60–70%. The concentrated hydrocarbon solutions of the products can be kept for up to 3 months at $-78°$ in the absence of light without appreciable decomposition.

Handling of Products

The methyltitanium trihalides can be obtained free of occluded hydrocarbon solvent by low-temperature vacuum distillation, but some thermal decomposition usually occurs and the resultant solids are difficult to handle. The compounds readily attack vacuum grease and decompose in the slightly warmer portions of the distillation apparatus.

The ampuled materials prepared in the syntheses described here contain sufficient solvent to form concentrated solutions on warming to ca. $-20°$. Any slight decomposition products can be removed by filtration, and the subsequent handling of the filtrate by pouring and distilling is done rapidly and accompanied by little thermal decomposition.

A convenient apparatus for this transfer from the ampul is shown in Fig. 11*b*. The frozen ampul, the neck of which is scored by a scratch mark, is attached to the frit arm by thick-walled rubber tubing, and the complete apparatus is evacuated. The scored portion of the neck extends inside the wider-bore frit arm. The apparatus is filled with nitrogen, and the ampul is allowed to warm until the methyltitanium trihalide has melted and dissolved in its occluded solvent. The ampul neck is broken at the score mark by slight pressure against the inside of the frit arm and the solution is filtered into the cooled receiver by application of vacuum.

Properties

The methyltitanium trihalides are deep-violet crystalline solids, which melt to yellow liquids (CH_3TiCl_3, 28.5°; CH_3TiBr_3, 2–3°).[3] The Ti—C stretching frequency of trichloromethyltitanium in the vapor phase occurs at 533 cm^{-1} (CH_3 compound) and at 518 cm^{-1} (CD_3 compound).[6] The visible spectrum of trichloromethyltitanium in *n*-hexane solution has been reported and discussed.[7] The ^1H nmr spectra of the methyltitanium trihalides in dichloromethane solution show sharp singlets at 7.09 (CH_3TiCl_3) and 7.45 (CH_3TiBr_3).[8]

Trichloromethyltitanium is a Lewis acid and reacts with both monodentate[9,10] (L) and bidentate[10,11] (B) ligands to form complexes of the types $CH_3TiCl_3 \cdot L$, $CH_3TiCl_3 \cdot 2L$, and $CH_3TiCl_3 \cdot B$. The six-coordinate complexes are much more stable thermally than the parent trichloromethyltitanium. Both trihaloalkyltitanium compounds also coordinate further to halide ions to form complexes of the types[8,12] $[(C_2H_5)_4N][(CH_3)_2Ti_2X_7]$, $[(C_2H_5)_4N]_2[(CH_3)_2Ti_2X_8]$, and $[(C_2H_5)_4N]_2[CH_3TiX_5]$ (X = Cl or Br). The Ti—C stretching frequencies in the chloro complexes are ca. 50 cm^{-1} lower than in the parent trichloromethyltitanium. Both oxygen[13] and sulfur dioxide[8] can be inserted into the Ti—C bonds.

References

1. C. Beerman and H. Bestian, *Angew. Chem.*, **71,** 618 (1959).
2. H. de Vries, *Recl. Trav. Chim. Pays-Bas,* **80,** 866 (1961).
3. K. H. Thiele, P. Zdunneck, and D. Baumgart, *Z. Anorg. Chem.*, **378,** 62 (1970).
4. D. F. Shriver, "The Manipulation of Air-sensitive Compounds," McGraw-Hill Book Company, New York, 1969.
5. R. J. H. Clark, "The Chemistry of Titanium and Vanadium," Elsevier Publishing Company, Amsterdam, 1968.
6. H. M. van Looy, L. M. Rodriguez, and J. A. Gabant, *J. Polymer Sci. (A)*, **4,** 1927 (1966).
7. C. Dijkgraaf and J. P. G. Rousseau, *Spectrochim. Acta,* **25A,** 1455 (1969).
8. R. J. H. Clark and M. A. Coles, *J. Chem. Soc., Dalton Trans.*, **1972,** 2454.
9. G. A. Razuvaev and L. M. Bobinova, *Dokl. Akad. Nauk S.S.S.R.*, **152,** 1363 (1963).
10. G. W. A. Fowles, D. A. Rice, and J. D. Wilkins, *J. Chem. Soc. (A)*, **1971,** 1920.
11. K. H. Thiele and K. Jacob, *Z. Anorg. Chem.*, **356,** 195 (1968); R. J. H. Clark and A. J. McAlees, *J. Chem. Soc. (A)*, **1970,** 2026; *Inorg. Chem.*, **11,** 342 (1972).
12. R. J. H. Clark and M. A. Coles, *Chem. Commun.*, **1971,** 1587.
13. R. J. H. Clark and A. J. McAlees, *J. Chem. Soc. (A)*, **1972,** 640.

35. ETHYLENEBIS(TRIPHENYLPHOSPHINE)PALLADIUM(0) AND RELATED COMPLEXES

Submitted by A. VISSER,* R. VAN DER LINDE,* and R. O. DE JONGH*
Checked by H. SELBECK,† M. MOLIN,† and G. WILKE†

Ethylene (*tert*-phosphine) complexes of zero-valent nickel[1] and platinum[2] have been known for years. Analogous palladium complexes can be synthesized along the same lines as those reported for the nickel compounds, using ethoxydiethylaluminum(III) as the reducing agent in the presence of ethylene. These palladium-ethylene complexes may serve as starting materials for oxidative addition reactions, since the ethylene ligand is loosely bonded.[3]

A. ETHYLENEBIS(TRIPHENYLPHOSPHINE)PALLADIUM(0)

$$Pd(acac)_2 + 2P(C_6H_5)_3 + C_2H_4 + 2Al(C_2H_5)_2(OC_2H_5) \longrightarrow Pd(C_2H_4)[P(C_6H_5)_3]_2 + \cdots$$

Procedure

■ **Caution.** *Alkylaluminum compounds are pyrophoric and react violently with water. These materials should be handled with care and can safely be destroyed by the addition of isopropyl alcohol and kerosene* (1 : 6 *volume ratio*).

The reaction is carried out in a 250-ml three-necked flask equipped with a magnetic stirrer, a 100-ml calibrated dropping funnel, and a three-way stopcock (Fig. 12). The apparatus is evacuated and filled with argon; 5.4 g (17.5 mmole) of bis(acetylacetonato)palladium(II)‡ and 9.4 g (36 mmole) of triphenylphosphine are introduced through the third neck. Then, from the dropping funnel, 20 ml of toluene§ and 100 ml of diethyl ether are added. During these operations, the apparatus is gently flushed with argon. A gas inlet tube, reaching beneath the liquid surface, is placed in the third neck, and ethylene¶ is bubbled through. At the same time, the flask is cooled to 0°. With constant

*Unilever Research, Vlaardingen, The Netherlands.
†Max-Planck-Institut für Kohlenforschung, 433 Mülheim/Ruhr, West Germany.

‡Prepared as described by P. G. Charles and M. D. Pawlikowski in *J. Phys. Chem.*, **62**, 440 (1958) for Ni $(acac)_2$. Palladium acetylacetonate and triphenylphosphine were crystallized and stored under argon.

§Solvents were dried with solid potassium hydroxide and distilled from sodium-potassium alloy and benzophenone.

¶Ethylene, lecture bottle from Baker Chemical Co., can be used without purification.

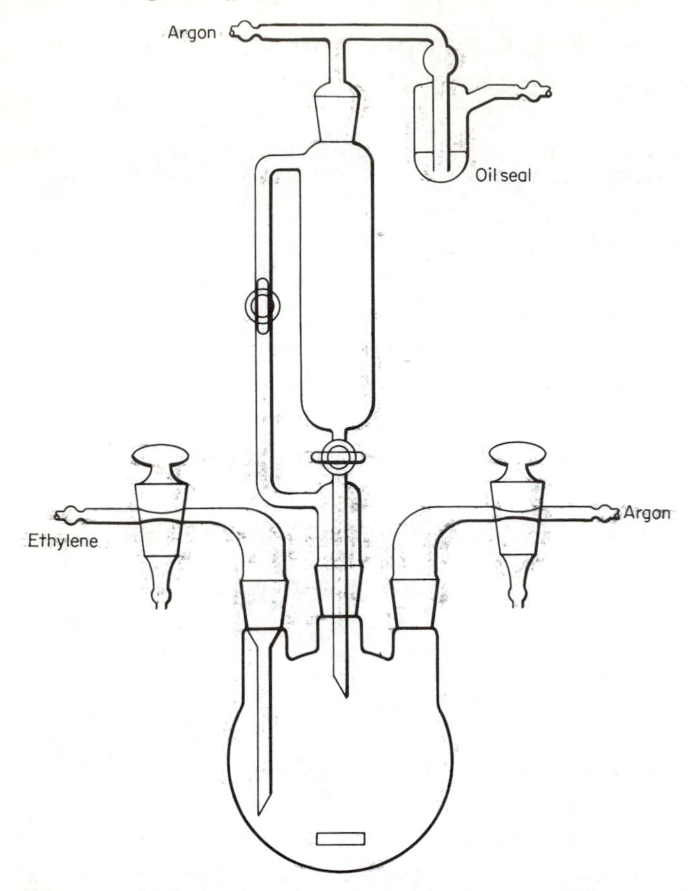

Fig. 12. Apparatus for the preparation of ethylenebis(triphenylphosphine)palladium(0). The inlet tube should be as large as possible to prevent plugging, but an inner diameter of 10 mm will suffice. The checkers find it more convenient to have the argon inlet attached to a stoppered neck in the flask.

stirring, 6.3 ml (5.3 g, 41 mmoles) of ethoxydiethylaluminum(III)* in 30 ml of diethyl ether is added from the dropping funnel over a 15-min period. The resulting reaction mixture is then stirred for 24 hr at room temperature while it is kept saturated with ethylene. The reaction mixture is then cooled again to 0° for 2 hr (stirring is continued) to complete crystallization. By means of a siphon and a slight overpressure of argon, the contents of the flask are transferred onto a frit cooled to 0°, as depicted in Fig. 13. The precipitate is

*Ethoxydiethylaluminum(III) can be prepared from commercial triethylaluminum by the addition of 1 mole of ethanol in benzene solution; b.p. 50–54° at 10^{-3} torr.

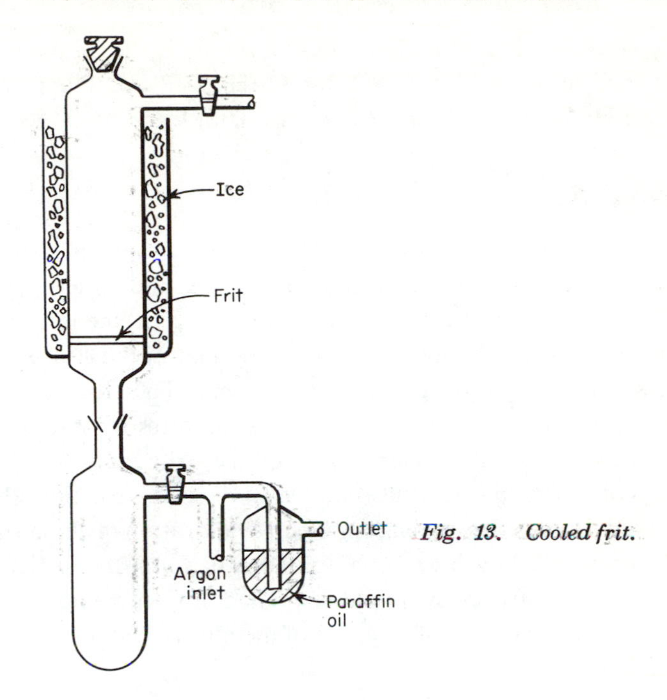

Fig. 13. Cooled frit.

thoroughly washed 3 times with ether saturated with ethylene at 0° and dried* *in vacuo* on the frit, first at 0°, then at room temperature. An 8.7-g sample of the white (occasionally grayish) air-sensitive complex is obtained (yield: 75%).

The purity can be checked conveniently by complexometric titration of palladium[4,5] after destruction of the complex with sulfuric and nitric acids and by volumetric determination of the ethylene upon displacement with triphenyl phosphite. *Anal.* Calcd. for $C_{38}H_{34}P_2Pd$: Pd, 16.1; C_2H_4, 4.26; C, 69.3; H, 5.2. Found: Pd, 16.2; C_2H_4, 4.06; C, 68.9; H, 4.9.

B. ETHYLENEBIS(TRICYCLOHEXYLPHOSPHINE)PALLADIUM(0) AND ETHYLENEBIS(TRI-*o*-TOLYL PHOSPHITE)PALLADIUM(0)

Procedure

These complexes can be synthesized by the same procedure with yields of 50 and 70%, respectively. For the tri-*o*-tolyl phosphite complex, the reaction mixture must be cooled to −30° to effect crystallization. *Anal.* Calcd. for

*The checkers recommend predrying by passing ethylene over the compound in order to prevent possible decomposition of the complex.

$C_{38}H_{70}P_2Pd$: Pd, 15.3; C_2H_4, 4.04; C, 65.6; H, 10.1. Found: Pd, 15.3; C_2H_4, 3.99; C, 65.8; H, 9.9. Calcd. for $C_{44}H_{46}O_6P_2Pd$: Pd, 12.7; C_2H_4, 3.34; C, 63.0; H, 5.6. Found: Pd, 12.7; C_2H_4, 3.32; C, 62.8; H, 5.4.

Properties

Ethylenebis(*tert*-phosphine)palladium(0) complexes are air-sensitive but can be kept for months, when dry and pure, under inert gas in a refrigerator. They are soluble in organic solvents like benzene and toluene; palladium may precipitate from the solutions, especially that of the triphenylphosphine complex, unless they are saturated with ethylene. Therefore, the complexes may be crystallized from toluene–diethyl ether mixtures saturated with ethylene.

The ethylenebis(tricyclohexylphosphine)palladium(0) and ethylenebis(tri-*o*-tolyl phosphite)palladium(0) complexes dissolve in diethyl ether when argon is passed through the suspension. Upon evaporation of these solutions *in vacuo,* the corresponding bis(*tert*-phosphine)palladium(0) complexes are obtained.[3]

Ethylenebis(triphenylphosphine)palladium, however, decomposes under this treatment with the formation of metallic palladium. For infrared data, see Table I.

TABLE I Infrared Data of L_2Pd-Ethylene Complexes[3]

L	Infrared absorptions tentatively assigned to complexed ethylene, cm^{-1}
Triphenylphosphine	* 1488, 1203, 388
Tricyclohexylphosphine	3048, 2980, 1483, 1202, 1195, 350
Tri-*o*-tolyl phosphite	*

*Absorptions obscured by bands of L.

References

1. G. Wilke and G. Hermann, *Angew. Chem.*, **74,** 693 (1962).
2. C. D. Cook and G. S. Jauhal, *Inorg. Nucl. Chem. Lett.*, **3,** 31 (1967).
3. R. van der Linde and R. O. de Jongh, *Chem. Commun.*, **1971,** 563.
4. G. Schwarzenbach, "Die komplexometrische Titration," p. 94, Ferdinand Enke Verlag, Stuttgart, 1956.
5. H. Flaska, *Microchim. Acta,* **41,** 226 (1953).

Chapter Five

MAIN-GROUP AND ACTINIDE COMPOUNDS

36. TETRAETHYLAMMONIUM, TETRAPHENYLARSONIUM, AND AMMONIUM CYANATES AND CYANIDES

Submitted by R. L. DIECK,* E. J. PETERSON,* A. GALLIART,* T. M. BROWN,* and T. MOELLER*
Checked by A. RUCKENSTEIN† and J. L. BURMEISTER†

The absence of readily available cyanates and cyanides which exhibit an appreciable degree of solubility in nonaqueous systems has hindered the preparation of hygroscopic coordination compounds containing these pseudohalides as ligands. It was therefore desirable to have a convenient method for the preparation of cyanate and cyanide salts which contain large counterions. The ion-exchange process described herein allows for the preparation of large quantities of tetraethylammonium and tetraphenylarsonium pseudohalides within a reasonable period of time. In addition, the method can be extended to the preparation of other tetraalkylammonium cyanates and cyanides. A more convenient and rapid method for the preparation of ammonium cyanate in high yields is also described.

A. TETRAETHYLAMMONIUM CYANATE AND CYANIDE

A recently developed general method of synthesis for the tetraalkylammonium cyanides[1] was extended subsequently to the cyanates.[2] Previous to

*Chemistry Department, Arizona State University, Tempe, Ariz. 85281.
†Chemistry Department, University of Delaware, Newark, Del. 19711.

the latter report, there were no preparative methods available for the synthesis of quaternary ammonium cyanates. Those described for the cyanides[3–6] employed reagents which were not readily available and either necessitated the use of highly toxic chemicals or required all operations to be performed under dry nitrogen. The ion-exchange method outlined below avoids these problems and limitations.

1. Tetraethylammonium cyanate

$$[(C_2H_5)_4N]Br + OCN^- \xrightarrow[\text{ion exchange}]{CH_3OH} [(C_2H_5)_4N]OCN + Br^-$$

Procedure

A 6- by 50-cm column terminating in a Teflon stopcock is packed with 1 lb of IRA-400 resin* and charged with 454 g (5.5 moles) of potassium cyanate dissolved in 4 l. of 50% aqueous methanol. The flow rate during the charging procedure should not exceed 2 drops/sec. After completion of this step, the column is rendered anhydrous by rinsing it with 5 l. of absolute methanol. An increased flow rate can be used for this step.

A solution of 43.3 g (0.21 mole) of tetraethylammonium bromide in 350 ml of anhydrous methanol is placed on the prepared anhydrous column and eluted to a volume of 2 l. with absolute methanol. Again a flow rate of 2 drops/sec is advised. The eluate is concentrated on a rotary evaporator (water aspirator) to a very viscous solution, which is then evaporated to dryness under dynamic vacuum (10^{-5} torr) at ambient temperature. The yield, based upon tetraethylammonium bromide, is 27 g (75%). Further purification was not found necessary, but if additional purification of the products is desired, it can be accomplished by dissolving the product in 150 ml of acetonitrile in a dry-box. The pure cyanate is precipitated by the addition of 200 ml of ethyl acetate. The product is isolated by filtration in a dry-box and dried under dynamic vacuum at ambient temperature. *Anal.* Calcd. for $C_9H_{20}N_2O$: C, 62.79; H, 11.63; N, 16.09; O, 9.40. Found: C, 62.7; H, 11.8; N, 16.3; O, 9.3.

Properties

Tetraethylammonium cyanate is a white solid which deliquesces in air but can be stored for an indefinite time in a dry inert atmosphere. It is soluble in

*Available from Mallinckrodt Chemical Works, P.O. Box 5439, St. Louis, Mo. 63160.

methanol, ethanol, acetonitrile, and 1,2-dichloroethane. The melting point is $258 \pm 1°$. The infrared spectrum, taken as a Nujol mull, has the following fundamental absorptions (in cm^{-1}): ν_{as} at 2155 (s) and δ at 624 (m). The ν_s and 2δ modes exhibit Fermi resonance at 1282 (m) and 1195 (m) cm^{-1}, since each vibrational mode is displaced or perturbed from its expected value because these two vibrational states are close in energy.

2. Tetraethylammonium cyanide

$$[(C_2H_5)_4N]\,Br + CN^- \xrightarrow[\text{ion exchange}]{CH_3OH} [(C_2H_5)_4N]CN + Br^-$$

Procedure

■ **Caution.** *Since the cyanide ion is extremely toxic, it should be handled with care. The solutions should not be made acidic for fear of liberating* HCN.

A column, identical to that described in part 1 of Sec. A, is packed with 1 lb of IRA-400 resin and charged with 454 g (9.26 moles) of sodium cyanide dissolved in 4 l. of 50% aqueous methanol. The column is then rinsed with 5 l. of anhydrous methanol. A solution of 44 g (0.21 mole) of tetraethylammonium bromide in 400 ml of anhydrous methanol is prepared and placed on the charged anhydrous column. The collection and purification of the tetraethylammonium cyanide are carried out exactly as described for the cyanate, except that purification of the cyanide is always necessary. After the product is isolated from the acetonitrile–ethyl acetate solution, it is dried under dynamic vacuum at ambient temperature for 24 hr. The yield, based on tetraethylammonium bromide, is 20 g (60%). *Anal.* Calcd. for $C_9H_{20}N_2$: C, 69.23; H, 12.82; N, 17.95. Found: C, 69.1; H, 12.7; N, 17.7.

Properties

The physical properties of tetraethylammonium cyanide are quite similar to those of the tetraethylammonium cyanate. The cyanide is soluble in methanol, ethanol, acetonitrile, and 1,2-dichloroethane. It also deliquesces in air and must be kept in a dry inert atmosphere. The melting point of this white solid is $254 \pm 1°$. The infrared spectrum of the cyanide, taken as a Nujol mull, has ν(CN) bands at 2145 (m) and 2058 (m) cm^{-1}, in addition to bands exhibited by the tetraethylammonium cation at 2980 (vs), 2950 (s), 1490 (vs), 1435 (vs), 1401 (vs), 1365 (vs), 1177 (vs), 1075 (m), 1058 (s), 1008 (vs), and 798 (vs) cm^{-1}.

B. TETRAPHENYLARSONIUM CYANATE AND CYANIDE

The tetraphenylarsonium cation is useful in coordination chemistry because of its large size. Tetraphenylarsonium cyanate dihydrate was prepared by Norbury and Sinha in 1968.[7] The preparation involved precipitation of the cyanate from an aqueous solution, but the method suffers from the disadvantage of low yields. Tetraphenylarsonium cyanide has been prepared from methanol solution with fairly good yields,[3] but the product is isolated as the monohydrate. The ion-exchange method developed for the tetraalkylammonium salts can be readily applied to the preparation of anhydrous tetraphenylarsonium cyanate and cyanide, thus eliminating many of the problems initially involved in the synthesis of these componds.

1. Tetraphenylarsonium cyanate

$$[(C_6H_5)_4As]Br + OCN^- \xrightarrow[\text{ion exchange}]{CH_3OH} [(C_6H_5)_4As]OCN + Br^-$$

Procedure

A 2- by 60-cm column terminating in a Teflon stopcock is packed with 50 g of IRA-400 resin and then charged with 50 g (0.61 mole) of potassium cyanate dissolved in 450 ml of 50% aqueous methanol. A flow rate of 2 drops/sec is employed during the charging process. After the charging is complete, the column is rendered anhydrous by rinsing it with 550 ml of anhydrous methanol, using an increased flow rate.

A solution of tetraphenylarsonium bromide* is prepared by dissolving 5 g (0.011 mole) of the bromide in 20 ml of anhydrous methanol. This solution is placed on the column and eluted to a total volume of 220 ml. Again, the flow rate from the column should not exceed 2 drops/sec. The eluate is evaporated to dryness under dynamic vacuum (10^{-5} torr) at ambient temperature. The yield, based on tetraphenylarsonium bromide, is 4.31 g (94%). If necessary, recrystallization is accomplished by dissolving the crude product in 20 ml of acetonitrile, followed by precipitation of the pure product by the addition of 75 ml of ethyl acetate. The cyanate is isolated by filtration, and the product is dried under dynamic vacuum at ambient temperature for 24 hr. Both recrystallization and filtration steps should be performed in a dry-box. *Anal.* Calcd. for $C_{25}H_{20}AsNO$: C, 70.59; H, 4.75; N, 3.29; As, 17.61. Found: C, 70.3; H, 4.7; N, 3.2; As, 17.6.

*Available from Strem Chemicals, Inc. Danvers, Mass. 01923.

Properties

Anhydrous tetraphenylarsonium cyanate is a white solid which is hygroscopic and must be stored in a dry inert atmosphere. It is soluble in ethanol, methanol, and acetonitrile. The compound melts, with decomposition, at ca. 224°. The infrared spectrum, taken as a Nujol mull, exhibits the following bands due to the cyanate fundamental absorptions (in cm^{-1}): ν(As) at 2140 (s) and δ at 622 (s). This spectrum agrees well with a recent spectral study of this compound.[8] Again, as with tetraethylammonium cyanate, Fermi resonance is exhibited in the spectrum of this compound at 1280 (m) and 1192 (m) cm^{-1}.

2. Tetraphenylarsonium cyanide

$$[(C_6H_5)_4As]Br + CN^- \xrightarrow[\text{ion exchange}]{CH_3OH} [(C_6H_5)_4As]CN + Br^-$$

Procedure

A 2- by 60-cm column is packed with 50 g of IRA-400 resin and charged with 50 g (1.03 moles) of sodium cyanide dissolved in 450 ml of 50% aqueous methanol, using a flow rate of approximately 2 drops/sec. The charged column is then rendered anhydrous by rinsing it with 550 ml of absolute methanol, using an increased flow rate. A solution of 5 g (0.012 mole) of tetraphenylarsonium bromide dissolved in 20 ml of absolute methanol is placed on the charged anhydrous column and eluted to a total volume of 220 ml. The collection and purification of the tetraphenylarsonium cyanide proceed exactly as described for the corresponding cyanate. The yield, based on tetraphenylarsonium bromide, is 4.24 g (96%). *Anal.* Calcd. for $C_{25}H_{20}AsN$: C, 73.35; H, 4.93; N, 3.42; As, 18.30. Found: C, 72.9; H, 5.1; N, 3.4; As, 18.7.

Properties

Tetraphenylarsonium cyanide is a hygroscopic white solid which must be stored in a dry inert atmosphere. The compound undergoes slow decomposition upon standing but can be repurified by the above recrystallization procedure. The compound melts, with decomposition, at ca. 219° and exhibits an appreciable degree of solubility in ethanol, methanol, and acetonitrile. The infrared spectrum, taken as a Nujol mull, exhibits bands due to ν(CN) at 2250 (w) and 2058 (w) cm^{-1}, in addition to those absorptions attributed to the tetraphenylarsonium cation.[8]

C. AMMONIUM CYANATE

$$[(C_2H_5)_4N]OCN + NH_4SCN \xrightarrow{CH_3CN} [(C_2H_5)_4N]SCN + NH_4OCN$$

The preparation of ammonium cyanate has been dealt with previously in this series.[9] The method is, however, somewhat hazardous and time-consuming. Furthermore, ammonium cyanate is not particularly stable, rearranging to urea upon standing. The method described below allows for the preparation of ammonium cyanate with a minimum amount of effort, thus avoiding the need to prepare large quantities at any one time. If necessary, the method can be scaled up to prepare whatever quantity is desired.

Procedure

In a dry-box, 31.2 g (0.41 mole) of anhydrous ammonium thiocyanate dissolved in 500 ml of anhydrous acetonitrile is combined with a solution of 70.5 g (0.41 mole) of tetraethylammonium cyanate in 200 ml of acetonitrile. The white precipitate which immediately forms is collected on a fine-fritted sintered-glass crucible by suction filtration and washed several times with acetonitrile. The product is dried for $\frac{1}{2}$ hr under a dynamic vacuum at ambient temperature. The yield, based on tetraethylammonium cyanate, is 19.7 g (80%). *Anal.* Calcd. for CH_4N_2O: C, 20.00; H, 6.71; N, 46.65; O, 26.64. Found: C, 20.2; H, 6.5; N, 46.5; O, 26.8.

Properties

Ammonium cyanate is a white solid, which rearranges to urea upon prolonged storage or heating. It is extremely soluble in water, slightly soluble in acetonitrile, ethanol, and chloroform, and insoluble in benzene and diethyl ether. The infrared spectrum and x-ray diffraction powder pattern are most useful for determining the absence of urea in the final product. Nujol mulls of pure ammonium cyanate exhibit absorptions at 3160 (s), 2190 (s), 1334 (m), 1243 (m), and 640 (m) cm^{-1} and are free of any infrared-active bands in the regions characteristic of urea, that is, 3456 and 1683 cm^{-1}.

References

1. J. Solodar, *Syn. Inorg. Metal-Org. Chem.*, **1,** 141 (1971).
2. A. Galliart and T. M. Brown, *Syn. Inorg. Metal-Org. Chem.*, **2,** 273 (1972).
3. S. Andreades and E. W. Zahnow, *J. Am. Chem. Soc.*, **91,** 4181 (1969).
4. O. W. Webster, W. Mahler, and R. E. Benson, *J. Am. Chem. Soc.*, **84,** 3678 (1962).

5. A. R. Norris, *Can. J. Chem.*, **45,** 2703 (1967).
6. V. Gutmann and H. Brady, *Z. Anorg. Allg. Chem.*, **361,** 213 (1968).
7. A. Norbury and A. Sinha, *J. Chem. Soc. (A)*, **1968,** 1598.
8. O. Ellestad, P. Klaeboe, E. Tucker, and J. Songstad, *Acta Chem. Scand.*, **26,** 1721 (1972).
9. R. Baird and R. Pinnell, *Inorganic Syntheses*, **13,** 17 (1972).

37. DIINDENYLMAGNESIUM

$$(C_2H_5)MgBr + C_9H_8 \longrightarrow (C_9H_7)MgBr + C_2H_6$$

$$2(C_9H_7)MgBr \xrightarrow{190^\circ} (C_9H_7)_2Mg + MgBr_2$$

Submitted by K. D. SMITH* and J. L. ATWOOD*
Checked by E. C. ASHBY† and H. S. PRASAD†

As attention is focused on organometallic compounds of the lanthanide and actinide elements,[1,2] synthetic intermediates in the production of these substances become increasingly important. Although most reactions involve the action of either an alkali metal organometallic[3] or a Grignard reagent[4] on an anhydrous metal chloride in donor solvent, for some purposes the pure Group II organometallic may be preferred. Dicyclopentadienylberyllium[5,6] and dicyclopentadienylmagnesium[7] have received considerable interest, and quite recently, diindenylmagnesium has been used in the preparation of triindenylsamarium[8] and triindenylscandium.[9]

Procedure‡

Magnesium turnings (5 g, 0.21 mole) are covered with 100 ml of sodium-dried diethyl ether in a 250-ml three-necked flask. One neck of the flask is fitted with a condenser, which is in turn connected to a mercury bubbler. Of the remaining two entrances to the flask, one is attached via a stopcock to a cylinder of high-purity N_2, and one is fitted with a pressure-equalizing funnel for addition of ethyl bromide (bromoethane). The vessel is then flushed with N_2, and 15 ml of ethyl bromide (0.20 mole) is slowly added with stirring. The solution is refluxed for 2 hr, at which time the ethylmagnesium bromide Grignard reagent should be milky white. Then, with rapid N_2 flow, the

*Chemistry Department, University of Alabama, Tuscaloosa, Ala. 35486.
†Chemistry Department, Georgia Institute of Technology, Atlanta, Ga. 30332.
‡Oxygen and water must be rigorously excluded from all stages of the preparation.

stopcock is removed from the condenser, and 22 ml of freshly distilled indene (0.19 mole) and 100 ml of toluene are added. The stopcock is replaced, and the reaction temperature is elevated so that the toluene solution refluxes vigorously. All diethyl ether is driven off by N_2 flowing slowly through the solution. After 2 hr, the N_2 flow is closed off, and the solution is allowed to reflux for 8 hr. Solvent is then removed under vacuum, and the flask is taken into a dry-box. The substance is powdered and transferred to a Schlenk sublimation apparatus,[10] removed from the dry-box, and thermolyzed under vacuum at 190°. The crude product should be resublimed to free the white crystalline diindenylmagnesium from a yellow oil contaminant. A yield of pure product of 6 g (25%) should be considered average. *Anal.** Calcd. for $(C_9H_7)_2Mg$: C, 84.94; H, 5.51; Mg, 9.55. Found: C, 85.5; H, 5.61; Mg, 9.77.

Properties

The white crystalline solid has no clear melting point. Decomposition begins at approximately 170°, but sublimation may be accomplished at 190° under reduced pressure with some loss of material. It is soluble in ethers and slightly so in aromatic hydrocarbons. The substance rapidly decomposes with the slightest exposure to either H_2O or O_2. Diindenylmagnesium crystallizes in the orthorhombic space group $P2_12_12_1$ with lattice constants $a = 21.494(9)$, $b = 12.375(5)$, $c = 10.394(4)$ A, and $\rho_{calcd} = 1.23$ cm^{-3} for 8 molecules per unit cell.

References

1. H. J. Gysling and M. Tsutsui, *Adv. Organomet. Chem.*, **9,** 361 (1970).
2. R. G. Hayes and J. L. Thomas, *Organomet. Chem. Rev. (A)*, **7,** 1 (1971).
3. U. Müller-Westerhoff and A. Streitwieser, *J. Am. Chem. Soc.*, **90,** 7364 (1968).
4. T. J. Marks and A. M. Seyam, *J. Am. Chem. Soc.*, **94,** 6545 (1972).
5. P. G. Laubereau and J. H. Burns, *Inorg. Nucl. Chem. Lett.*, **6,** 59 (1970).
6. F. Baumgartner, E. O. Fischer, B. Kannelakopulos, and P. Laubereau, *Angew. Chem., Int. Ed.*, **5,** 134 (1966).
7. R. S. P. Coutts and P. C. Wailes, *J. Organomet. Chem.*, **25,** 117 (1970).
8. J. L. Atwood, J. H. Burns, and P. G. Laubereau, *J. Am. Chem. Soc.*, **95,** 1830 (1973).
9. K. D. Smith and J. L. Atwood, unpublished work, 1973.
10. D. F. Shriver, "The Manipulation of Air-sensitive Compounds," p. 148, McGraw-Hill Book Company, New York, 1969.

*Analysis done by Schwarzkopf Microanalytical Laboratory, Woodside, N.Y. 11377.

38. FLUOROMETHYLSILANES

$(CH_3)_xSiF_{4-x}$ $(x = 1, 2, 3)$

Submitted by A. P. HAGEN* and L. L. McAMIS*
Checked by D. J. SEPELAK† and C. H. VAN DYKE†

Fluoromethylsilanes have been synthesized from the corresponding chloromethylsilanes using a variety of fluorinating agents including ZnF_2,[1] anhydrous HF,[2] and SbF_3 with $SbCl_5$ as a catalyst.[3,4] Each procedure requires complex apparatus in addition to a vacuum system for handling a dangerous substance. The present procedure, with a very simple reaction system, results in excellent yields of a product which is easily separated from unreacted starting material.

Each synthesis is carried out in a glass pressure vessel.‡ This type of reactor was made from a 4-mm quick-opening angle valve§ and a length of heavy-walled borosilicate glass tubing (18 mm i.d., 26 mm o.d.) cut to give the desired volume. This vessel has been found to be suitable for reaction mixtures having an autogenous pressure of up to 50 atm. (■ **Caution.** *It is essential that heavy-walled glass tubing be used to construct the reaction vessel, in order to guard against explosion at high pressures. The vessel when under pressure should be carefully manipulated behind a safety shield, and the operator should be wearing a leather glove.*) Volatile reactants are purified by normal vacuum-line procedures and then condensed directly into the reactor.[5]

A. TRIFLUOROMETHYLSILANE

$$CH_3SiCl_3 + SbF_3 \rightarrow CH_3SiF_3 + SbCl_3$$

Procedure

In a nitrogen-filled glove bag an excess (ca. 10 g, 0.056 mole) of antimony trifluoride¶ is placed in a 30-ml glass reactor, which is then evacuated and pumped for 24 hr to provide a final drying at room temperature.

*Chemistry Department, University of Oklahoma, Norman, Okla. 73069. Financial support from a National Science Foundation grant is gratefully acknowledged.
†Department of Chemistry, Carnegie-Mellon University, Pittsburgh, Pa. 15213.
‡See D. L. Morrison and A. P. Hagen, *Inorganic Syntheses,* **13,** 65 (1972), fig. 2.
§Fischer and Porter, Warminster, Pa. 18974, cat. 795–005–004.
¶Available from Ozark-Mahoning Co., Tulsa, Okla. 74119.

Commercial trichloromethylsilane,* (ca. 6 g, 0.04 mole) is placed in a glass tube which can be attached to the vacuum system through a stopcock. The CH_3SiCl_3 is then cooled to liquid-nitrogen temperature (−196°), and the tube is evacuated. The stopcock is then closed, and the nitrogen bath is removed. When the sample has melted, it is solidified at −196° and evacuated again. This process is repeated until the sample is free of noncondensable impurities (air). The tube is then allowed to warm slowly while the contents distill slowly through a −47° trap (hexyl alcohol slush) into a −196° trap. All transfers in the vacuum line are performed without pumping. The material which passes into the −196° trap is then distilled again through a −47° trap into a −196° trap. The process is continued until no more material stops in the −47° trap. The material in the −47° trap is discarded after each distillation. This process may be represented as follows:†

RT ∼ −47° ∼ −196°
↓
RT ∼ −47° ∼ −196°
↓
Repeated until no material stops in −47° trap

The material which stopped in the −196° trap is then warmed slowly to room temperature while distilling through a −130° trap (*n*-pentane slush) into a −196° trap. The material (CH_3SiCl_3) which stops in the −130° trap is cooled to −196° and again warmed to room temperature and allowed to distill through a −130° trap into a −196° trap. This process may be represented as follows:

RT ∼ −130° ∼ −196°
↓
RT ∼ −130° ∼ −196°
↓
Repeated until no material passes into −196°trap

The material (HCl) in the −196° trap is discarded. The CH_3SiCl_3 in the −130° trap will have an infrared spectrum and vapor pressure consistent with the literature (see Table I).

The purified CH_3SiCl_3 is then transferred into the SbF_3-filled reactor which has been cooled to liquid-nitrogen temperature. The Teflon stopcock is then closed, and the reaction vessel is surrounded by an ice-water bath held in a beaker. The reactor is now most conveniently left overnight, permitting the ice to melt so that the reactor is held at room temperature for a few hours.

*Available from PCR, Inc., Gainesville, Fla. 32601.
†RT stands for room temperature.

TABLE I Physical Properties

Substance	Infrared spectrum, Ref.	b.p., °C	m.p., °C	Vapor pressure* $\log P$ (mm) $= A/T + B \log T + C$			Ref.
				A	B	C	
CH_3SiCl_3	6	+65.7	−77.8	−1457	0	7.410	3
CH_3SiF_3	7	−30.2	−72.8	−1238.1	0	7.9752	3
$(CH_3)_2SiCl_2$	6	+70.0	−76.1	−1645	0	7.6738	4
$(CH_3)_2SiF_2$	8	+ 2.7	−87.5	−1809.76	−3.9471	19.0738	4
$(CH_3)_3SiCl$	6	+57.3	−57.7	−1578	0	7.6559	4
$(CH_3)_3SiF$	9	+16.4	−74.3	−1405	0	7.7326	4

*For CH_3SiCl_3, $T = (t + 256)$.

The reactor is then opened, and the volatile components (unreacted CH_3-$SiCl_3$ and product CH_3SiF_3) are condensed into two −196° traps in series.

The substances which stopped in the −196° traps are combined and then warmed slowly to room temperature while distilling through a −96° trap (toluene slush) into a −196° trap. This is represented as

$$\text{RT} \sim -96° \sim -196°$$

Unreacted CH_3SiCl_3 stops in the −96° trap, and the product CH_3SiF_3 is in the −196° trap. The product is then warmed slowly to room temperature while distilling through a −130° trap into a −196° trap. This is to remove any trace of HCl, which would be in the −196° trap. The product CH_3SiF_3 in the −130° trap should have an infrared spectrum and molecular weight consistent with literature values. This procedure gives a yield of 88% based on the initial amount of silane.*

B. DIFLUORODIMETHYLSILANE AND FLUOROTRIMETHYLSILANE

$$3(CH_3)_2SiCl_2 + 2SbF_3 \rightarrow 3(CH_3)_2SiF_2 + 2SbCl_3$$
$$3(CH_3)_3SiCl + SbF_3 \rightarrow 3(CH_3)_3SiF + SbCl_3$$

Procedure

The reaction is carried out in the same type of glass pressure reactor as described previously in the synthesis of trifluoromethylsilane. In a nitrogen-filled

*If a low yield of CH_3SiF_3 is obtained after the overnight reaction, the checkers suggest that the materials be recondensed back into the vessel containing SbF_3 and allowed to react for an additional 18–24 hr.

glove bag an excess (ca. 10 g, 0.056 mole) of antimony trifluoride is placed in the reactor, which is then evacuated and pumped for 24 hr to provide a thorough final drying at room temperature.

Commercial dichlorodimethylsilane (ca. 6 g, 0.06 mole) is purified by following the procedure described in Sec. A.

The purified $(CH_3)_2SiCl_2$ is then transferred into the SbF_3-filled reactor, which has been cooled to liquid-nitrogen temperature. The Teflon stopcock is then closed, and the reaction vessel is surrounded by an ice-water bath held in a beaker. The reactor is now most conveniently left overnight, permitting the ice to melt so that the reactor is held at room temperature for a few hours. The reactor is then opened, and the volatile components are condensed into two $-196°$ traps in series.

The substances which stop in the $-196°$ traps are combined and then warmed slowly to room temperature while distilling through a $-78°$ trap (Dry Ice and acetone) into a $-196°$ trap. This is represented as

$$\text{RT} \sim -78° \sim -196°$$

Unreacted $(CH_3)_2SiCl_2$ stops in the $-78°$ trap, and the product $(CH_3)_2SiF_2$ is in the $-196°$ trap. The product is then warmed slowly to room temperature while distilling through a $-130°$ trap into a $-196°$ trap. The product $(CH_3)_2SiF_2$ is in the $-130°$ trap and should have an infrared spectrum consistent with that in the literature (see Table I). This procedure gives a yield of 65% based on the initial amount of silane.

This same procedure beginning with $(CH_3)_3SiCl$ results in a nearly quantitative yield of $(CH_3)_3SiF$.

Properties

The fluoromethylsilanes are readily soluble in ether and are less reactive with water than their chloro analogs. Upon exposure to moist air, CH_3SiF_3 fumes,[2] but no observable reaction takes place with $(CH_3)_2SiF_2$ and $(CH_3)_3SiF$. All three compounds are hydrolyzed by aqueous $0.5M$ sodium hydroxide. These compounds can be heated to and held at 300° for 24 hr without undergoing any decomposition or rearrangement reactions.

Some of the physical properties of the compounds are listed in Table I.

References

1. A. E. Newkirk, *J. Am. Chem. Soc.*, **68,** 2736 (1946).
2. W. H. Pearlson, T. J. Brice, and J. H. Simons, *J. Am. Chem. Soc.*, **67,** 1769 (1945).

3. H. S. Booth and W. F. Martin, *J. Am. Chem. Soc.*, **68,** 2655 (1946).
4. H. S. Booth and J. F. Suttle, *J. Am. Chem. Soc.*, **68,** 2658 (1946).
5. A suitable vacuum line is depicted in W. L. Jolly and J. E. Drake, *Inorganic Syntheses*, **7,** 34 (1963).
6. A. L. Smith, *J. Chem. Phys.*, **21,** 1997 (1953).
7. R. L. Collins and J. R. Nielsen, *J. Chem. Phys.*, **23,** 351 (1955).
8. H. Kriegsmann, *Z. Elektrochem.*, **62,** 1033 (1958).
9. H. Kriegsmann, *Z. Anorg. Allg. Chem.*, **294,** 113 (1958).

39. URANIUM HEXACHLORIDE

(Hexachlorouranium)

$$UF_6 + 2BCl_3 \longrightarrow UCl_6 + 2BF_3$$

Submitted by T. A. O'DONNELL* and P. W. WILSON†
Checked by H. J. HURST†

Although there are a number of reported methods for preparing uranium hexachloride, most of them can be grouped under two types.[1,2] (1) Uranium hexachloride can be prepared by further chlorination of lower uranium chlorides. This type includes those preparations in which a uranium oxide is the starting material, since a lower uranium chloride is normally formed as an intermediate in these chlorination reactions. (2) Uranium hexachloride is formed in the thermal decomposition of uranium pentachloride. Neither method yields uranium hexachloride in a very pure form. Uranium/chlorine ratios of 1:5.8 to 5.9 are normally the best encountered.

The procedure described below, which uses a metathetical reaction between uranium hexafluoride and boron trichloride (trichloroborane),[3] is the only preparation available which gives pure uranium hexachloride. It is also a significant improvement on previous methods because of the ease with which it can be conducted and its excellent yields (approaching 100%).

■ **Caution.** *Uranium hexafluoride is toxic and must be handled with care.*

Procedure

The apparatus used is shown diagrammatically in Fig. 14. The vacuum line is made from $\frac{3}{8}$-in. nickel or Monel tubing. The metal valves can be of the bel-

*Department of Inorganic Chemistry, University of Melbourne, Parkville, Victoria 3052, Australia.

†Australian Atomic Energy Commission, Lucas Heights, New South Wales 2232, Australia.

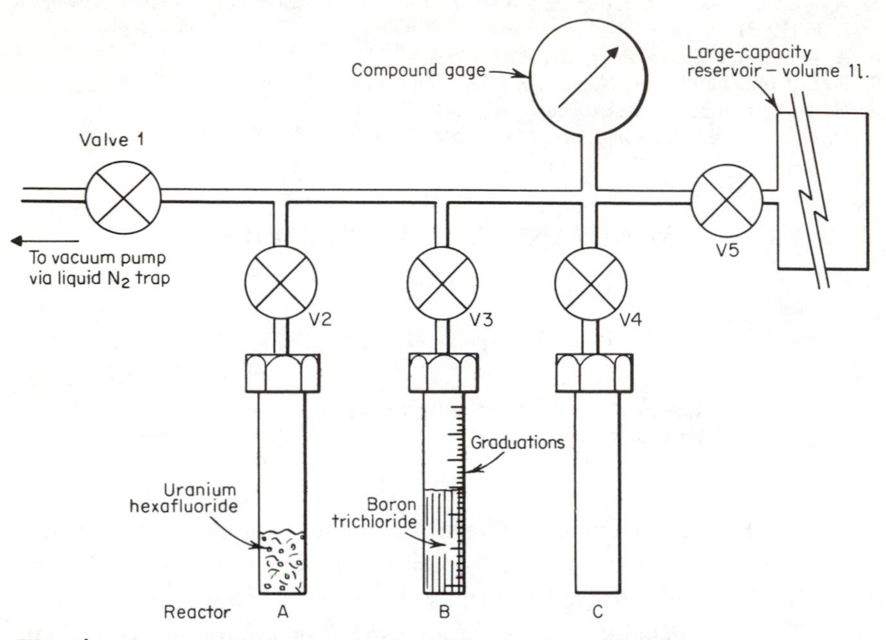

Fig. 14. Apparatus for the preparation of uranium hexachloride.

lows, diaphragm, or Teflon-backed variety. They should be resistant to fluoride corrosion (Whitey 1KS6 valves are suitable). The gage shown is used only as a safety feature and can be of the Bourdon tube type (compound gage, range 30 in. and 30 psi, made from nickel or Monel; such gages are made by the Ashcroft Co., Stratford, Conn.). The large-capacity reservoir should be made from nickel or aluminum and have a volume of approximately 1 l. The reactor tubes are made from molded polytrifluorochloro ethylene (Kel-F). They have a capacity of 20 cc and can be obtained from the Argonne National Laboratory, Argonne, Ill. This system is typical of those used for handling fluorides and has been fully described elsewhere.[4]

The preparation is started by evacuating the reactors, vacuum line, and reservoir (the contents of reactors *A* and *B* are cooled to $-196°$ before evacuating). All valves are then closed, and the reagents are warmed to room temperature. Valves 2 and 4 are opened, and 2 g of uranium hexafluoride (prepared as described by O'Donnell et al.[3]) is sublimed into reactor *C* by cooling reactor *C* to $-196°$. It is best to preweigh the required quantity of uranium hexafluoride into reactor *A* from some other, larger storage container. If this is done, all the uranium hexafluoride in reactor *A* is sublimed into reactor *C*, leaving reactor *A* empty. Valve 2 is then closed, and valve 3 is opened. At least 2.7 g of boron trichloride is distilled into reactor *C*. This amount represents a twofold excess. The transfer can be conveniently effected if reactor *B* is calibrated in 1-ml graduations and at least 1.1 ml of boron trichloride is transferred. Valve 3 is then closed, and valve 5 is opened. The liquid-nitrogen bath around reactor *C* is removed and replaced by another bath whose temperature is less than 10° (a toluene slush bath is best). Reaction starts as soon as the boron trichloride melts (at −107°). [■ **Caution.** *If the reaction were carried out in a vacuum line of small volume, a dangerous increase in pressure might occur at this stage, as boron trifluoride is evolved. Using a 1-l. reservoir for a 2-g preparation ensures that the pressure never exceeds 1 atm (the Kel-F tubes are not recommended for use above 3 atm).*]

After about 30 min the reaction ceases. At this stage the boron trifluoride and the excess boron trichloride must be removed. One way is to open valve 1 and pump them into the vacuum system (which should have a liquid-nitrogen scavenger trap). It is a considerable advantage of the preparative method that all the reagents and products, except uranium hexachloride, are volatile.

If for some reason excess boron trichloride is not used and unreacted uranium hexafluoride remains, it also is volatile and is removed from the uranium hexachloride at this stage. The residue is essentially pure uranium hexachloride. The uranium hexachloride is in its purest form immediately following the preparation. Attempts to sublime the compound invariably decrease its purity, as there is always some thermal decomposition.

The yield is essentially quantitative, based on UF_6 used. *Anal.* Calcd. for UCl_6: 52.8; Cl, 47.2. Found: U, 52.9; Cl, 47.1.

Properties

Uranium hexachloride is a black solid melting at 177.5°. Since it is hygroscopic and reacts vigorously with water, it should be handled only in dryboxes. The crystal structure has been determined:[5] hexagonal symmetry, space group D_{3d}^{3}-$C\bar{3}$ (m, $n = 3$), with an almost perfect octahedron of chlorine atoms around each uranium atom. Uranium hexachloride can be sublimed at 75–100° at low pressures, but normally some thermal decomposition results. The ultraviolet-visible spectrum of gaseous uranium hexachloride has been determined.[6] No fine structure was observed in the spectrum. Because previously available preparative methods were inadequate, there has been very little study of the chemistry of uranium hexachloride. It reacts with hydrogen fluoride to give uranium pentafluoride and slowly with uranium hexafluoride to give uranium tetrafluoride and chlorine.[2,3]

Extension of Method

The scale of the preparation can easily be increased if certain precautions are noted. A large amount of heat is released in the metathetical reaction ($\Delta G = -502$ kJ/mole). A secondary reaction[3] can occur during the preparation

$$2UF_6 + UCl_6 \longrightarrow 3UF_4 + 3Cl_2$$

Normally the rate of this secondary reaction is slow, and UCl_6 is not contaminated with UF_4, but in larger-scale preparations, if heat is not adequately dissipated, the rate of the secondary reaction can become considerable. In these circumstances the UCl_6 will be contaminated with UF_4. For larger-scale preparations apparatus must therefore be designed to dissipate rapidly the heat released. It should also be noted that with this preparative method it is often convenient to prepare the UCl_6 *in situ,* e.g., within a spectroscopic cell.[6] This can completely eliminate handling and consequent contamination of the UCl_6.

This procedure has been used to prepare many other chlorides, e.g., tungsten hexachloride,[7] vanadium tetrachloride,[8] and the pentachlorides of niobium and tantalum.[8] However, it is normally much easier to prepare chlorides than fluorides. This method will therefore find application only where this generalization is not applicable. In particular, it will find application where the chloride is thermally unstable under normal chlorination conditions and must be prepared at or below room temperature. This method has been used to advantage for the preparation of $ReCl_6$[9] and the new compound osmium pentachloride.[10]

References

1. J. J. Katz and E. Rabinowitch, "The Chemistry of Uranium," National Nuclear Energy Series, Div. VIII, Vol. 5, p. 497, McGraw-Hill Book Company, New York, 1951.
2. K. W. Bagnall, "Halogen Chemistry," V. Gutman (ed.), Vol. 3, p. 303, Academic Press, Inc., London, 1967.
3. T. A. O'Donnell, D. F. Stewart, and P. W. Wilson, *Inorg. Chem.,* **5,** 1438 (1966).
4. J. H. Canterford and T. A. O'Donnell, "Technique of Inorganic Chemistry," H. B. Jonassen and A. Weissberger (eds.), Vol. VII, p. 273, John Wiley & Sons, Inc., New York, 1968.
5. W. H. Zachariasen, *Acta Cryst.,* **1,** 285 (1948).
6. H. J. Hurst and P. W. Wilson, *Spectrosc. Lett.,* **5,** 275 (1972).
7. T. A. O'Donnell and D. F. Stewart, *Inorg. Chem.,* **5,** 1434 (1966).

8. J. H. Canterford and T. A. O'Donnell, *Aust. J. Chem.*, **21**, 1421 (1968).
9. J. H. Canterford and A. B. Waugh, *Inorg. Nucl. Chem. Lett.*, **7**, 395 (1971).
10. R. C. Burns and T. A. O'Donnell, submitted for publication in *Inorg. Nucl. Chem. Lett.*

40. CHLOROTRIS(η-CYCLOPENTADIENYL)COMPLEXES OF URANIUM(IV) AND THORIUM(IV)

$$UCl_4 + 3C_5H_5Tl \xrightarrow{DME} (\eta\text{-}C_5H_5)_3UCl + 3TlCl$$

$$ThCl_4 + 3C_5H_5Tl \xrightarrow{DME} (\eta\text{-}C_5H_5)_3ThCl + 3TlCl$$

Submitted by T. J. MARKS,* A. M. SEYAM,*,† and W. A. WACHTER*,‡
Checked by G. W. HALSTEAD§ and K. N. RAYMOND§

The compounds $(\eta\text{-}C_5H_5)_3UCl$ and $(\eta\text{-}C_5H_5)_3ThCl$ are useful precursors for the synthesis of a large number of organoactinides,[1] such as tris(cyclopentadienyl)metal alkoxides,[2] tetrahydroborates,[3,4] alkyls and aryls,[5–9] halides,[10] amides,[11] and other derivatives.[12] The chlorouranium compound was first synthesized by Reynolds and Wilkinson[13] in 1956 from uranium tetrachloride and sodium cyclopentadienide in tetrahydrofuran. Vacuum sublimation, in our hands, invariably gives the pure product in low yield. The thorium analog was originally prepared in a similar manner,[14] except that the rapid rate of attack by thorium tetrachloride on tetrahydrofuran requires that dimethoxyethane be used as solvent. A pure product is again obtained only upon sublimation, and in low yield.

We describe here improved syntheses of $(\eta\text{-}C_5H_5)_3UCl$ and $(\eta\text{-}C_5H_5)_3ThCl$. The newer procedures have a number of attractive features. Cyclopentadienylthallium is used as the cyclopentadienylating reagent.[3,15] Unlike cyclopentadienylsodium, it is air-stable and can be stored and handled without special precautions. In addition, it generally gives cleaner reactions (less reduction) and is readily separated from the reaction solution by careful filtration.[16] Dimethoxyethane is used as a solvent for both preparations. The above innovations yield products upon solvent removal which are sufficiently pure for most synthetic work; sublimation, which is accompanied by extensive thermal decomposition, is unnecessary.

*Department of Chemistry, Northwestern University, Evanston, Ill. 60201.
†UNESCO Fellow, on leave from the University of Jordan.
‡NSF Predoctoral Fellow.
§Department of Chemistry, University of California, Berkeley, Calif. 94720.

In both syntheses, air and moisture are excluded from reagents at all times. Hence, all operations are performed under dry, prepurified nitrogen, and all glassware is oven-dried, or, when possible, flame-dried. All solvents are distilled from sodium-potassium alloy–benzophenone under nitrogen. For general discussions of techniques and apparatus employed in the synthesis of air-sensitive organometallics, the reader is referred to the excellent books by Shriver[17] and by Eisch and King.[18]

A. CHLOROTRIS(η-CYCLOPENTADIENYL)URANIUM(IV)

Procedure

A 1-l. three-necked flask is charged with 21.6 g (0.057 mole) of uranium tetrachloride* and a large magnetic stirring bar in a glove bag or glove box.† The flask is fitted with a gas inlet valve and stoppers and is removed to the bench top. With a strong flow of nitrogen gas through the flask, it is fitted with a condenser also equipped with a gas inlet. Next, the vessel is filled with 700 ml of dimethoxyethane, freshly distilled; with strong stirring, 46.0 g (0.171 mole) of cyclopentadienylthallium‡ is added. (■ **Caution** *This compound is toxic and should be handled in a hood.*) The milky-tan suspension is stirred at room temperature for 24 hr. After this time, the mixture will be suction-filtered by pouring it through a glass connecting tube into a fritted Schlenk funnel.§ The frit assembly rests on the top of a 1-l. three-necked flask fitted with a nitrogen inlet valve and a valve connected to a vacuum line. The sintered disk is covered with a layer of diatomaceous earth. The apparatus is evacuated and filled with nitrogen several times before filtration is performed. The reaction mixture is slowly poured into the Schlenk frit, while gentle suction is applied. The filtration residue is then washed with 50 ml of dimethoxyethane, and suction is applied again. Next, the dark red-brown filtrate is stripped of solvent via trap-to-trap distillation to yield a golden-brown solid as the product.

*The uranium tetrachloride(anhydrous) was purchased from Rocky Mountain Research, Denver, Colo. 80204. It can be synthesized by the procedure of J. A. Hermann and J. F. Suttle, *Inorganic Syntheses,* **5,** 143 (1957).

†This compound is hygroscopic and should be protected from the atmosphere. Storage in a Schlenk tube is recommended.

‡Cyclopentadienylthallium was purchased from Research Organic/Inorganic Chemical Co., Sun Valley, Calif. 91352. It can be easily synthesized by the method of C. C. Hunt and J. R. Doyle, *Inorg. Nucl. Chem. Lett.,* **2,** 283 (1966) or (recommended by the checkers) that of H. Meister, *Angew. Chem.,* **69,** 533 (1957).

§See reference 17, p. 150, for a diagram of a typical setup. Similar glassware can be purchased from Ace Glass, Inc., Vineland, N.J. 08360, and from Kontes Glass Co., Vineland, N.J. 08360.

Under a nitrogen flow, this is scraped to the bottom of the flask and washed with 50 ml of freshly distilled *n*-hexane. The washings are removed with a syringe, and the product is dried 8 hr under high vacuum. Yield 20.0 g (75%). *Anal.* Calcd. for $C_{15}H_{15}UCl$: C, 38.45; H, 3.23. Found: C, 38.3; H, 3.31.

Properties

Chlorotris(η-cyclopentadienyl)uranium(IV) is an oxygen-sensitive brown solid. It can be handled in air for brief periods of time with minimal oxidation, which is evidenced by darkening of the color. The compound is soluble in ethereal and aromatic solvents but only sparingly soluble in aliphatic hydrocarbons. Solutions are exceedingly air-sensitive. The ^1H nmr spectrum in benzene exhibits a sharp singlet 9.6 ppm to high field of the solvent (τ12.4). The infrared spectrum (Nujol mull) exhibits typical π-cyclopentadienyl bands at 1013 (m) and 784 (s) cm^{-1}. Oxidation is evidenced by the appearance of the antisymmetric ν(OUO) stretch of the uranyl group at 930 cm^{-1}.

B. CHLOROTRIS(η-CYCLOPENTADIENYL)THORIUM(IV)

Procedure

In the same manner as in Sec. A, a 1-l. three-necked flask is charged with 16.0 g (0.043 mole) of anhydrous thorium tetrachloride.* Next, the flask is evacuated and gently flamed with a torch.† After cooling, a magnetic stirring bar and 250 ml of freshly distilled dimethoxyethane are added to the flask under a strong flush of nitrogen. The flask is then fitted with a condenser equipped with a gas inlet, and, with rapid stirring, 36.0 g (0.133 mole) of cyclopentadienylthallium is added. The mixture is refluxed for 2 days by heating in an oil bath on a hot plate with magnetic stirring. The solvent is then stripped from the tan suspension by trap-to-trap distillation, and the residue is dried under high vacuum for 1 hr.

The slate-gray residue is next transferred, in a glove bag or glove box, to an evacuable Soxhlet extractor.‡ After the apparatus has been evacuated and filled

*This reagent was purchased from Research Organic/Inorganic Chemical Co., Sun Valley, Calif. 91352.

†Samples of thorium tetrachloride are sometimes contaminated with thionyl chloride, which can be removed by this procedure.

‡The apparatus employed was similar to Kontes unit K-212800 (Kontes Glass Co., Vineland, N.J. 08360) except that both receiving flask and condenser were attached by greaseless O-ring joints. The receiving flask was equipped with a gas inlet and an O-ring stopper. All gas inlets were fitted with Teflon needle valves.

with nitrogen several times, 60 ml of freshly distilled benzene is introduced via syringe into the receiving flask. Next, all inlet valves are closed, and the benzene is frozen in liquid nitrogen. The apparatus is subsequently evacuated to 10^{-3} mm, all valves are closed, and the benzene is thawed. This cycle is repeated, and then extraction is allowed to take place under high vacuum by closing the valves, allowing the apparatus to warm to room temperature, and introducing cool water into the condenser.* In some cases it may be necessary to warm the receiving flask to ca. 40°. After 2 days, the benzene is removed from the apparatus by trap-to-trap distillation. The resulting flesh-colored extracted solid is scraped to the bottom of the receiving flask and washed with three 10-ml portions of freshly distilled hexane. The washings are removed with a syringe. The product is then dried 8 hr under high vacuum. Yield, 13.0 g (65%) of light-flesh-colored product. *Anal.* Calcd. for $C_{15}H_{15}$-ThCl: C, 38.90; H, 3.27. Found: C, 38.0; H, 3.00. A purer, white crystalline product in considerably lower yield can be obtained by vacuum sublimation at 175°.

Properties

Chloro(η-cyclopentadienyl)thorium(IV) is a white solid, less oxygen-sensitive than the uranium analog but considerably more moisture-sensitive. Exposure to air causes the appearance of an ocher color. The compound is less soluble in all organic solvents than the uranium analog; solutions are extremely air-sensitive. The ^{1}H nmr spectrum in benzene-d_6 exhibits a sharp singlet at τ3.81. The infrared spectrum (Nujol mull) shows π-cyclopentadienyl bands at 1016 (m), 812 (sh), and 788 (s) cm^{-1}.

References

1. For excellent reviews of the subject see H. Gysling and M. Tsutsui, *Adv. Organomet. Chem.*, **9**, 361 (1970); R. G. Hayes and J. L. Thomas, *Organomet. Chem. Rev. (A)*, **7**, 1 (1971); B. Kanellakopulos and K. W. Bagnall in "MTP International Review of Science, Inorganic Chemistry," Ser. 1, Vol. 7, p. 299, H. J. Emeleus and K. W. Bagnall (eds.), University Park Press, Baltimore, 1972.
2. R. von Ammon, R. D. Fischer, and B. Kanellakopulos, *Chem. Ber.*, **105**, 45 (1972).
3. M. L. Anderson and L. R. Crisler, *J. Organomet. Chem.*, **17**, 345 (1969).
4. T. J. Marks, W. J. Kennelly, J. R. Kolb, and L. A. Shimp, *Inorg. Chem.*, **11**, 2540 (1972).
5. T. J. Marks and A. M. Seyam, *J. Am. Chem. Soc.*, **94**, 6545 (1972).

*Soxhlet extraction can also be carried out at atmospheric pressure in a less sophisticated apparatus. The product may be slightly less pure but is suitable for most preparative purposes.

6. A. E. Gabala and M. Tsutsui, *J. Am. Chem. Soc.,* **95,** 91 (1973).
7. G. Brandi, M. Brunelli, G. Lugli, and A. Mazzei, *Inorg. Chim. Acta,* **7,** 319 (1973).
8. T. J. Marks, A. M. Seyam, and J. R. Kolb, *J. Am. Chem. Soc.,* **95,** 5529 (1973).
9. T. J. Marks and W. A. Wachter, *J. Am. Chem. Soc.*, in press.
10. R. D. Fischer, R. von Ammon, and B. Kanellakopulos, *J. Organomet. Chem.,* **25,** 123 (1970).
11. T. J. Marks, J. R. Kolb, and A. R. Newman, manuscript in preparation.
12. T. J. Marks, P. S. Poskozim, and A. M. Seyam, unpublished results.
13. L. T. Reynolds and G. Wilkinson, *J. Inorg. Nucl. Chem.,* **2,** 246 (1956).
14. N. TerHarr and M. Dubeck, *Inorg. Chem.,* **3,** 1649 (1964).
15. J. Leong, K. O. Hodgson, and K. N. Raymond, *Inorg. Chem.,* **12,** 1329 (1973).
16. R. B. King, *Inorg. Chem.,* **9,** 1936 (1970).
17. D. F. Shriver, "The Manipulation of Air-sensitive Compounds," McGraw-Hill Book Company, New York, 1969.
18. J. J. Eisch and R. B. King (eds.), "Organometallic Syntheses," Vol. 1, Part I, Academic Press, Inc., New York, 1965.

Chapter Six

LIGANDS USED TO PREPARE METAL COMPLEXES

(*Assembled through the cooperation of Professor Devon W. Meek*)

41. TRIMETHYLPHOSPHINE

$$3CH_3Li + PCl_3 \xrightarrow{Et_2O} (CH_3)_3P + 3LiCl$$

Submitted by R. T. MARKHAM,* E. A. DIETZ, Jr.,* and D. R. MARTIN*
Checked by O. STELZER† and R. SCHMUTZLER†

Trimethylphosphine is a useful ligand which commonly is prepared by the reaction of a phosphorus trihalide with a methyl Grignard reagent and isolated as the silver iodide complex, $[AgIP(CH_3)_3]_4$.[1,2] Numerous references to the preparation of $P(CH_3)_3$ have appeared, some without giving the yields,[3] while others report yields of approximately 40%.[4,5] It has been stated that 60% recovery of $P(CH_3)_3$ is possible if the reaction is conducted at low temperatures, but no experimental data were given.[6] Previous syntheses also required an excess of alkylating agent to optimize the yields, while the recoveries were calculated on the amount of deficient phosphorus trihalide. For a meaningful comparison of methods, the quantity of $P(CH_3)_3$ reported should be based upon the amount of alkylating agent employed.

In some instances the substitution of methyllithium for a methyl Grignard reagent offers significant advantages in product isolation and product yields.

*Chemistry Department, The University of Texas at Arlington, Arlington, Tex. 76019.
The support of The Robert A. Welch Foundation is gratefully acknowledged.

†Lehrstuhl B für anorganische Chemie, der Technischen Universität, Pockelstrasse 4, 33 Braunschweig, West Germany.

The present availability of this reagent makes it a desirable reactant for the synthesis presented here.

Procedure

■ **Caution.** *Trimethylphosphine is a volatile toxic material which ignites spontaneously in air.*

To a three-necked 1-l. flask equipped with an addition funnel, stirrer, and Dry Ice condenser is added $\frac{1}{3}$ mole of PCl_3 and 300 ml of dry $(C_2H_5)_2O$. The outlet of the Dry Ice condenser leads to a trap containing 300 ml of saturated KI solution which is $1.1M$ in AgI to prevent any possible loss of $P(CH_3)_3$. With the flask cooled to $-78°$ and under a dry N_2 purge, 1 mole of a $2.2M$ ethereal solution of CH_3Li is added slowly ($\frac{3}{4}$ hr) to the stirred PCl_3 solution. After addition of the CH_3Li, the reaction flask is allowed to warm to 0°, and 300 ml of H_2O is added carefully to dissolve the white precipitate of LiCl. The H_2O-Et_2O mixture is transferred rapidly to a separatory funnel, and the H_2O layer is removed. The ethereal solution then is added to the 300 ml of KI-AgI solution; vigorous agitation of this mixture produces the white precipitate of $[AgIP(CH_3)_3]_4$. The precipitate is collected on a filter* and washed thoroughly with a total of 300 ml of saturated KI solution, 500 ml of water, and 400 ml of diethyl ether. The final product is dried under vacuum (ca. 10^{-4} torr) at room temperature. Melting point: 134–136° (literature,[5] 131–133°).

From this procedure 62 g of complex is recovered, representing a 60% yield based upon either the CH_3Li or PCl_3. The $P(CH_3)_3$ contained in the silver iodide complex is liberated by placing 10 g of the complex in a 250-ml round-bottomed flask connected with a 18/9 ball-and-socket glass joint to a vacuum manifold consisting of a manometer, trap, and vacuum source. The flask is evacuated at room temperature and then, by manipulation of stopcocks, connected to a trap at $-196°$. The flask is then heated by raising an oil bath, preheated to 200°, around it. After $\frac{1}{2}$ hr at 200°,† 2.25 g (29.6 mmoles) of very pure $P(CH_3)_3$ is isolated. For the two reactions, the overall yield of $P(CH_3)_3$ from PCl_3 or CH_3Li is 55%.

The usefulness of this preparation is demonstrated by the good yields of product which can be obtained even when the reaction is conducted at room temperature. For example, when $\frac{1}{3}$ mole of PCl_3 and 1 mole of CH_3Li were

*The checkers recommend collecting the finely divided precipitate on a Büchner funnel with filter paper.

†The checkers find that the complex should be heated for 1 hr and that the temperature can exceed 200°.

allowed to react at 25°, 27 g of silver complex was produced. This amount of complex is comparable to many of the yields from the syntheses using methyl Grignard reagents.

Properties

The purity of the trimethylphosphine produced by this reaction was checked by vapor-pressure data. The experimental value at 0° was 159 torr, which falls within the range of previous reports of from 158[7] to 161[4] torr.

The molecular weight was found by vapor-density measurement to be 77.3 g/mole (calcd. 76.1 g/mole). The ^1H nmr[8] and infrared[9] data also were identical to those previously reported. For ^1H nmr, $\delta_{TMS} = -0.89$ ppm, $J_{HCP} = 2.7$ Hz (1–1 doublet).

References

1. F. G. Mann and A. F. Wells, *J. Chem. Soc.*, **1938,** 702.
2. R. Thomas and K. Eriks, *Inorganic Syntheses,* **9,** 59 (1967).
3. See literature review in reference 2.
4. A. B. Burg and R. I. Wagner, *J. Am. Chem. Soc.*, **75,** 3872 (1953).
5. J. G. Evans, P. L. Goggin, R. J. Goodfellow, and J. G. Smith, *J. Chem. Soc. (A)*, **1968,** 464.
6. M. A. A. Beg and H. C. Clark, *Can. J. Chem.*, **38,** 119 (1960).
7. B. Silver and S. Luz, *J. Am. Chem. Soc.*, **83,** 786 (1961).
8. J. F. Nixon and R. Schmutzler, *Spectrochim. Acta,* **22,** 565 (1966).
9. J. Goubeau, R. Baumgärtner, W. Koch, and U. Müller, *Z. Anorg. Allg. Chem.*, **337,** 174 (1965).

42. TERTIARY PHOSPHINES

$$(C_6H_5)_3P + 2Li \xrightarrow{\text{THF}} (C_6H_5)_2PLi + C_6H_5Li$$

$$C_6H_5Li + C_4H_9Cl \longrightarrow C_6H_6 + LiCl + CH_2{=}C(CH_3)_2$$

$$(C_6H_5)_2PLi + RX \xrightarrow{\text{THF}} (C_6H_5)_2PR + LiX$$

Submitted by V. D. BIANCO* and S. DORONZO*
Checked by K. J. REIMER,† A. G. SHAVER,† P. FIESS,† and H. C. CLARK†

*Istituto di Chimica Generale ed Inorganica, Università degli Studi di Bari, Bari, Italy.
†Chemistry Department, University of Western Ontario, London 72, Ontario, Canada.

Many methods[1–31] are reported for the preparation of tertiary phosphines Ph_2PR (R = alkyl or aryl group), but only a few of them can be used conveniently because of difficulties in working procedures or disappointing yields. The method involving the reaction between lithium diphenylphosphide and alkyl (or aryl) halides is undoubtedly the most convenient both for the availability of starting materials and for the simplicity of working procedures. This method is of general applicability to a wide range of phosphines of the type Ph_2PR, where R is CH_3, C_2H_5, n-C_4H_9, C_6H_5—CH_2, C_6H_{11}, or $(CH_2)_2N(C_2H_5)_2$, and offers high yields particularly when the phenyllithium formed in the cleavage of Ph_3P by lithium metal is selectively destroyed by means of *t*-butyl chloride.

■ **Caution.** *Phosphines are toxic compounds. All the manipulations should be carried out in an efficient fume hood.*

Since the phosphines are easily oxidized compounds, all reactions must be carried out under an atmosphere of dry, oxygen-free nitrogen, and all reaction products must be manipulated in careful absence of air, by standard vacuum-line techniques.

The tetrahydrofuran is dried and purified by shaking with sodium hydroxide pellets, refluxing 24 hr over sodium metal, and distilling in an inert atmosphere from lithium tetrahydridoaluminate or sodium diphenylketyl immediately before use. Other solvents are dried and purified in the usual way.[32] Alkyl or aryl halides are also freshly distilled.

The lithium shavings are prepared from a piece of lithium (99.2% grade, Schuchardt, München, West Germany) weighing slightly more than the amount required, which is cut into small pieces; the pieces are pounded with a clean hammer into thin sheets about 0.5 mm thick. Then the sheets are cut into thin strips about 1 mm wide which are transferred to a beaker containing anhydrous ether. The required amount of lithium is weighed under dry ether and allowed to fall directly into the reaction flask. The lithium thus retains its bright luster. The lithium may also be pressed into wires of about 0.5-mm diameter by using a proper press (e.g., TOM PRESS NPL 1S/40, producing wires of lithium of 0.6–0.4-mm diam.); these wires are then cut into small pieces. During the preparation of the shavings, oxidation of the lithium must be avoided. Pieces or ribbon of lithium should not be used since these react with difficulty with triphenylphosphine.

Since the procedures employed are essentially the same, only the preparation of methyldiphenylphosphine is reported in detail as a typical procedure for the preparation of phosphines.

A. METHYLDIPHENYLPHOSPHINE[33]

Procedure

A 1-l., three-necked, round-bottomed flask is equipped with a nitrogen inlet, pressure-equalizing dropping funnel, magnetic stirrer, and reflux condenser. The flask is charged with 250 ml of dry, oxygen-free, freshly distilled THF and 40 g (0.15 mole) of triphenylphosphine; then 2.1 g (0.30 mole) of lithium shavings is added. The mixture is stirred at room temperature (20–25°) until the lithium is dissolved.* If unreacted lithium remains after the reaction has gone to completion, it may be removed before continuing the preparation. To the deep-red solution 14.1 g (0.15 mole) of freshly prepared *t*-butyl chloride[34] is added dropwise with stirring (the color of the reaction mixture clarifies, becoming red-orange). Then, 20.5 g (0.14 mole) of CH_3I in 20 ml of THF is added dropwise, the temperature being maintained between 0 and 5° with an ice-NaCl bath; the color disappears, and LiI precipitates. After the addition of CH_3I is complete, the mixture is stirred at room temperature for 4 hr. The reflux condenser is now replaced (under nitrogen flow) by a distilling head equipped with a thermometer and a Liebig condenser, and the THF is almost completely distilled off at atmospheric pressure. The distilling apparatus is then removed, deoxygenated water (80 ml) is added to the residue to dissolve lithium iodide, and 80 ml of purified and deoxygenated benzene is subsequently added. The mixture is transferred to a separatory funnel, the flask is washed with two 15-ml portions of benzene, and the washings are added to the mixture in the separatory funnel. The mixture is shaken vigorously, the organic layer is separated, and the aqueous layer is extracted twice with benzene (2 × 50 ml). The organic fractions are collected and dried over Na_2CO_3 (4 hr). The solution, filtered from the desiccant, is fractionated at atmospheric pressure to eliminate the benzene, and the residue is fractionated under reduced pressure to yield 24 g (80% yield)† of pure phosphine as a colorless liquid (b.p.

*The reaction between lithium shavings and triphenylphosphine in anhydrous THF for the reported preparations goes to completion within 12 hr at room temperature. However, complete reaction may be reached within 8 hr by warming the mixture to 45–50°. The time of this initial step depends on several factors, e.g., the purity of the lithium, the use of pieces of lithium instead of shavings or wires, or the use of lithium metal partially oxidized on its surface. Inefficient stirring or, more important, using wet THF not freshly distilled in the absence of air can also alter the time required.

†The yields are largely affected by several factors, e.g., some of those reported in the note above with regard to the time of reaction between Li and Ph_3P. Very important factors are also an efficient extraction of the phosphine from the mixture by means of the appropri-

120–122°/1.5 torr) (literature,[10] 120–122°/1.5 torr). *Anal.* Calcd. for $C_{13}H_{13}P$: C, 78.0; H, 6.5; P, 15.5. Found: C, 77.9; H, 6.7; P, 15.3.

B. ETHYLDIPHENYLPHOSPHINE

Procedure

In a similar reaction apparatus, 123 g (0.46 mole) of Ph_3P and 6.52 g (0.94 mole) of lithium shavings are allowed to react in 750 ml of dry, oxygen-free, freshly distilled THF. When the reaction is complete,* the mixture is treated with 43.3 g (0.94 mole) of freshly prepared *t*-butyl chloride; 50.5 g (0.46 mole) of ethyl bromide in 50 ml of THF is then added dropwise, with stirring, to the solution of lithium diphenylphosphide, at a temperature below 5°. When the addition of ethyl bromide is complete, the mixture is stirred at room temperature for 4 hr. The solvent is almost completely removed by distillation, deoxygenated water (200 ml) is added, and the phosphine is extracted with 200 ml of purified and deoxygenated diethyl ether. The organic layer is separated, and the aqueous layer is extracted twice with diethyl ether (2 × 60 ml). The organic fractions collected are dried over Na_2SO_4 (4 hr) and then fractionated to give 88 g (89% yield) of phosphine as a colorless liquid (b.p. 138–141°/1 torr) (literature,[10] 129–132°/0.7 torr). *Anal.* Calcd. for $C_{14}H_{15}P$: C, 78.5; H, 7.0; P, 14.5. Found: C, 78.2; H, 6.8; P, 14.56.

C. *n*-BUTYLDIPHENYLPHOSPHINE

Procedure

This phosphine is prepared by using 52.5 g (0.2 mole) of Ph_3P (in 300 ml of dry, oxygen-free THF) and 2.78 g (0.4 mole) of lithium shavings. Phenyllithium is destroyed by using 18.5 g (0.2 mole) of freshly prepared *t*-butyl chloride, and then 27.40 g (0.2 mole) of *n*-butyl bromide in 30 ml of THF is added dropwise to the solution of lithium diphenylphosphide at a temperature below 5°. When the addition of *n*-butyl bromide is complete, the reaction mixture is refluxed for 30 min and worked up as reported above, using 100 ml

*See first footnote on p. 157.

ate solvent and the dimensions of the fractionating apparatus, which, if inadequate, may cause a considerable loss of phosphine. The checkers were only able to obtain the following yields: $MePh_2P$ (65%), *n*-$BuPh_2P$ (70%), and $(C_6H_{11})Ph_2P$ (51%).

of deoxygenated water and extracting the phosphine with diethyl ether (100 and 2 × 50 ml). The yield (82%) is 40 g of phosphine (b.p. 88–90°/0.1 torr) (literature,[15] 87–92°/0.1 torr). *Anal.* Calcd. for $C_{16}H_{19}P$: C, 79.35; H, 7.85; P, 12.80. Found: C, 79.20; H, 7.74; P, 12.74.

D. CYCLOHEXYLDIPHENYLPHOSPHINE

Procedure

The phosphine is prepared according to the given procedure starting from 16.1 g (0.061 mole) of Ph_3P (in 100 ml of dry, oxygen-free THF) and 0.85 g (0.125 mole) of lithium shavings; 5.65 g (0.061 mole) of freshly prepared *t*-butyl chloride is used to destroy the phenyllithium, and 9.95 g (0.061 mole) of cyclohexyl bromide in 10 ml of THF is added to the solution of lithium diphenylphosphide at a temperature below 5°. When the addition of cyclohexyl bromide is complete, the mixture is refluxed for 30 min; lithium bromide is removed on a filter, and the THF is almost completely removed by distillation at atmospheric pressure. The residue is then fractionated, and the oily fraction boiling at 180°/2.5 torr is crystallized from oxygen-free ethanol. Ten grams (62% yield) of phosphine (white crystals) are obtained (m.p. 60–61°) (literature,[16] 60–61°). *Anal.* Calcd. for $C_{18}H_{21}P$: C, 80.5; H, 7.9; P, 11.60. Found: C, 80.3; H, 7.77; P, 11.59.

E. BENZYLDIPHENYLPHOSPHINE

Procedure

The preparation is carried out similarly to that of cyclohexyldiphenylphosphine, using 95 g (0.362 mole) of Ph_3P (in 700 ml of dry, oxygen-free THF), 5.10 g (0.736 mole) of lithium shavings, 33.5 g (0.362 mole) of freshly prepared *t*-butyl chloride, and 45.7 g (0.362 mole) of benzyl chloride (in 50 ml of THF). The reaction mixture is then refluxed for 2 hr. After the distillation of the solvent and the fractionation of the residue, the oily fraction boiling at 203–206°/1.5 torr is crystallized from oxygen-free ethanol and yields 75 g (75% yield) of phosphine as a white crystalline product (m.p. 141–142°, decomposes); (literature,[17] 142–143°). *Anal.* Calcd. for $C_{19}H_{17}P$: C, 82.5; H, 6.15; P, 11.25. Found: C, 82.6; H, 6.2; P, 11.13.

F. 2-(DIPHENYLPHOSPHINO)TRIETHYLAMINE

Procedure

■ **Caution.** *The compound* $(C_2H_5)_2NCH_2CH_2Br$ *is a "nitrogen mustard" and should be handled with great care.*

To the solution of lithium diphenylphosphide prepared from 21.8 g (0.083 mole) of Ph_3P (in 110 ml of dry, oxygen-free THF) and 1.20 g (0.166 mole) of lithium shavings, 28.50 g (0.158 mole) of $(C_2H_5)_2N(CH_2)Br$* in 30 ml of THF is added dropwise under vigorous stirring at a temperature below 5°. The reaction mixture is refluxed for 20 min and then worked up as reported above, using 50 ml of deoxygenated water and extracting the phosphine with 120 ml of benzene. The yield is 18.9 g (80%) of phosphine as colorless liquid (b.p. 146°/1 torr) (literature,[22] 185°/3 torr). *Anal.* Calcd. for $C_{18}H_{24}NP$: C, 75.8; H, 8.4; N, 4.9; P, 10.9. Found: C, 75.5; H, 8.2; N, 5.0; P, 10.8.

Properties

Phosphines are very reactive substances. They are easily subject to oxidation by atmospheric oxygen, and the phosphine oxides $Ph_2P(O)R$ are easily obtained by reaction with hydrogen peroxide or potassium permanganate (in ether). The phosphines may be characterized by reaction with alkyl halides to obtain the corresponding alkylphosphonium salts. The phosphines are soluble in THF, benzene, diethyl ether, dioxane, acetone, and ethanol (excluding Ph_2BenzylP and Ph_2CyclohexylP) and are insoluble in water.

References

1. L. Horner, P. Beck, and H. Hoffmann, *Chem. Ber.*, **92**, 2088 (1969).
2. L. Maier, *J. Inorg. Nucl. Chem.*, **24**, 1073 (1962).
3. D. Seyferth and J. M. Burlitch, *J. Org. Chem.*, **28**, 2463 (1963).
4. G. Aksnes and L. J. Brudvik, *Acta Chem. Scand.*, **17**, 1616 (1963).
5. H. Schindlbauer and V. Hilzensauer, *Monatsh. Chem.*, **96**, 961 (1965).
6. L. Maier, *Helv. Chim. Acta*, **46**, 2667 (1963).
7. R. A. Zingaro and B. E. McGlothlin, *J. Chem. Eng. Data*, **8**, 226 (1963).

*The amine can be prepared as the hydrobromide salt $Br(CH_2)_2N(C_2H_5)_2 \cdot HBr$ by the published method.[35] It is also commercially available at Fluka AG, Buchs, Switzerland; Schuchardt, München, West Germany; and K and K Laboratories, Inc., Plainview, N.Y. 11803. The hydrobromide salt (45 g) is dissolved in water (25 ml) and treated with 15 ml of 50% aqueous KOH. The oily layer is separated using a separatory funnel and stored over KOH pellets in a Dry Ice–acetone bath. A yield of 24.8 g (80%) of $(C_2H_5)_2N(CH_2)_2Br$ is obtained.

8. L. Horner and A. Mentrup, *Ann. Chem.*, **65**, 646 (1961); Ger. Pat. 1,114,190 (1961); *Chem. Abstr.*, **57**, 2256*f* (1962).
9. F. G. Mann, B. P. Tong, and V. P. Wystrach, *J. Chem. Soc.*, **1963**, 1155.
10. S. T. D. Gough and S. Trippett, *J. Chem. Soc.*, **1961**, 4263.
11. S. Trippett, *J. Chem. Soc.*, **1963**, 19.
12. F. A. Hart, *J. Chem. Soc.*, **1960**, 3324.
13. H. Normant, T. Cuvigny, J. Normant, and B. Angelo, *Bull. Soc. Chim. Fr.*, **1965**, 3446.
14. W. Kuchen and H. Buchwald, *Chem. Ber.*, **92**, 227 (1959); *Angew. Chem.*, **69**, 307 (1957).
15. M. M. Rauhut and A. M. Semsel, U.S. Pat. 3,099,691 (1963); *Chem. Abstr.*, **60**, 555*b* (1964).
16. K. Issleib and H. Volker, *Chem. Ber.*, **94**, 392 (1961).
17. M. C. Browning, J. R. Mellor, D. J. Morgan, S. A. J. Pratt, L. E. Sutton, and L. M. Venanzi, *J. Chem. Soc.*, **1962**, 693.
18. R. B. Fox, *J. Am. Chem. Soc.*, **72**, 4147 (1950).
19. S. Trippett, *J. Chem. Soc.*, **1961**, 2813.
20. D. Jerchel and J. Kimming, *Chem. Ber.*, **83**, 277 (1950).
21. A. M. Aguiar, H. J. Greenberg, and K. E. Rubenstein, *J. Org. Chem.*, **28**, 2091 (1963).
22. K. Issleib and R. Rieschel, *Chem. Ber.*, **98**, 2086 (1965).
23. G. R. Dobson, R. C. Taylor, and T. D. Walsh, *Inorg. Chem.*, **6**, 1929 (1967).
24. D. Wittenberg and H. Gilman, *J. Org. Chem.*, **23**, 1063 (1958).
25. A. M. Aguiar, J. Beisler, and A. Mills, *J. Org. Chem.*, **27**, 1001 (1962).
26. A. M. Aguiar, H. J. Greenberg, and K. E. Rubenstein, *J. Org. Chem.*, **28**, 2091 (1963).
27. H. Schindlbauer, *Monatsh. Chem.*, **96**, 2051 (1965).
28. D. Jerchel and J. Kimming, *Chem. Ber.*, **83**, 277 (1950).
29. J. Chatt and F. A. Hart, *J. Chem. Soc.*, **1960**, 1378.
30. A. M. Aguiar and D. Daigle, *J. Am. Chem. Soc.*, **86**, 2299 (1964).
31. R. A. Baldwin and R. M. Washburn, *J. Org. Chem.*, **30**, 3860 (1965).
32. A. I. Vogel, "A Textbook of Practical Organic Chemistry," 3d ed., Longmans, Green & Co., Ltd., London, 1956.
33. The synthesis of CH_3PPh_2 by the reaction of Ph_2PCl with CH_3MgX is described in *Inorganic Syntheses*, **15**, 129 (1973).
34. Reference 32, p. 276.
35. F. Cortese, "Organic Syntheses," Collective Vol. II, p. 91, John Wiley & Sons, Inc., New York, 1950.

43. DIPHENYLPHOSPHINE

$$(C_6H_5)_3P + 2Li \xrightarrow{THF} (C_6H_5)_2PLi + C_6H_5Li$$

$$(C_6H_5)_2PLi + H_2O \longrightarrow (C_6H_5)_2PH + LiOH$$

Submitted by V. D. BIANCO* and S. DORONZO*
Checked by J. CHAN† and M. A. BENNETT†

*Istituto di Chimica Generale ed Inorganica, Università degli Studi di Bari, Bari, Italy.
†Research School of Chemistry, Australian National University, Canberra, A.C.T. 2600, Australia.

Secondary aromatic phosphines can be prepared from tertiary phosphines by cleavage with sodium in liquid ammonia,[1] and the detailed preparation of diphenylphosphine by this method has been reported.[2] Diphenylphosphine has also been prepared by the reaction of chlorodiphenylphosphine with alkali metals[3] or with lithium tetrahydroaluminate.[4] This phosphine has been also obtained from diphenyltrichlorophosphorane[5] or tetraphenyldiphosphine-disulfide[6] with lithium aluminum hydride. A faster and easier method of preparation, which gives equally high yields, consists in the cleavage of triphenylphosphine with lithium metal in tetrahydrofuran, followed by hydrolysis of lithium diphenylphosphide with water to generate the phosphine.[7]

■ **Caution.** *Diphenylphosphine is a bad-smelling, toxic compound. All the manipulations should be carried out in an efficient fume hood.*

Procedure

The preparation must be carried out under dry, oxygen-free nitrogen and the reaction products must be manipulated in the absence of air, by standard vacuum-line techniques.

The tetrahydrofuran (THF) is dried and purified by shaking with sodium hydroxide pellets, refluxing 24 hr over sodium metal, and distilling immediately before use from lithium aluminum hydride (lithium tetrahydroaluminate) or from sodium diphenylketyl. Dry diethylether and deoxygenated water are also used. (■ **Caution.** *Serious explosions may occur when impure THF is purified if it contains peroxides* [see *Inorganic Syntheses,* **12:** 317 (1970)].) A 500-ml, three-necked, round-bottomed flask is equipped with a nitrogen inlet, pressure-equalizing dropping funnel, magnetic stirrer, and reflux condenser. The flask is charged with 20 g (0.0764 mole) of triphenylphosphine in 200 ml of dry THF, and 1.10 g (0.159 mole) of lithium shavings* is then added. The mixture is stirred at room temperature until the lithium is dissolved. The flask is cooled in an ice-water bath, and the deep-red solution is treated, under stirring, with 100 ml of deoxygenated water to hydrolyze the lithium diphenylphosphide. During this treatment the solution clarifies, becoming almost colorless, and LiOH partially precipitates. At this point, the reflux condenser and the dropping funnel are removed (under nitrogen flow), the main neck is stoppered, and 80 ml of dry diethyl ether is added to the mixture, with vigorous shaking to extract the phosphine with the organic solvent. The mixture is now transferred (under nitrogen atmosphere) to a separatory

*For the preparation of lithium shavings and reaction time with triphenylphosphine, see Synthesis 42.

funnel, the upper organic layer containing the phosphine is separated, and the aqueous layer is shaken again with another 30 ml of diethylether. The collected organic fractions (stored under nitrogen) are now washed in a separatory funnel with dilute (1:15) HCl (2 × 40 ml) and then with water until the acidity is completely removed. The ether solution is dried over Na_2CO_3 (4 hr), the desiccant is removed on a filter, and the solvent is distilled almost completely using a 500-ml flask equipped with a distilling head, thermometer, and Liebig condenser. The residue is transferred (under nitrogen) to a small flask for distillation and fractionated under reduced pressure. The yield is 11.5 g (81 %) of diphenylphosphine, b.p. 112–113° at 1.5 torr. *Anal.* Calcd. for $C_{12}H_{11}P$: C, 77.42; H, 5.91; P, 16.67. Found: C, 77.0; H, 5.84; P, 16.5.

Properties

Diphenylphosphine is toxic; it has a strong unpleasant odor and the vapor should not be inhaled. Contact with the skin may cause a rash. $d_4^{16} = 1.07$. $\nu(PH) = 2280 \pm 10\ cm^{-1}$; H is $\tau 4.80$, doublet, J_{PH} 218 Hz. In performing nmr spectroscopy, the C_6D_6 solutions should be handled in an inert atmosphere, to avoid rapid oxidation.

References

1. W. Müller, dissertation, Tübingen, 1957; K. Issleib and H. O. Fröhlich, *Z. Naturforsch.*, **14b**, 349 (1959); R. G. Hayter and F. S. Humiec, *Inorg. Chem.*, **2**, 306 (1963); W. Hewertson and H. R. Watson, *J. Chem. Soc.*, **1962**, 1490.
2. W. Gee, R. A. Shaw, and B. C. Smith, *Inorganic Syntheses*, **9**, 19 (1967).
3. L. Horner, P. Beck, and H. Hoffmann, *Chem. Ber.*, **92**, 2088 (1959).
4. W. Kuchen and H. Buchwald, *Chem. Ber.*, **91**, 2871 (1958).
5. L. Horner, P. Beck, and H. Hoffmann, *Chem. Ber.*, **91**, 1583 (1958).
6. K. Issleib and A. Tzschach, *Chem. Ber.*, **92**, 704 (1959).
7. D. Wittenberg and H. Gilman, *J. Org. Chem.*, **23**, 1063 (1958).

44. TRI(PHENYL-d_5)PHOSPHINE

$$C_6D_6 + Br_2 \xrightarrow{C_5H_5N} C_6D_5Br + DBr$$

$$C_6D_5Br + Mg \xrightarrow{Et_2O} C_6D_5MgBr$$

$$3C_6D_5MgBr + PCl_3 \xrightarrow{Et_2O} (C_6D_5)_3P + 3MgBrCl$$

Submitted by V. D. BIANCO* and S. DORONZO*
Checked by S. DOWNIE† and M. A. BENNETT†

Intramolecular aromatic substitution in transition metal complexes is an area of research where there is considerable current interest, and tri(phenyl-d_5)-phosphine is a ligand often used in such investigations.[1] This phosphine can be prepared[2] from phenyl-d_5-magnesium bromide by its reaction with PCl_3 in diethylether, as described below.

Procedure

1. Bromobenzene-d_5

A 250-ml, two-necked, round-bottomed flask is equipped with a reflux condenser, pressure-equalizing dropping funnel, and magnetic stirrer. At the top of the condenser is attached a tube connected to an inverted funnel dipping just below the surface of about 200 ml of water in a beaker. The flask is charged with 10 g (0.119 mole) of C_6D_6 (Fluka AG, Buchs, Switzerland; Schuchardt, Munich, West Germany; Ryvan Chemical Co. Ltd., Southampton, England) and 0.5 g of pyridine; 23.2 g (0.145 mole) of bromine is placed in the dropping funnel. The flask is cooled to 0° (ice-water bath), and the bromine is added dropwise, with stirring, to the solution of benzene-d_6. Evolution of DBr begins immediately, and the solution becomes red. When the addition of bromine is complete, the reaction mixture is stirred 1 hr at 25–30°. The mixture is then heated to 65–70° (water bath) and kept at this temperature until the DBr is completely evolved (12 hr).‡ After the complete evolution of DBr, the condenser and the dropping funnel are removed, and the reaction mixture is washed in a separatory funnel successively with water (2 × 80 ml), with a 6% solution of NaOH (3 × 80 ml),‡§ and finally with water again several times

*Istituto di Chimica Generale ed Inorganica, Università degli Studi di Bari, Bari, Italy.
†Research School of Chemistry, Australian National University, Canberra, A.C.T. 2600, Australia.

‡The evolution of DBr is tested with moist litmus paper.

§After treatment with NaOH, the solution clarifies, becoming colorless or pale yellow.

until the alkalinity is completely removed. The organic layer is dried over $MgSO_4$ (4 hr), filtered from the desiccant, and fractionated. The fraction boiling at 155–157°/760 torr is collected; 10.4 g (54% yield) of C_6D_5Br is obtained.

2. Tri(phenyl-d$_5$)phosphine

A 250-ml, three-necked, round-bottomed flask is equipped with a reflux condenser, pressure-equalizing dropping funnel, nitrogen inlet, and magnetic stirrer. The entire system is flushed with nitrogen to remove the moist air. To the dropping funnel is added 10.3 g (0.064 mole) of C_6D_5Br, and the flask is charged with 1.54 g (0.064 mole) of magnesium turnings and 26 ml of dry diethyl ether. About 0.5 ml of C_6D_5Br is then added during slow stirring, and after the reaction has started,* the remaining C_6D_5Br is slowly added over a period of 45 min (continuous moderate stirring). After the addition is complete, the solution of Grignard compound is stirred at room temperature for an additional 30 min. The dropping funnel is now charged with a solution of 2.91 g (0.021 mole) of PCl_3 in 18 ml of diethyl ether, and the flask is cooled to $-10°$ (ice–conc. HCl bath). The solution of PCl_3 is then added dropwise to the Grignard compound, with stirring. When the addition of PCl_3 is complete, the reaction mixture is allowed to warm up at room temperature for $\frac{1}{2}$ hr to ensure complete reaction. The flask is now cooled again to 0°, and the stirred mixture is treated with H_2O (15 ml). At this point the reflux condenser and the dropping funnel are removed (under nitrogen flow), the main neck is stoppered, and the reaction mixture is transferred (under nitrogen atmosphere) to a separatory funnel. The phosphine is extracted four times with diethyl ether (50, 30, 20, and 10 ml), and the collected organic fractions (stored under nitrogen atmosphere) are dried over Na_2CO_3 (4 hr). The solution is filtered from the desiccant, and the solvent is removed under vacuum at room temperature. The residue is transferred to a 100-ml flask equipped with a distilling head, thermometer, and Liebig condenser and is heated at 280° (flaming directly) to eliminate the secondary product $C_6D_5—C_6D_5$. After the removal† of this product, the residual colorless liquid is allowed to cool to room temperature, and a white solid separates.‡ The flask is now opened to the atmosphere, and the solid is scraped out and transferred to a vacuum sublimation apparatus. The $(C_6D_5)_3P$ is sublimed, using a bath temperature of 140–160° and a pressure of

*Some crystals of iodine as catalyst may be used.

†The checkers find that if the $C_6D_5—C_6D_5$ is not removed, the final impure product melts in the range 113–130°.

‡If at room temperature the residue appears to be semisolid, the solidification can be completed by cooling in an ice-water bath.

0.5 torr. Yield: 4.07 g (69.5% based on C_6D_5Br) of white $(C_6D_5)_3P$, melting at 76–77°. *Anal.* Calcd. for $C_{18}D_{15}P$: C, 77.98; P, 11.19. Found: C, 77.85; P, 11.10.

Properties

Tri(phenyl-d_5) phosphine is a white, crystalline nonhygroscopic compound, practically insoluble in water and soluble in ethanol, acetone, benzene, and diethylether.

References

1. G. W. Parshall, *Acc. Chem. Res.*, **3,** 139 (1970).
2. M. A. Bennett and D. L. Milner, *J. Am. Chem. Soc.*, **91,** 6983 (1969).
3. E. Lindner, H. D. Ebert, and A. Haag, *Chem. Ber.*, **103,** 1872 (1970).

45. METHYL DIFLUOROPHOSPHITE

$$PF_3 + HOCH_3 + C_5H_5N \rightarrow F_2POCH_3 + C_5H_5N{\cdot}HF$$

Submitted by L. F. CENTOFANTI* and L. LINES†
Checked by R. J. CLARK‡

Methyl difluorophosphite has been prepared by the fluorination of CH_3POCl_2 with SbF_3.[1] A more satisfactory route is the direct reaction of PF_3, alcohol, and a base.[2]

■ **Caution.** *Since many of the fluoro derivatives of phosphorus oxoacids are known to be extremely toxic, all products of this reaction should be handled with care.*

Procedure

Standard high-vacuum techniques are used throughout.[3,4] Phosphorus trifluoride can be obtained from Ozark-Mahoning Co.

In a typical experiment, a 2000-ml bulb equipped with a stopcock is attached to the vacuum system by a standard-taper connection. The bulb is thor-

*Monsanto Industrial Chemical Corp., St. Louis, Mo. 63166.
†Olin Research Center, New Haven, Conn. 06504.
‡Department of Chemistry, Florida State University, Tallahassee, Fla. 32306.

oughly evacuated. Approximately equimolar amounts of PF_3 (140 mmoles), CH_3OH (140 mmoles), and pyridine (12.6 g, ca. 150 mmoles) are condensed into the bulb, and the stopcock is closed. The mixture is allowed to warm slowly to ambient temperature. An immediate reaction is indicated by the formation of white clouds and solids. Calculations would indicate a bulb pressure of about 2 atm during and after reaction. The pressure in the bulb is always found to be less than this value (usually about 1 atm). The described vessel will easily hold 2 atm. If the reaction is scaled up, metal vessels should be substituted for the glass bulb.

After $2\frac{1}{2}$ hr at room temperature, the mixture is frozen at $-196°$ and allowed to warm slowly, while its volatile contents are distilled through U tubes maintained at $-64°$ (chloroform slush), $-140°$ (cooled 30–60° petroleum ether), and $-196°$ (liquid nitrogen). White solids remain behind in the reaction vessel. An unidentified oil at $-64°$ and 5.2 mmoles of PF_3 at $-196°$ are discarded. The desired PF_2OCH_3 (84 mmoles) is found at $-140°$. The yield, based on consumed PF_3, is 62%.

The procedure can also be used for the preparation of the corresponding ethyl (57% yield), isopropyl (50% yield,) *t*-butyl (24% yield), and trifluoroethyl (33% yield) esters. For the *t*-butyl and trifluoroethyl esters, the reaction times are increased to 12 and 4 hr, respectively.

Properties

Methyl difluorophosphite is a colorless gas having the following characteristics:[1] b.p. = $-15.5°$ (760 torr); f.p. = $-117.3°$. The vapor pressure can be expressed by

$$\log P_{\text{mm}} = 7.734 - \frac{1250.5}{T}$$

Infrared and nmr parameters have been reported in the literature.[2,5] The gas-phase spectrum[2] exhibits the following peaks: 3001 (ms), 2948 (m), 2924 (mw,sh) (CH stretch); 1397 (mw), 1177 (w,sh), 1137 (mw), 1097 (m,sh), 1047 (vs), 967 (s), 820 (vs), 792 (vs), 557 (w).

References

1. D. R. Martin and P. J. Pizzolato, *Inorganic Syntheses,* **4,** 141 (1953).
2. L. Lines and L. F. Centofanti, *Inorg. Chem.,* **12,** 2111 (1973).
3. R. T. Sanderson, "Vacuum Manipulation of Volatile Compounds," John Wiley & Sons, Inc., New York, 1958.
4. D. F. Shriver, "The Manipulation of Air-sensitive Compounds," McGraw-Hill Book Company, New York, 1969.
5. A. Cowley and M. Damasco, *J. Am. Chem. Soc.,* **93,** 6815 (1971).

46. BI-, TRI-, AND TETRADENTATE PHOSPHORUS-SULFUR LIGANDS

Submitted by D. W. MEEK,* G. DYER,* and M. O. WORKMAN*
Checked by P. S. K. CHIA† and S. E. LIVINGSTONE†

Recent investigations in the purposeful design and syntheses of polydentate ligands containing heavy donor atoms have led to transition metal complexes with unusual coordination numbers and stereochemistries.[1–4] Chelating phosphine, arsine, stibine, and thioether ligands are useful for studies in the areas of homogeneous catalysis, pentacoordination,[1–4] and oxidative-addition reactions.[5,6] Subtle variations in the nature of a polydentate ligand often lead to marked changes in the electronic and magnetic properties of a resultant transition metal complex;[7] thus, viable synthetic routes for new ligands are needed to facilitate studies on the effects of changing the donor basicity, chelate chain length, and mixed sets of donor atoms.

The syntheses of the three phosphorus-sulfur ligands I, II, and III are

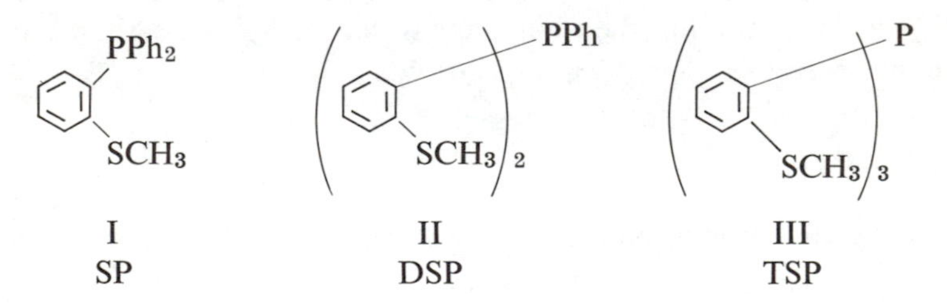

accomplished by treating diphenylphosphinous chloride, Ph_2PCl, phenylphosphonous dichloride, $PhPCl_2$, and phosphorus trichloride, PCl_3, respectively, with the lithium reagent derived from *o*-bromothioanisole.[2,8–11] These syntheses illustrate the synthetic route and procedure that can lead to a large number of chelating ligands.[2,8,9,12] For example, the corresponding arsenic-sulfur ligands[2] can be made analogously, and other substituents can be placed on the sulfur and/or phosphorus atoms by choosing the appropriate intermediates.

■ **Caution.** *These reactions should be carried out in a purified nitrogen atmosphere and in a good hood. Although these three phosphorus-sulfur ligands are solids and possess very little odor when pure, the intermediates have very unpleasant odors and should be assumed to be toxic.*

*Chemistry Department, The Ohio State University, Columbus, Ohio 43210.
†Chemistry Department, University of New South Wales, Kensington, New South Wales 2033, Australia.

Preparation of the Intermediates

$$C_6H_7NS + Na \xrightarrow{C_2H_5OH} C_6H_6NSNa + \frac{1}{2}H_2$$
$$C_6H_6NSNa + CH_3I \longrightarrow C_7H_9NS + NaI$$

1. ***o*-Methylthioaniline,** (benzene ring with NH_2 and SCH_3 ortho)

Procedure

Livingstone's method for alkylating *o*-mercaptoaniline is used.[10] *o*-Mercaptoaniline (150 g, 1.2 moles), obtained from Eastman, is dissolved in 625 ml of absolute ethanol, and the solution is placed in a 2-l., three-necked, round-bottomed flask fitted with a nitrogen gas inlet and outlet. Sodium spheres (27 g, 1.17 moles) are cleaned by washing with benzene and added slowly to the *o*-mercaptoaniline solution at such a rate that the temperature is maintained just below the reflux temperature of ethanol. A nitrogen atmosphere is maintained over the reaction mixture to reduce the fire hazard from the hydrogen gas being evolved. The reaction mixture is stirred mechanically with a sealed ground-glass stirrer. When all the sodium metal has been added, methyl iodide (165 g, 1.17 moles) is added in small quantities, with cooling. The solution becomes dark upon addition of the methyl iodide. After complete addition of methyl iodide, the mixture is heated at the reflux temperature for 40 min. Then the reaction mixture is cooled to room temperature and poured into 3 l. of distilled water. The brown oily product is extracted with ether, and the ether phase is dried overnight with anhydrous sodium sulfate. The ether is removed by distillation; subsequent vacuum distillation of the residual red-brown oil yields ca. 138 g (83%) of clear colorless *o*-methylthioaniline (b.p., 120–122°/15 torr). The product is light-sensitive and turns dark brown on long exposure. The compound should be stored in a brown bottle, preferably in the dark.

2. ***o*-Bromothioanisole,** (benzene ring with Br and SCH_3 ortho)

$$C_7H_9NS + NaNO_2 + 2H_2SO_4 \xrightarrow{<5°} [C_7H_7N_2S]HSO_4 + 2H_2O + NaHSO_4$$
$$[C_7H_7N_2S][HSO_4] + KBr \xrightarrow[Cu_2Br_2]{65°} C_7H_7BrS + KHSO_4 + N_2$$

Procedure

o-Methylthioaniline is converted to *o*-bromothioanisole by the following method: 80 g (0.57 mole) of *o*-methylthioaniline is mixed with 550 ml of water in a 3-l. Erlenmeyer flask. The resulting mixture is cooled to 10° in an ice bath, and 200 g of concentrated sulfuric acid is added in small portions. A white solid forms initially and dissolves on addition of more sulfuric acid. After complete addition of the 200 g of sulfuric acid, the mixture is cooled to 0–5° in an ice-salt bath. A solution of sodium nitrite (45 g, 0.65 mole, in 200 ml of water) is prepared and added to the resulting sulfuric acid mixture in small portions, *while keeping the temperature below* 5°. A bright yellow color develops quickly in solution and becomes darker orange as the reaction proceeds. After complete addition of the sodium nitrite solution, the *diazo* solution is stored in an ice bath while the next reagent solution is prepared.

Cuprous bromide (138 g, 0.50 mole) is mixed with potassium bromide (150 g, 1.25 moles). This mixture is suspended in 200 ml of water and heated to 65°. Then the already prepared diazo solution is added in small portions, *while keeping the solution temperature between* 65 *and* 80°. Nitrogen gas is evolved quite vigorously, but the reaction stops when the temperature falls below 65°. After complete addition of the diazo solution, the mixture is heated until no more nitrogen gas is evolved. The mixture is then cooled to 50°, and 1000 ml of concentrated hydrochloric acid is added. The resulting solution is placed in a 3-l., three-necked, round-bottomed flask, and the product is *steam-distilled.* This steam distillation requires about 15 hr to remove most of the product on this scale, but larger-scale runs may be set up to steam-distill on a continuous basis for 2–3 days. The large amount of condensate is washed with sodium hydroxide solution (20% by weight) and extracted with ether to give a yellow-orange solution. The ether extract solution is dried with anhydrous sodium sulfate overnight; then the ether solvent is removed by distillation at atmospheric pressure, and the product is vacuum-distilled (b.p. 79°/3 torr) to give 94 g (80%) of a clear yellow liquid.

3. ***In Situ* Preparation of [2-(Methylthio)phenyl]lithium,**

$$C_7H_7BrS + n\text{-}C_4H_9Li \ n\text{-} \xrightarrow{C_6H_{14}} C_7H_7LiS + n\text{-}C_4H_9Br$$

Procedure

The product from Part 2, *o*-bromothioanisole (77 g, 0.38 mole), is placed in a dry 1-l. three-necked flask fitted with a water-cooled condenser, a strong mechanical stirrer, a 250-ml pressure-equalizing dropping funnel, and a nitrogen inlet and outlet. Ether (50 ml) is added, and the system is purged with dry nitrogen for 30 min. The ether solution is cooled to 5° in an ice bath, and a solution of *n*-butyllithium* (24.3 g, 0.38 mole) in hexane (235 ml total volume) is added dropwise with stirring. After one-half the butyllithium is added, a white solid begins to form. When addition of the butyllithium is complete, the resulting mixture is stirred at 0° for 1 hr.

Preparation of the Ligands

Synthesis of each of three phosphorus-sulfur chelating ligands makes use of the lithium reagent from Part 3 and follows a similar procedure; the specific details for each ligand are given below.

A. [*o*-(METHYLTHIO)PHENYL]DIPHENYLPHOSPHINE, o-$C_6H_4(PPh_2)(SCH_3)$

$$C_7H_7LiS + (C_6H_5)_2PCl \longrightarrow C_{19}H_{17}PS + LiCl$$

Procedure

Liquid *o*-bromothioanisole (38.5 g, 0.19 mole) in sodium-dried ether (200 ml) is treated dropwise over a period of 2 hr with *n*-butyllithium (125 ml of a 1.5*N* solution in hexane, 0.19 mole) at 0°, under nitrogen with vigorous stirring. After the resultant mixture has been stirred for 1 hr more at 0°, diphenylphosphinous chloride (42 g, 0.19 mole), which has been vacuum-distilled, is added in 125 ml of ether over 3 hr. The resulting white precipitate is then hydrolyzed with 0.2*N* hydrochloric acid (125 ml). After a period of stirring to ensure that all the inorganic salts have dissolved, the white crystalline ligand is collected on a filter, washed with water, ethanol, and ether, and dried in a desiccator (yield: 43 g, 74%), (m.p., 101–102°). *Anal.* Calcd. for $C_{19}H_{17}PS$: C, 74.00; H, 5.56; S, 10.38. Found: C, 74.3; H, 5.76; S, 10.6.

*If *n*-butyllithium is unavailable commercially, it can be prepared easily from lithium metal and freshly distilled *n*-butyl bromide by published methods, e.g., R. G. Jones and H. Gilman, *Org. Reactions,* **6**, 339 (1951).

Properties

[*o*-(Methylthio)phenyl]diphenylphosphine (SP) is a white solid that is not sensitive to air, has only a faint odor when pure, and is quite soluble in organic solvents such as acetone, alcohol, and benzene. An Hnmr spectrum of SP in dichloromethane solution has a sharp singlet at $\tau 7.67$ (CH_3 protons) and a multiplet centered at $\tau 2.70$ (C_6H_5 protons).[11] The ligand forms stable complexes with a large number of transition metals, e.g., square-planar $M(SP)X_2$ and square-pyramidal $[M(SP)_2X]^+$ complexes are prevalent.[9,11,12] Some of the complexes undergo *S*-demethylation readily to give complexes of the corresponding phosphine-mercaptide ligand.[9,11]

B. BIS[*o*-(METHYLTHIO)PHENYL]PHENYLPHOSPHINE,

$$\left(C_6H_4(SCH_3) \right)_2 PPh$$

$$2C_7H_7LiS + C_6H_5PCl_2 \longrightarrow C_{20}H_{19}PS_2 + 2LiCl$$

Procedure

o-Bromothioanisole (77 g, 0.38 mole) is treated as in Sec. A; then phenylphosphonous dichloride (34 g, 0.19 mole) is dissolved in 100 ml of ether and added dropwise with stirring. The solution becomes slightly warm but shows very little change in physical appearance. After complete addition of the phenylphosphonous dichloride solution, the two-phase mixture is stirred at room temperature for 2 hr. Any excess lithium reagent is hydrolyzed by adding 40 ml of water and then 75 ml of 0.2*N* hydrochloric acid. The white solid is collected on a filter, washed with a small amount of cold ethanol, and dried. The ether-hexane layer is separated from the aqueous phase, dried over anhydrous sodium sulfate, and evaporated to dryness to give an additional portion of white solid. Total yield is ca. 60 g (89%) of bis[*o*-(methylthio)phenyl]-phenylphosphine. Recrystallization from 1-butanol gives white needles, m.p., 115–116°. Slightly impure samples melted ca. 99–104°. *Anal.* Calcd. for $C_{20}H_{19}PS_2$: C, 67.79; H, 5.36; P, 8.75; S, 18.07. Found: C, 67.9; H, 5.51; P, 8.53; S, 18.0.

Properties

Bis[*o*-(methylthio)phenyl]phenylphosphine (DSP) is a white solid that is not appreciably air-sensitive, is soluble in common organic solvents such as alcohol, acetone, and benzene, and does not have a strong odor when pure. An Hnmr spectrum of DSP in d-chloroform solution has a sharp singlet (CH_3) at $\tau 7.62$ and a multiplet (phenyl) centered at $\tau 2.56$.[11] The ligand forms stable four- and five-coordinate complexes with several transition metal ions,[8,9,11] for example, Ni(II), Pd(II), Pt(II), and Co(II). The x-ray structural determination of $Ni(DSP)I_2$ showed a square-pyramidal arrangement of ligands around the nickel.[13]

C. TRIS[*o*-(METHYLTHIO)PHENYL]PHOSPHINE, $\left(C_6H_4{-}SCH_3 \right)_3 P$

$$3C_7H_7LiS + PCl_3 \longrightarrow C_{21}H_{21}PS_3 + 3LiCl$$

Procedure

A solution of *o*-bromothioanisole (77 g) in 400 ml of ether is treated with an equimolar quantity (250 ml of a 1.5*N* solution) of *n*-butyllithium in hexane over 2 hr at 0° under a nitrogen atmosphere. Phosphorus trichloride (17.5 g in 250 ml of ether) is added dropwise via a pressure-equalizing dropping funnel over 3 hr, and then the reaction mixture is hydrolyzed with 250 ml of 0.2*N* hydrochloric acid. The resultant white precipitate is washed thoroughly with water, ethanol, and ether and then dried (yield: 45 g, 89%). The compound is recrystallized from *n*-butanol. *Anal.* Calcd. for $C_{21}H_{21}PS_3$: C, 62.97; H, 5.28; S, 24.02. Found: C, 62.9; H, 5.43; S. 24.0.

Properties

Tris[*o*-(methylthio)phenyl]phosphine functions as a tetradentate tripodlike ligand that forms stable trigonal-bipyramidal complexes with Ni(II).[2,14] With heavy metals the ligand can function either as a bidentate or as a tridentate.[2]

References

1. L. M. Venanzi, *Angew. Chem., Int. Ed.*, **3**, 453 (1964).
2. G. Dyer and D. W. Meek, *Inorg. Chem.*, **4**, 1398 (1965).
3. C. A. McAuliffe and D. W. Meek, *Inorg. Chim. Acta*, **5**, 270 (1971).
4. L. Sacconi, *J. Pure Appl. Chem.*, **17**, 95 (1968).
5. J. Halpern, *Acc. Chem. Res.*, **3**, 386 (1970), and references contained therein.
6. T. E. Nappier, Jr., D. W. Meek, R.M. Kirchner, and J. A. Ibers, *J. Am. Chem. Soc.*, **95**, 4194 (1973).
7. L. Sacconi, *J. Chem. Soc.* (*A*), **1970**, 248.

8. M. O. Workman, G. Dyer, and D. W. Meek, *Inorg. Chem.*, **6,** 1543 (1967).
9. S. E. Livingstone and T. N. Lockyer, *Inorg. Nucl. Chem. Lett.*, **3,** 35 (1967).
10. S. E. Livingstone, *J. Chem. Soc.*, **1956,** 437.
11. P. G. Eller, J. M. Riker, and D. W. Meek, *J. Am. Chem. Soc.*, **95,** 3540 (1973).
12. G. Dyer and D. W. Meek, *J. Am. Chem. Soc.*, **89,** 3983 (1967).
13. D. W. Meek and J. A. Ibers, *Inorg. Chem.*, **8,** 1915 (1969).
14. L. P. Haugen and R. Eisenberg, *Inorg. Chem.*, **8,** 1072 (1969).

47. TETRADENTATE TRIPOD LIGANDS CONTAINING NITROGEN, SULFUR, PHOSPHORUS, AND ARSENIC AS DONOR ATOMS

Submitted by R. MORASSI* and L. SACCONI*
Checked by C. A. McAULIFFE† and W. LEVASON†

Neutral polydentate ligands are widely used for the synthesis of coordination compounds of transition metal ions.[1–13] When different synthetic routes are possible for a given ligand, those which give the highest yields with the least number of steps and minimize the number of possible isomeric or by-products will obviously be preferred. The above criteria may be exemplified by a series of neutral tetradentate ligands of the tripod or umbrella type with a nitrogen atom as the top donor; these may be described as 2-substituted triethylamine derivatives (I):

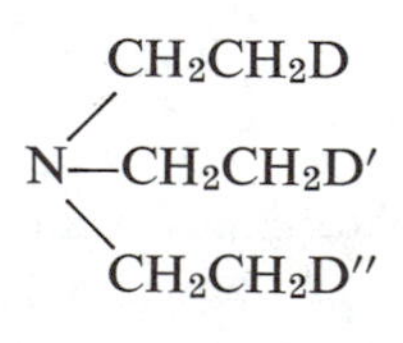

I

where D, D′, and D″ stand for any of the following donor groups: $N(C_2H_5)_2$, OCH_3, S—R (R=CH_3, C_2H_5, *i*-C_3H_7, *t*-C_4H_9), $P(C_6H_5)_2$, and $As(C_6H_5)_2$.

Structurally, the ligands can be classified according to their symmetry and their relationship with *N,N*-diethylethylenediamine and 1,1,7,7-tetraethyldiethylenetriamine, as shown in Table I.

From a synthetic point of view, ligands of the same type are often derived

*Istituto di Chimica Generale, Laboratorio C.N.R., Università di Firenze, Firenze 50132, Italy.

†Chemistry Department, University of Manchester Institute of Science and Technology, Manchester M60 1QD, England.

TABLE I

No.	Ligand Type	Terminal Ligand Atoms
1.	$N(D_3)$	$D = D' = D''$
2.	$N(ND_2)$	$D'' = N(C_2H_5)_2$, $D = D' \neq D''$
3.	$N(N_2D)$	$D = D' = N(C_2H_5)_2$, $D'' \neq D, D'$
4.	$N(DD'_2)$	$D \neq D' = D''$, $D, D', D'' \neq N(C_2H_5)_2$
5.	$N(ND'D'')$	$D = N(C_2H_5)_2$, $D \neq D' \neq D''$

from common intermediates. All the above ligands were obtained in satisfactory yields by a proper combination of three types of reaction: (1) chlorination of alcoholic groups by standard methods ($SOCl_2$); (2) base-promoted alkylation of primary or secondary amino groups; (3) nucleophilic displacement of a halogen atom by strong (nucleophilic) bases (alkoxy, alkylthio, phosphido, or arsenido groups). A number of complexes with transition metal ions have been isolated and characterized.[1–13]

We report here the syntheses of the symmetric type 1 ligands tris[2-(diphenylphosphino)ethyl]amine, tris[2-(diphenylarsino)ethyl]amine, and tris[2-(methylthio)ethyl]amine.

Ligands of the Type $N(D_3)$

The most convenient synthetic route to these ligands is the one-step reaction

$$N(CH_2CH_2Cl)_3 + KD \text{ (or NaD)} \longrightarrow N(CH_2CH_2D)_3 + KCl \text{(or NaCl)}$$

Tris(2-chloroethyl)amine is readily available by chlorination of triethanolamine.[14]

The above reaction is successful only when strongly nucleophilic D groups are employed to maintain mild reaction conditions which minimize polymerization of the chloroamine.

This type of reaction was reported for tris(2-methoxyethyl)amine[15] and for the tetraamine tris(2-diethylaminoethyl)amine (Et_6tren).[16] The latter was also obtained by alkylation of *N,N*-diethylethylenediamine with $(C_2H_5)_2NCH_2CH_2Cl$.[17] The corresponding methyl derivative tris(2-dimethylaminoethyl)amine (Me_6tren) was prepared by methylation of tris(2-aminoethyl)amine (tren).[18] Similarly, methylation of tris(2-thiolethyl)amine was used to prepare tris(2-methylthioethyl)amine ($D = SCH_3$).[8] The method reported here (Sec. C) seems more efficient, in that it avoids the intermediate synthesis of $N(CH_2CH_2SH)_3$ and is based on the reaction above with $NaSCH_3$; the sodium methylmercaptide is either prepared in advance[19] or *in situ* by the reaction of CH_3SH with $NaOCH_3$.

A. TRIS[2-(DIPHENYLPHOSPHINO)ETHYL]AMINE

$$N(CH_2CH_2Cl)_3 + 3KP(C_6H_5)_2 \cdot 2 \text{ dioxan} \rightarrow N[CH_2CH_2P(C_6H_5)_2]_3 + 3KCl$$

■ **Caution.** *These syntheses must be carried out in an efficient hood, because the chloroamine is a nitrogen mustard analog and the phosphines and mercaptans are also toxic.*

Procedure

A 1-l. four-necked flask is equipped with a glass stirrer, nitrogen inlet tube, 250-ml dropping funnel with pressure-equalizing side arm, and reflux condenser with drying tube ($CaCl_2$). All glassware must be previously dried. Anhydrous tetrahydrofuran (THF)* (250 ml) is placed in the flask and deaerated by bubbling a slow stream of dry nitrogen for 15 min and stirring (flushing with dry nitrogen is continued throughout the whole reaction). To this solvent 120.3 g (0.3 mole) of dry $KP(C_6H_5)_2 \cdot 2$ dioxane is rapidly added. (■ **Caution.** *This salt, which is prepared by alkali metal cleavage of* PPh_3,[20] *is commonly stored under an inert atmosphere. Any contact with air must be avoided, as it causes immediate hydrolysis and oxidation and may result in spontaneous ignition. The presence of hydrolysis and oxidation products can also cause considerable difficulty in crystallization of the ligand.*) During addition the mixture is stirred until a clear orange solution is obtained. A solution of 20.4 g (0.1 mole) of $N(CH_2CH_2Cl)_3$[14] in 50 ml of anhydrous THF is placed in the dropping funnel and deaerated with a stream of dry nitrogen gas. (■ **Caution.** *Tris-(2-chloroethyl)amine has strong vesicant properties, and any contact with the skin must be avoided. The use of rubber gloves is imperative.*) The funnel is stoppered, and the solution is slowly added dropwise with stirring to the orange solution of the phosphide for 30–40 min at room temperature. The solution becomes gradually opalescent and changes rather sharply from orange to a milky yellowish color near the end of the addition. The mixture is then refluxed for 15 min and, after cooling, is poured into 3 times its volume of water, whereupon a pale-yellow oil separates. The mixture is ice-cooled until the oil crystallizes completely; it is then filtered and washed with water and then with ethanol. Recrystallization from dimethylformamide-ethanol gives colorless needles. The yield is about 80%. *Anal.* Calcd. for $C_{42}H_{42}NP_3$: C, 77.15; H, 6.47; N, 2.14; P, 14.15. Found: C, 77.4; H, 6.5; N, 2.0; P, 14.2.

*The use of absolutely dry THF is essential. It must be stored over solid KOH, then distilled over sodium and benzophenone, and finally treated with $LiAlH_4$ and redistilled.

Properties

The ligand is a colorless solid, m.p. 101–102°. It is air-stable, and fairly soluble in THF, dimethylformamide, acetone, and chlorinated hydrocarbons, slightly soluble in diethylether and petroleum ether, and practically insoluble in water and in cold ethanol.

B. TRIS[2-(DIPHENYLARSINO)ETHYL]AMINE

$$N(CH_2CH_2Cl)_3 + 3KAs(C_6H_5)_2 \cdot 2\ \text{dioxane} \rightarrow N[CH_2CH_2As(C_6H_5)_2]_3 + 3KCl$$

Procedure

The reaction apparatus and procedure are the same as for Sec. A, except that $KP(C_6H_5)_2 \cdot 2$ dioxane is replaced by an equimolar amount (135 g) of $KAs(C_6H_5)_2 \cdot 2$ dioxane.* Typical yields are about 80%. *Anal.:* Calcd. for $C_{42}H_{42}NAs_3$: C, 64.20; H, 5.39; As, 28.61. Found: C, 64.1; H, 5.7; As, 28.5.

Properties

The ligand is a colorless, air-stable solid, m.p. 95–96°. The solubility properties are similar to those in Sec. A.

C. TRIS[2-(METHYLTHIO)ETHYL]AMINE

$$N(CH_2CH_2Cl)_3 + 3NaSCH_3 \longrightarrow N(CH_2CH_2SCH_3)_3 + 3NaCl$$

Procedure

■ **Caution.** *The methylmercaptan used in this synthesis is a poisonous gas with an obnoxious odor. All procedures must be carried out in an efficient fume hood. The apparatus must be connected to a series of washing bottles charged with chromic acid mixture or other absorbing solutions.*

A 1-l. three-necked flask is equipped with a glass stirrer, a reflux condenser with drying tube ($CaCl_2$), and a 250-ml dropping funnel with external cooling bath (see Fig. 15). All glassware must be previously dried. Anhydrous methanol (250 ml) is placed in the flask, followed by 23 g (1 mole) of sodium added

*This salt is prepared by alkali metal cleavage of $AsPh_3$.[21] Its properties are similar to those of $KPPh_2 \cdot 2$ dioxane, and it must be handled similarly.

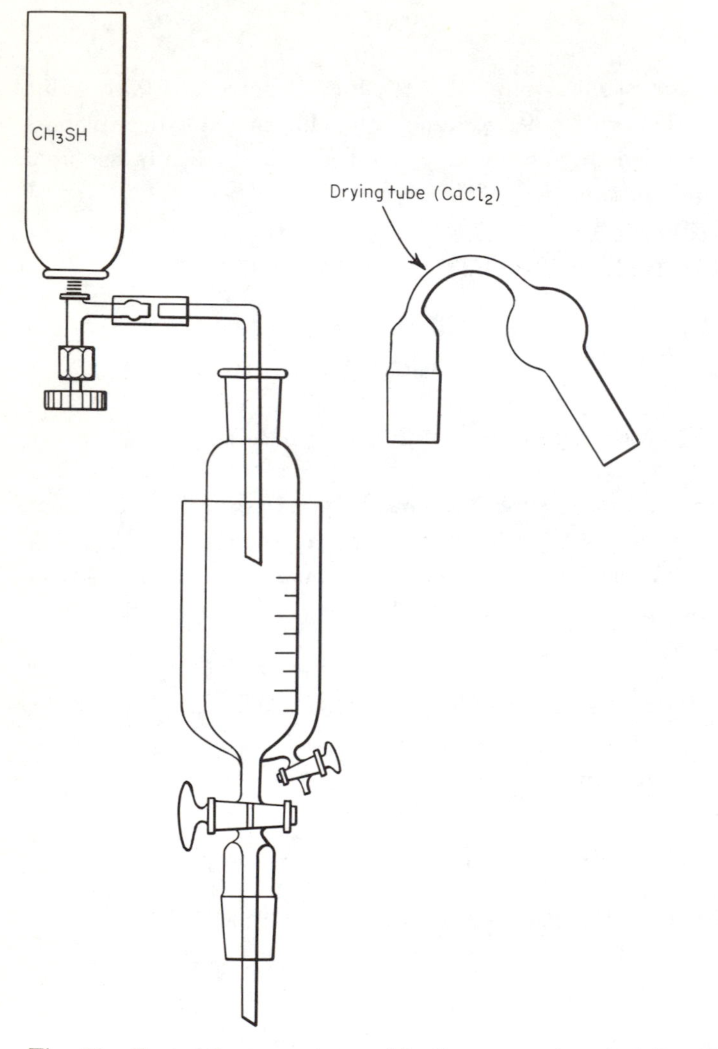

Fig. 15. Part of the apparatus used for the preparation of tris(2-methylthioethyl)amine.

in small portions with stirring. When all the sodium has reacted, the flask is allowed to cool at room temperature. The dropping funnel is cooled externally with ice, and 53.5 ml (48 g, 1 mole) of CH_3SH is poured into the funnel, as shown in Fig. 15. This is diluted with 50 ml of precooled anhydrous methanol, and the solution is added dropwise with stirring to the sodium methoxide solution during a 1-hr period. The cooling bath is then drained off, and a solution of 68 g (0.33 mole) of $N(CH_2CH_2Cl)_3$[14] in 70 ml of anhydrous methanol is placed in the dropping funnel. The sodium mercaptide solution is heated to

50–60°, and the chloroamine solution is *slowly* added dropwise with stirring for 4 hr. The mixture is finally refluxed for 1 hr. After the mixture has cooled, the NaCl which separates is filtered off, and the solution is concentrated to small volume under reduced pressure. The residue is diluted with diethylether, filtered again, concentrated, and finally vacuum-distilled. The ligand is collected at 135–140°/2 torr. The yield is about 30%. *Anal.* Calcd. for $C_9H_{21}NS_3$: C, 45.14; H, 8.84. Found: C, 45.3; H, 9.0.

Properties

The ligand is a straw-yellow oil with a smell like that of hydrogen sulfide. It is soluble in most organic solvents and insoluble in water. The color slowly darkens on standing.

Cognate Preparations

With the same apparatus and procedure, similar syntheses can be performed by replacing CH_3SH with C_2H_5SH, *i*-C_3H_7SH, and *t*-C_4H_9SH, and using the corresponding alcohol as solvent.[9] In these cases, the dropping funnel need not be cooled since these mercaptans are liquid at room temperature.

Properties

The ligands are all pale-yellow liquids with a smell like that of hydrogen sulfide; the color slowly darkens on standing. *Tris(2-ethylthioethyl)amine,* b.p. 138–140°/0.5 torr; yield: about 60%. *Anal.* Calcd. for $C_{12}H_{27}NS_3$: C, 51.19; H, 9.67; N, 4.97; S, 34.14. Found: C, 50.9; H, 9.5; N, 5.0; S, 33.9. *Tris(2-isopropylthioethyl)amine,* b.p. 135–136°/0.2 torr; yield: about 60%. *Anal.* Calcd. for $C_{15}H_{33}NS_3$: C, 55.80; H, 10.30; N, 4.34; S, 29.76. Found: C, 55.4; H, 10.4; N, 4.4; S, 29.1. *Tris(2-t-butylthioethyl)amine,* b.p. 154–156°/0.3 torr; yield: about 70%. *Anal.* Calcd. for $C_{18}H_{39}NS_3$: C, 59.20; H, 10.76; N, 3.84; S, 26.30. Found: C, 59.8; H, 11.2; N, 3.9; S, 26.0.

References

1. L. Sacconi and I. Bertini, *J. Am. Chem. Soc.*, **90,** 5443 (1968).
2. L. Sacconi, I. Bertini, and F. Mani, *Inorg. Chem.*, **7,** 1417 (1968).
3. L. Sacconi and R. Morassi, *Inorg. Nucl. Chem. Lett.*, **4,** 449 (1968).
4. L. Sacconi and R. Morassi, *J. Chem. Soc.* (*A*), **1969,** 2904.
5. L. Sacconi and R. Morassi, *ibid.*, **1970,** 575.
6. R. Morassi and L. Sacconi, *J. Chem. Soc.* (*A*), **1971,** 492.

7. R. Morassi and L. Sacconi, *ibid.*, 1487.
8. M. Ciampolini, J. Gelsomini, and N. Nardi, *Inorg. Chim. Acta,* **2,** 343 (1968).
9. G. Fallani, R. Morassi, and F. Zanobini, *Inorg. Chim. Acta,* **12,** 147 (1975).
10. M. Bacci and S. Midollini, *Inorg. Chim. Acta,* **5,** 220 (1971).
11. M. Bacci, R. Morassi, and L. Sacconi, *J. Chem. Soc. (A),* **1971,** 3686.
12. F. Mani and L. Sacconi, *Inorg. Chim. Acta,* **4,** 365 (1970).
13. R. Morassi, I. Bertini, and L. Sacconi, *Coord. Chem. Rev.,* **11,** 343 (1973).
14. J. P. Mason and D. J. Gasch, *J. Am. Chem. Soc.,* **60,** 2816 (1938).
15. S. M. Ali, F. M. Brewer, J. Chadwick, and G. Garton, *J. Inorg. Nucl. Chem.,* **9,** 124 (1959).
16. H. Lettre and W. Riemenschneider, *Ann.,* **575,** 18 (1952).
17. R. H. Mizzoni, M. A. Hennessey, and C. R. Scholz, *J. Am. Chem. Soc.,* **76,** 2414 (1954).
18. M. Ciampolini and N. Nardi, *Inorg. Chem.,* **5,** 41 (1966).
19. W. R. Kirner, *J. Am. Chem. Soc.,* **50,** 2451 (1928).
20. K. Issleib and A. Tzschach, *Chem. Ber.,* **92,** 1118 (1959).
21. A. Tzschach and W. Lange, *Chem. Ber.,* **95,** 1360 (1962).

48. DIMETHYL(PENTAFLUOROPHENYL)PHOSPHINE AND DIMETHYL(PENTAFLUOROPHENYL)ARSINE LIGANDS

Submitted by J. L. PETERSON* and D. W. MEEK*
Checked by D. P. SHAH† and S. O. GRIM†

A polyfluorophenyl group has been used in the design and synthesis of potential ligands to modify the electronic properties of the phosphorus donor.[1–4] A fluorinated aromatic group, which is more electronegative than the corresponding unfluorinated group,[3,5] causes a decrease in the σ-donor properties of the fluoro ligand‡ compared with the unfluorinated analog.[1–4] The introduction of a fluoro group into a potential ligand also permits the use of ^{19}F nmr for characterization of ligands and the resultant complexes.

*Chemistry Department, The Ohio State University, Columbus, Ohio 43210.
†Department of Chemistry, University of Maryland, College Park, Md. 20742.

‡The term *fluoro ligand* will be used to refer to a ligand in which some of the hydrogen atoms are replaced by fluorine;[2,3] the term *perfluoro ligand* will be reserved for those ligands in which all the hydrogen atoms are replaced by fluorine, for example, PF_3, $P(CF_3)_3$, or

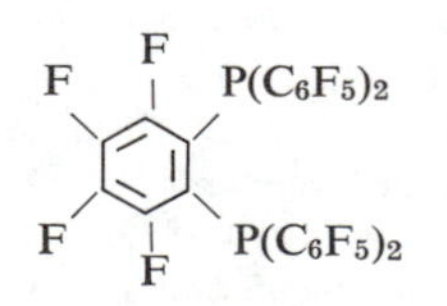

Dimethyl(pentafluorophenyl)phosphine, $C_6F_5P(CH_3)_2$, has been prepared by the reaction of methylmagnesium bromide and (pentafluorophenyl)phosphonous dichloride, $C_6F_5PCl_2$, the latter having been prepared from (pentafluorophenyl)magnesium bromide and phosphorus trichloride.[6] Fild et al.[6] reported a second route to this phosphine, namely, by treating (pentafluorophenyl)magnesium bromide and dimethylphosphinous chloride, $(CH_3)_2PCl$. The latter method gives a much more efficient use of C_6F_5Br, but it requires $(CH_3)_2PCl$, which is expensive and much less readily available than PCl_3.

The analogous arsine, dimethyl(pentafluorophenyl)arsine, $C_6F_5As(CH_3)_2$, has not been previously reported; however, the necessary precursors, (pentafluorophenyl)arsonous dichloride, $C_6F_5AsCl_2$,[7] or dimethylarsinous iodide, $(CH_3)_2AsI$, are known.

Dimethyl(pentafluorophenyl)phosphine and dimethyl(pentafluorophenyl)-arsine form stable complexes with transition metals, for example, Pt(II), Pd(II), Ni(II), Au(I), and Co(II),[1,9] and they are particularly interesting ligands for stereochemical studies, as both ^1H and ^{19}F nmr spectra are sensitive indicators of the coordination environment of the ligand. These pentafluorophenyl ligands can be prepared by the following procedures in approximately 15-hr working time with an elapsed time of 30 hr.

■ **Caution.** *These preparations should be conducted in a purified nitrogen atmosphere and in well-ventilated hoods. Dimethylphosphinous chloride is spontaneously flammable in air. These ligands and the intermediates dimethylphosphinous chloride,* $(CH_3)_2PCl$, *and dimethylarsinous iodide,* $(CH_3)_2AsI$, *should be handled very carefully, as they should be considered quite toxic.*

A. DIMETHYL(PENTAFLUOROPHENYL)PHOSPHINE

$$C_6F_5Br + Mg \rightarrow C_6F_5MgBr$$

$$C_6F_5MgBr + (CH_3)_2PCl \rightarrow C_6F_5P(CH_3)_2 + MgBrCl$$

Procedure

A 250-ml, three-necked, round-bottomed flask (to which is attached at the bottom a coarse-sintered-glass frit, a Teflon stopcock, and a standard-taper male joint) is equipped with a 100-ml pressure-equalizing addition funnel, a reflux condenser, and a Tru-bore glass mechanical stirrer with a Teflon blade.[10] The system is flushed with dry nitrogen via a gas inlet in the addition funnel and a gas outlet in the condenser, leading to a mineral-oil bubbler. Magnesium turnings (3.5 g, 0.144 mole) and 100 ml of sodium-dried diethyl ether are added to the flask, followed by addition of 28.2 g (0.114 mole) of freshly distilled bromopentafluorobenzene* to the addition funnel.

The flask is cooled to 0°, and the bromopentafluorobenzene is added dropwise with stirring. After addition is completed (about 45 min), the mixture is warmed to room temperature and stirred for $1\frac{1}{2}$ hr. The resulting solution is filtered directly into a nitrogen-filled 500-ml, three-necked, round-bottomed flask equipped with a reflux condenser and a mechanical stirrer. After the filtration is completed, the Grignard flask[10] is replaced by a 100-ml pressure-equalizing addition funnel containing 9.98 g of freshly distilled (b.p. 73–76°) dimethylphosphinous chloride,† $(CH_3)_2PCl$, in 50 ml of dry diethyl ether. The Grignard solution is cooled to 0°, and $(CH_3)_2PCl$ is added dropwise with vigorous stirring. The resulting mixture is then refluxed for 2 hr before hydrolysis at 0° with 100 ml of saturated NH_4Cl solution. The pH of the aqueous phase is adjusted to about 6 (pHydrion paper) by dropwise addition of 6*M* HCl, and the two phases are separated. After drying the nonaqueous phase (top layer) over anhydrous sodium sulfate and removing the ether by distillation at atmospheric pressure, the residue is vacuum-distilled to give $(CH_3)_2PC_6F_5$; b.p. 42–45°/1.5 torr (literature, 48°/0.4).[6] The yield is 12.4 g or 55%.

Properties

Dimethyl(pentafluorophenyl)phosphine is a colorless, air-sensitive liquid with a very unpleasant odor. The ^1H nmr spectrum of a dichloromethane solution of $(CH_3)_2PC_6F_5$ consists of a doublet of triplets that is centered at $\tau 8.54$ with $J_{PH} = 5.2$ Hz and $J_{FH} = 1.0$ Hz (literature, $\tau 8.43$; $J_{PH} = 5.4$ Hz and $J_{FH} = 1.2$ Hz as a neat liquid).

A methyl iodide derivative can be prepared by treating 1.5 g of dimethyl(pentafluorophenyl)phosphine with 2.0 g of CH_3I in 25 ml of absolute ethanol at reflux temperature for 4 hr. On cooling the mixture to room temperature, *hygroscopic* needles are obtained; these are recrystallized from an acetone-benzene mixture to give 1.0 g of colorless needles. The ^1H nmr spectrum of the methyl iodide derivative is a doublet of triplets that is centered at $\tau 7.16$ with $J_{PH} = 14.8$ Hz and $J_{FH} = 1.7$ Hz.

*Bromopentafluorobenzene may be purchased from Pierce Chemical Co., Rockford, Ill. 61105, or PCR, Inc., Gainesville, Fla. 32601.

†Dimethylphosphinous chloride may be purchased from either Orgmet, Inc., Haverhill, Mass. 01830, or Maybridge Chemical Co., Ltd., Tintagel, Cornwall, England. It can also be prepared by the method of G. W. Parshall, *Inorganic Syntheses,* **15,** 192 (1974).

B. DIMETHYL(PENTAFLUOROPHENYL)ARSINE

$$C_6F_5Br + Mg \rightarrow C_6F_5MgBr$$

$$C_6F_5MgBr + (CH_3)_2AsI \rightarrow C_6F_5As(CH_3)_2 + MgBrI$$

Procedure

The (pentafluorophenyl)magnesium bromide Grignard reagent is prepared as in Sec. A and then cooled to 0° in an ice-water bath and stirred at 0° as 23.2 g (0.10 mole) of dimethylarsinous iodide* is added dropwise; after the addition is completed, the resultant mixture is warmed and refluxed for 2 hr. The resulting mixture is cooled in an ice bath and hydrolyzed by dropwise addition of 50 ml of saturated NH_4Cl solution. The two phases are separated, and the organic phase (top layer) is dried for 12 hr over anhydrous sodium sulfate. The solvent is removed by distillation at atmospheric pressure. Subsequent vacuum-distillation of the remaining residue gives 15.2 g (55%)† of $C_6F_5As(CH_3)_2$, b.p. 54°/1.0 torr.

Properties

Dimethyl(pentafluorophenyl)arsine is a colorless, malodorous liquid which has a ^1H nmr spectrum in dichloromethane consisting of a triplet centered at $\tau 8.51$ with $J_{FH} = 1.1$ Hz. A methyl iodide derivative of the arsine can be prepared by refluxing 1.7 g of the arsine with 15 ml of methyl iodide for 2 hr. After evaporation of the solvent and recrystallization of the product from a dichloromethane-benzene mixture, a colorless compound which melts at 167–168° is obtained. The ^1H nmr spectrum of the methyl iodide derivative is a triplet at $\tau 7.16$ ($J_{FH} = 1.4$ Hz) in dichloromethane, and the infrared spectrum (Nujol mull) contains strong peaks at 1080 and 980 cm^{-1} characteristic of the C_6F_5 group. *Anal.* (done on the methyl iodide derivative). Calcd. for $C_9H_9F_5AsI$: C, 26.09; H, 2.17; I, 30.68. Found: C, 25.5; H, 2.21; I, 30.1.

References

1. E. C. Alyea and D. W. Meek, *J. Am. Chem. Soc.*, **91**, 5761 (1969).
2. P. G. Eller, J. M. Riker, and D. W. Meek, *J. Am. Chem. Soc.*, **95**, 3540 (1973).
3. P. G. Eller and D. W. Meek, *Inorg. Chem.*, **11**, 2518 (1972).

*Dimethylarsinous iodide was prepared from cacodylic acid and potassium iodide by the procedure given in reference 8.

†The checkers obtained 62%.

4. N. V. Duffy, A. J. Layton, R. S. Nyholm, D. Powell, and M. L. Tobe, *Nature,* **212,** 117 (1966).
5. M. G. Hogben and W. A. G. Graham, *J. Am. Chem. Soc.,* **91,** 283 (1969), and references contained therein; M. G. Hogben, R. S. Gay, A. J. Oliver, J. A. J. Thompson, and W. A. G. Graham, *ibid.,* 291.
6. M. Fild, O. Glemser, and O. Hollenberg, *Naturwiss.,* **52,** 590 (1965).
7. M. Green and D. Kirkpatrick, *Chem. Commun.,* **1967,** 57.
8. R. D. Feltham, A. Kasenally, and R. S. Nyholm, *J. Organomet. Chem.,* **7,** 285 (1965).
9. J. L. Peterson, Ph.D. dissertation, The Ohio State University, Columbus, December 1973.
10. G. Kordosky, B. R. Cook, J. Cloyd, Jr., and D. W. Meek, *Inorganic Syntheses,* **14,** 14 (1973); see flask *A* in fig. 2.
11. M. G. Barlow, M. Green, R. N. Haszeldine, and H. G. Higson, *J. Chem. Soc.* (*B*), **1966,** 1025.

49. TRIS(*o*-DIMETHYLARSINOPHENYL)ARSINE AND TRIS(*o*-DIMETHYLARSINOPHENYL)STIBINE

Submitted by W. LEVASON* and C. A. McAULIFFE*
Checked by A. I. PLAZA† and S. O. GRIM†

The syntheses of a range of quadridentate ligands with threefold symmetry of the type $E'(RL)_3$ (E = P, As, Sb; R = o-C_6H_4, —$CH_2CH_2CH_2$—, or —CH_2-CH_2—; L = PR'_2, AsR'_2, SR′, SeR′) have been reported in the last decade. Interest in these ligands stems from their ability to form a large number of essentially trigonal-bipyramidal complexes with a wide variety of transition metal ions.[1] The preparation of two ligands of this class, tris(*o*-dimethylarsinophenyl)arsine[2] and tris(*o*-dimethylarsinophenyl)stibine,[3] is described here.

■ **Caution.** *o-Bromophenyldichloroarsine and o-bromophenyldimethylarsine are malodorous and toxic and should be handled with care in an efficient fume hood. All the arsenic-containing intermediates and the ligands themselves are toxic.*

*Chemistry Department, University of Manchester Institute of Science and Technology, Manchester M60 1QD, England.
†Department of Chemistry, University of Maryland, College Park, Md. 20742.

o-Bromophenyldimethylarsine[4]

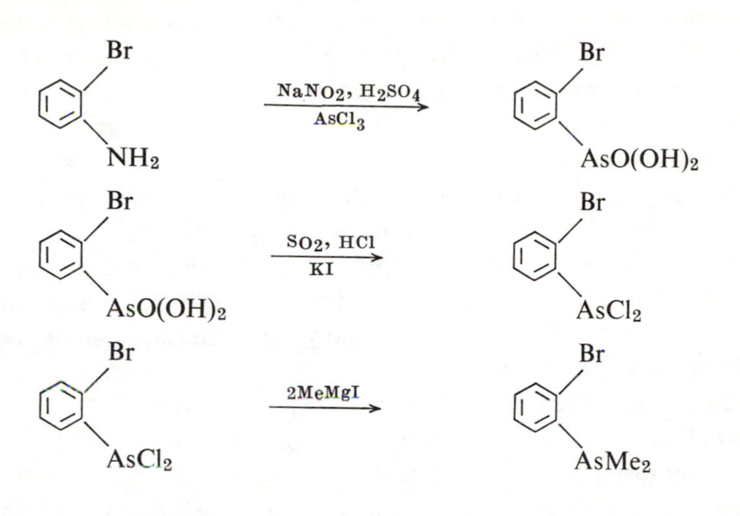

Procedure

o-Bromoaniline (100 g, 0.58 mole) is dissolved in a mixture of glacial acetic acid (500 ml) and propionic acid (50 ml) (the latter lowers the freezing point of the mixture) in a 1-l. bolt-head flask fitted with a *strong* paddle stirrer and stirrer motor. To the well-stirred solution concentrated sulfuric acid (35 ml) is added slowly over about 30 min. (■ **Caution.** *Much heat is generated.*) A thick white precipitate forms, and stirring is aided by swirling the flask. Arsenic trichloride (130 ml) is added, and the precipitate partially dissolves on stirring. The mixture is cooled to ca. 0° in an ice-salt bath, whereupon the amine sulfate precipitates. To the stirred suspension solid sodium nitrite (50 g, 0.5 mole) is added in small quantities over a 4-hr period, maintaining the temperature below 3°. Evolution of more than traces of nitrogen dioxide at this stage is indicative of too high a temperature and/or too rapid addition of the sodium nitrite. When the addition is complete, the mixture is allowed to warm up overnight.

Copper(I) bromide (0.5 g) is added to the stirred mixture, and the temperature is raised by heating on a steam bath. The yellow solution turns brown, and at ca. 60° nitrogen is evolved along with some nitrogen dioxide. (■ **Caution.** NO_2 *is highly toxic.*) The resulting brown solution is heated and stirred for 5 hr and allowed to stand overnight. A considerable quantity of sodium salt separates out overnight; this is filtered off, using a coarse-sintered glass funnel, and washed with glacial acetic acid (3 × 300 ml). The combined filtrate and washings are mixed with concentrated hydrochloric acid (500 ml) containing potassium iodide (2 g), and a rapid stream of sulfur dioxide is bubbled into the mixture for 3 hr. During this time a deep red-brown oil separates. This oil is extracted into carbon tetrachloride (500 ml), and the lower carbon tetrachloride layer is separated and dried over anhydrous sodium sulfate. The reddish-brown solution is filtered to remove the drying agent and rotary-evaporated to ca. 100 ml.

This oily liquid is distilled under reduced pressure and a nitrogen leak. Because of occasional blockages in the liquid-nitrogen traps it is advisable to use two traps in parallel followed by a third in series, permitting one trap to be emptied without losing continuity in the distillation. The fraction boiling 128–136° at 2.5 torr is collected. This is a yellow liquid when hot but on cooling is a red viscous substance. Yield: 114 g (65%).

A 2-l., three-necked, round-bottomed flask fitted with a 500-ml pressure-equalized dropping funnel, mechanical stirrer, and condenser and containing magnesium turnings (20 g, 0.83 mole) and sodium-dried ether (500 ml) is flushed with dry nitrogen. Methyl iodide (120 g, 0.83 mole) is added dropwise over 2 hr. (■ **Caution.** *This Grignard reaction should be initiated correctly, if necessary by addition of a few crystals of iodine, otherwise it will "take off" at a later stage.*) To the grayish solution of methylmagnesium iodide produced, a solution of *o*-bromophenyldichloroarsine (120 g, 0.4 mole) in dry ether (400 ml) is added dropwise with vigorous stirring; it is cooled if necessary. When the addition is complete, the yellow-green solution or suspension produced is stirred for 1 hr further under reflux to complete the reaction, cooled, and cautiously hydrolyzed by dropwise addition of a saturated aqueous ammonium chloride solution (ca. 70 g in 500 ml of water). The upper ether layer is separated under nitrogen and dried over anhydrous sodium sulfate. The pale-yellow ether solution is decanted from the drying agent, and the ether is distilled off at atmospheric pressure under nitrogen. The dark oily residue is transferred again under nitrogen to a 250-ml flask and distilled in vacuum, the yellow oil distilling at 85°/0.8 torr being collected as product. Yield: 85 g (78%) The *o*-bromophenyldimethylarsine so produced oxidizes readily in air with separation of a white solid, arsine oxide, and should be stored under nitrogen.

A. TRIS(*o*-DIMETHYLARSINOPHENYL)ARSINE

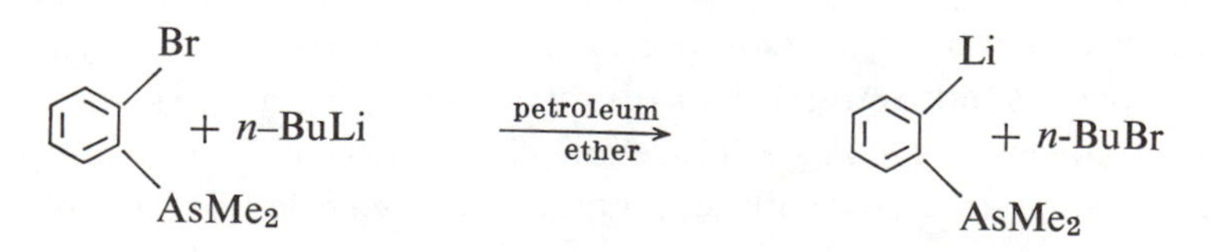

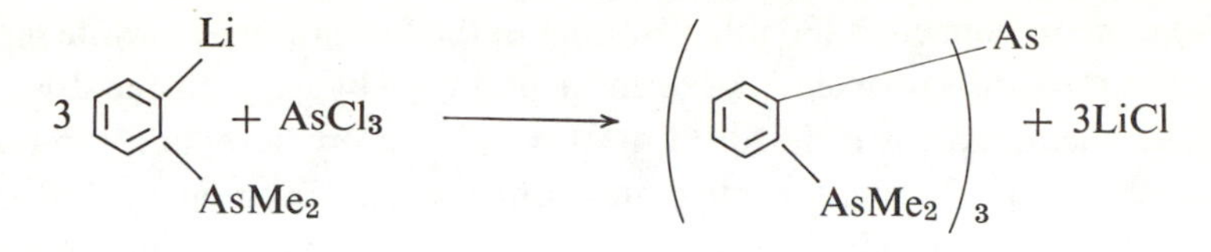

Procedure

o-Bromophenyldimethylarsine (42 g, 0.16 mole) is dissolved in petroleum ether (b.p. 40–60°, 300 ml) previously purified by shaking with concentrated sulfuric acid and distilling into a flask containing freshly extruded sodium wire. To this solution in a 1-l. three-necked flask fitted with a 250-ml pressure-equalized dropping funnel, mechanical stirrer, and condenser and flushed with dry nitrogen, a solution of *n*-butyllithium in petroleum ether (196 ml, 0.905*M*, 0.177 mole) is added dropwise. The mixture is stirred under reflux for 2 hr, the lithiated compound separating as yellow crystals during this time. A solution of arsenic trichloride (10.6 g, 0.06 mole) in petroleum ether (100 ml) is added dropwise to the refluxing mixture, and the heating is continued until the solution is decolorized (1–2 hr). The mixture is allowed to cool to room temperature and hydrolyzed by dropwise addition of 0.05*N* hydrochloric acid (200 ml). The upper organic layer is separated, dried over sodium sulfate, and evaporated, leaving a viscous oil. The oil is dissolved in a mixture of ethanol (400 ml) and methanol (200 ml), allowed to stand at ambient temperature until crystals appear, and then cooled to 0°. Yield: 15 g (ca. 45%). *Anal.* Calcd. for $C_{24}H_{30}As_4$: C, 46.7; H, 4.9; As, 48.5. Found: C, 46.7; H, 5.0; As, 48.2.

Properties

Tris(*o*-dimethylarsinophenyl)arsine forms colorless crystals, m.p. 115°. The ^1H nmr spectrum in $CDCl_3$ shows a single methyl resonance at $\tau 8.85$. It is readily soluble in dichloromethane and is air-stable for several hours.

B. TRIS(*o*-DIMETHYLARSINOPHENYL)STIBINE

Procedure

The preparation of this ligand is essentially similar to that of the quadridentate arsine described in Sec. A. To the lithiated species produced from *o*-bromophenyldimethylarsine (42 g) and *n*-butyllithium (192 ml, 0.91*M*, 0.187 mole) prepared as above, is added antimony trichloride (13.8 g, 0.06

mole) in petroleum ether (80 ml), resulting in the formation of a white precipitate. The mixture is refluxed for 4 hr more and then hydrolyzed and dried as in Sec. A. The solvent is distilled off at atmospheric pressure, and the residue is heated to ca. 100° under 1 torr to remove unchanged *o*-bromophenyldimethylarsine. The white product is tris(*o*-dimethylarsinophenyl)stibine. Yield: 16 g (ca: 44%). *Anal.* Calcd. for $C_{24}H_{30}As_3Sb$: C, 43.5; H, 4.5. Found: C, 43.4; H, 4.8.

Properties

The ligand is a white solid, m.p. 135°, which shows a single methyl resonance at τ8.80 ($CDCl_3$ solution) in the ^1H nmr spectrum.

References

1. B. Chiswell, "Aspects of Inorganic Chemistry," Vol. 1, Macmillan Company, Ltd., London, 1973.
2. O. St. C. Headley, R. S. Nyholm, C. A. McAuliffe, L. Sindellari, M. L. Tobe, and L. M. Venanzi, *Inorg. Chim. Acta*, **4**, 93 (1970).
3. L. Baracco and C. A. McAuliffe, *J. Chem. Soc., Dalton Trans.*, **1972**, 948.
4. E. R. H. Jones and F. G. Mann, *J. Chem. Soc.*, **1955**, 4472; R. D. Cannon, B. Chiswell, and L. M. Venanzi, *J. Chem. Soc.*, **1967**, 1277.

50. *cis*-2-DIPHENYLARSINOVINYLDIPHENYLPHOSPHINE AND 2-DIPHENYLARSINOETHYLDIPHENYLPHOSPHINE

By W. LEVASON* and C. A. McAULIFFE*
Checked by R. C. BARTH† and S. O. GRIM†

The coordination chemistry of diphosphine and diarsine ligands has been intensively investigated in recent years.[1] Very much less work has been carried out on the corresponding mixed ligands, the intermediate arsine-phosphine bidentates, owing to a considerable extent to the lack, until recently, of suitable synthetic routes to such ligands. Two routes are now available: the nucleophilic substitution of organohalogen compounds, two examples of which

*Department of Chemistry, University of Manchester Institute of Science and Technology, Manchester M60 1QD, England.
†Department of Chemistry, University of Maryland, College Park, Md. 20742.

are described below, and the base-catalyzed addition of secondary arsines across the carbon-carbon double bonds of vinylphosphines reported by R. B. King and co-workers.[2]

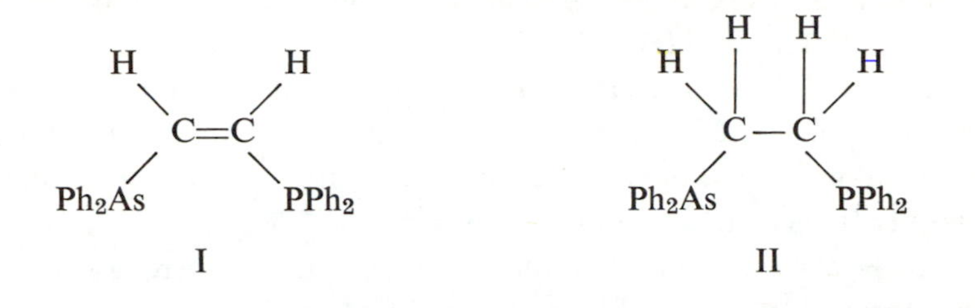

A. *cis*-2-DIPHENYLARSINOVINYLDIPHENYLPHOSPHINE[3]

$$2Li + AsPh_3 \xrightarrow{THF} LiAsPh_2 + LiPh$$

$$LiPh + t\text{-}BuCl \longrightarrow LiCl + PhH + (CH_3)_2C{=}CH_2$$

$$LiAsPh_2 + cis\text{-}ClCH = CHCl \longrightarrow LiCl + cis\text{-}Ph_2AsCh{=}CHCl$$

$$cis\text{-}Ph_2AsCH{=}CHCl + LiPPh_2 \longrightarrow cis\text{-}Ph_2AsCH{=}CHPPh_2 + LiCl$$

■ **Caution.** *Lithium diphenylarsenide and lithium diphenylphosphide are strong bases and very corrosive, and contact of them with the skin should be avoided. Hydrolysis of both compounds is vigorous and generates the malodorous and toxic secondary arsine or phosphine. Hence these experiments should be conducted in a fume hood. The products and the intermediates are also toxic and should be handled with due care. All manipulations should be done under nitrogen.*

Procedure

Triphenylarsine (61.2 g, 0.2 mole) is dissolved in tetrahydrofuran (400 ml) previously dried by distillation under nitrogen from sodium benzophenone ketyl. The apparatus consists of a 1-l. three-necked flask fitted with a nitrogen inlet, mechanical stirrer, and reflux condenser. Lithium rod (4.2 g, 0.6 mole) cut into small pieces (ca. $\frac{1}{2}$-cm cubes are suitable) is added, and the mixture is stirred under reflux for 4 hr to produce a deep-red solution. The solution is filtered through a glass-wool plug under nitrogen to remove the excess lithium, and *t*-butyl chloride (12.9 g, 0.14 mole) is added dropwise with cooling and stirring to destroy the phenyllithium.* (■ **Caution.** *Considerable heat is gen-*

*The fission of the Ph—P and Ph—As bonds occurs to the extent of ca. 60–70%. Yields and quantities of reactants were calculated on the 70% figure.

erated.) The resulting deep orange-red solution is transferred under nitrogen to a 500-ml pressure-equalized dropping funnel. This funnel is fitted into one neck of a 1-l. three-necked flask, also fitted with a nitrogen inlet, a mechanical stirrer, and reflux condenser and containing *cis*-1,2-dichloroethylene (79.8 g, 0.8 mole) in dry THF (200 ml). The flask is flushed with nitrogen and cooled to 0°, and the solution is vigorously stirred while the lithium diphenylarsenide solution is added dropwise over 2 hr. The temperature must be maintained at ca. 0° and the mixture vigorously stirred to avoid formation of the diarsine. Immediate decolorization of the arsenide solution occurs, and finally a pale-yellow solution is produced. This solution is stirred for 1 hr further and then hydrolyzed by dropwise addition of saturated aqueous ammonium chloride solution (200 ml). The THF layer is separated and dried over anhydrous sodium sulfate overnight. The THF is then rotary-evaporated to leave a white solid (or sometimes a pale-brown oil). This product is recrystallized from ethanol to yield white *cis*-2-chlorovinyldiphenylarsine; m.p. 98°; yield: ca. 38 g (ca. 65% based on $C_2H_2Cl_2$). *Anal.* Calcd. for $C_{14}H_{12}AsCl$: C, 57.9; H, 4.1. Found: C, 57.7; H, 4.1. The ^1H nmr spectrum exhibits a doublet at $\tau 3.5$.

A solution of lithium diphenylphosphide* is prepared (in a manner analogous to that used for the diphenylarsenide described above) from triphenylphosphine (44 g, 0.17 mole), lithium (4 g, 0.58 mole), and THF (400 ml); the phenyllithium again is destroyed by addition of *t*-butyl chloride (11.2 g, 0.12 mole) (■ **Caution.** *Considerable heat is generated.*) The red solution is added dropwise under nitrogen to a stirred solution of *cis*-2-chlorovinyldiphenylarsine† (35 g, 0.12 mole), instant discharge of the red color occurring. The mixture is stirred for 1 hr and then hydrolyzed by dropwise addition of saturated aqueous ammonium chloride solution (200 ml). The THF layer is separated and dried over anhydrous sodium sulfate overnight. After filtration the THF is rotary-evaporated to leave a brown solid. Repeated recrystallizations (usually two or three) from ethanol-dichloromethane yield white needles of the ligand; m.p., 105°. Yield: 35 g (68%). *Anal.* Calcd. for $C_{26}H_{22}AsP$: C, 69.9; H, 4.9. Found: C, 68.9; H, 4.8.

*$LiPPh_2$ can also be prepared[5] from lithium and Ph_2PCl or Ph_2PH, but the method described above using PPh_3 has been found satisfactory and uses the more readily available triphenylphosphine. Similar comments apply to the use of $AsPh_3$.

†It should be noted that the reactions described above cannot be performed in the reverse order, i.e., preparation of the chlorophosphine followed by reaction with $LiAsPh_2$. The reaction of lithium diphenylphosphine with *cis*-1,2-dichloroethylene, even with a large excess of the latter, yields only *cis*-1,2-bisdiphenylphosphinoethylene and not the chlorophosphine, and while 2-chloroethyldiphenylphosphine can be prepared, its isolation is more difficult, and yields are lower than for the corresponding arsine.[6]

Properties

cis-2-Diphenylarsinovinyldiphenylphosphine is a white, air-stable solid, m.p. 105°, which readily dissolves in dichloromethane and in hot alcohols. The ^1H nmr spectrum exhibits a doublet at $\tau 2.50$. It has characteristic infrared absorption at 1585 (s) cm^{-1} (C═C); 1060 and 1155 (w) cm^{-1} (olefinic C—H in-plane deformation), 1000 (vs) cm^{-1} (olefinic C—H out-of-plane deformation), and 1430 (vs) and 1480 (vs) cm^{-1} (asymmetrical C—H deformation). The absence of arsine oxide or phosphine oxide in the product can be checked by the absence of strong absorption in ν(AsO) and ν(PO) regions.

B. 2-DIPHENYLARSINOETHYLDIPHENYLPHOSPHINE[4]

$$LiAsPh_2 + ClCH_2CH_2Cl \xrightarrow{THF} LiCl + Ph_2AsCH_2CH_2Cl$$

$$LiPPh_2 + Ph_2AsCH_2CH_2Cl \xrightarrow{THF} LiCl + Ph_2AsCH_2CH_2PPh_2$$

■ **Caution.** *See comments at the beginning of Sec. A.*

Lithium diphenylarsenide and lithium diphenylphosphide solutions in THF are prepared as described in Sec. A.

Lithium diphenylarsenide in THF (quantities as in Sec. A) is added dropwise under nitrogen to a vigorously stirred solution of 1,2-dichloroethane (200 g, 2 moles, tenfold excess) in 200 ml of THF at 0°. Cooling and stirring vigorously avoids the formation of substantial quantities of the diarsine. After 1 hr of stirring, the solution is hydrolyzed, and the organic layer is dried over anhydrous sodium sulfate. The solvent is distilled off at atmospheric pressure, and the brown oil remaining is distilled *in vacuo*; b.p. 144–148°/2 torr. The yellow-brown oil produced is 2-chloroethyldiphenylarsine, which should be stored under nitrogen since it readily oxidizes to the corresponding arsine oxide in air. Yield: 37 g.

A solution of lithium diphenylphosphide prepared as in Sec. A from triphenylphosphine (39.3 g), lithium (2.3 g), and *t*-butyl chloride (10 g) in 250 ml of THF is added dropwise at room temperature under nitrogen to a stirred solution of 2-chloroethyldiphenylarsine (30 g, 0.1 mole) in dry THF (200 ml), and the resulting fawn-colored solution is stirred for 1 hr. It is then worked up as in Sec. A.

Recrystallization of the product from ethanol or ethanol-dichloromethane yields the ligand, ca. 19 g (65%). *Anal.* Calcd. for $C_{26}H_{24}PAs$: C, 70.5; H, 5.4. Found: C, 70.8; H, 5.6.

Properties

2-Diphenylarsinoethyldiphenylphosphine is a white crystalline solid, m.p. 120°, soluble in dichloromethane and hot alcohols. It is stable in air. The ^1H nmr spectrum exhibits a singlet at $\tau 7.89$ and a doublet at $\tau 7.93$. The absence of As=O and P=O vibration in the infrared spectrum is a useful check for the absence of oxidized impurities.

References

1. W. Levason and C. A. McAuliffe, *Adv. Inorg. Chem. Radiochem.*, **15,** 173 (1972); E. C. Alyea, "Aspects of Inorganic Chemistry," Vol. 1, p. 311, Macmillan & Co., Ltd., London, 1973.
2. R. B. King, *Acc. Chem. Res.*, **5,** 177 (1972).
3. K. K. Chow, M. T. Halfpenny, and C. A. McAuliffe, *J. Chem. Soc., Dalton Trans.*, **1973,** 147.
4. K. K. Chow and C. A. McAuliffe, *Inorg. Chim. Acta,* **14,** 5 (1975).
5. L. Maier, *Prog. Inorg. Chem.*, **5,** 27 (1963).
6. K. K. Chow, Ph.D. thesis, University of Manchester, England, 1972.

51. [2-(ISOPROPYLPHENYLPHOSPHINO)ETHYL]-DIPHENYLPHOSPHINE[1]

$$(C_6H_5)_2P(i\text{-}C_3H_7) + 2Li \xrightarrow{\text{THF}} LiP(C_6H_5)(i\text{-}C_3H_7) + C_6H_5Li$$

$$LiP(C_6H_5)(i\text{-}C_3H_7) + C_6H_5Li + t\text{-}C_4H_9Cl \longrightarrow C_6H_6 + CH_2{=}C(CH_3)_2 + LiCl + LiP(C_6H_5)(i\text{-}C_3H_7)$$

$$LiP(C_6H_5)(i\text{-}C_3H_7) + (C_6H_5)_2PCH{=}CH_2 \longrightarrow (C_6H_5)(i\text{-}C_3H_7)PCH_2CHLiP(C_6H_5)_2$$

$$(C_6H_5)(i\text{-}C_3H_7)PCH_2CHLiP(C_6H_5)_2 + H_2O \longrightarrow (C_6H_5)(i\text{-}C_3H_7)PCH_2CH_2P(C_6H_5)_2 + LiOH$$

Submitted by J. DEL GAUDIO* and S. O. GRIM*
Checked by T. G. ATTIG†

Unsymmetrical bistertiary phosphines[1] are a special class of the general group of polytertiary phosphines[2] which have received considerable attention for their varied coordination behavior. The compounds described below are especially interesting for (1) coordination studies in which slightly differing do-

*Department of Chemistry, University of Maryland, College Park, Md. 20742.
†Department of Chemistry, University of Western Ontario, London 72, Ontario, Canada.

nor properties of phosphorus, either steric or electronic or both, caused by alkyl substitution for phenyl, can be determined in the same molecule; (2) chiral studies, since the one phosphorus atom has three different groups attached plus the lone pair of electrons; (3) ^{31}P nmr studies, since the two phosphorus atoms are nonequivalent and phosphorus—phosphorus coupling will be directly observable in the nmr spectra; and (4) routine coordination chemistry.

■ **Caution.** *Tertiary phosphines are toxic and have obnoxious odors. Reactions, distillations, and other operations should be carried out in an efficient fume hood.*

*Procedure**

Into a 2-l. three-necked flask, fitted with a water condenser topped with a nitrogen inlet, a mechanical stirrer, and a powder-addition funnel, are placed 700 ml of tetrahydrofuran (THF), freshly distilled from sodium-benzophenone,† and 7.0 g (1 g atom) of lithium wire‡ which is finely cut with scissors over the powder funnel so that it drops directly into the flask as it is flushed with N_2. Diphenylisopropylphosphine§ (50 g, 0.22 mole) is added, and the powder funnel is replaced by a stopper. The mixture is stirred at ambient temperatures for about 5–6 hr. The reaction is only mildly exothermic, and the color becomes deep red as the reaction proceeds. The solution is then filtered through glass wool (to remove the excess lithium) into another three-necked flask fitted with a nitrogen inlet and stirrer.¶ The filtering funnel is replaced by an addition funnel. In order to destroy the phenyllithium, 20 g (0.22 mole) of dry $t\text{-}C_4H_9Cl$** in 100 ml of THF is added to the flask dropwise with stirring at room temperature over a 1-hr period, and then the mixture is stirred for 1 hr more.

*The procedure is quite lengthy. At least 2 days should be allowed for its completion.

†When a small amount of benzophenone (ca. 1 g) is added to THF (ca. 500 ml), which has been crudely dried with sodium, the deep-blue color which results is due to the sodium benzophenone ketyl and indicates that the solvent is dry.[3]

‡The checkers used lithium ribbon pounded into a foil and then finely cut with scissors.

§Diphenylisopropylphosphine and diphenylvinylphosphine were prepared by Grignard reactions from chlorodiphenylphosphine (Aldrich Chemical Co.).

¶The transfer can be made in several ways. If three nitrogen leads are used (one on the pouring flask, one on the receiving flask, and one over the funnel), the transfer can be made rapidly by using a powder funnel with glass wool. Or the solution can be poured through a flexible plastic connecting tube with standard-taper joints. Alternatively, the solution can be forced by nitrogen through an inverted U-glass "siphon" containing a wisp of glass wool.

**The $t\text{-}C_4H_9Cl$ was stored over molecular sieves to assure that it was dry and free of HCl. The stoichiometry is important here since excess $t\text{-}C_4H_9Cl$ can also destroy lithium isopropylphenylphosphide.

The mixture is cooled to $-23°$ with a Dry Ice–CCl_4 slush bath, and diphenylvinylphosphine (47 g, 0.22 mole) in 150 ml of dry THF is added dropwise with rapid stirring over $4\frac{1}{2}$ hr.* When addition is complete, the mixture is stirred for 1 hr more at $-23°$; then it is allowed to warm to room temperature and is stirred for 2 hr. The red solution is then hydrolyzed by the dropwise addition of deoxygenated saturated NH_4Cl solution. Sufficient NH_4Cl solution is added to cause the salts to clump and settle out of solution. The clear yellow solution is decanted under N_2 into a one-necked round-bottomed flask, the residue is washed with THF, and the mixture is decanted again. The solution and washings are concentrated by vacuum-evaporation to give a viscous oil.† Low-boiling fractions are removed by heating the oil at about 80° (0.1 torr) for 1 hr. A 10-ml aliquot of the crude product is placed in a Hickman still‡ (Fisher, 75-mm diameter by 60-mm height) with a receiving "cow" adapter. The Hickman still is heated with an oil bath. After some lower-boiling fractions are removed, the pure compound is collected with the oil-bath temperature between 130 and 155° (4×10^{-4} torr). The yield for each aliquot is 3 g of product, which corresponds to about a 50% yield over-all.

Properties

The compound, $(C_6H_5)_2PCH_2CH_2P(i\text{-}C_3H_7)C_6H_5$, is an air-sensitive oil. Its ^{31}P nmr spectrum is a pair of doublets at 12.8 (ppm) and 2.6 (ppm) with $J_{PCCP} = 30$ Hz. The ^1H nmr spectrum is quite complex since the methyl groups (centered at $\tau 9.0$) are nonequivalent due to the asymmetric phosphorus. Likewise the ethylene protons are nonequivalent and give a broad complex signal centered about $\tau 8.1$.

Other compounds prepared in the same manner are given in Table I.

*Temperature control is quite important. Very low to no yields are obtained at temperatures as low as $-78°$ or as high as 0°.

†A ^{31}P nmr spectrum at this point indicates that about 55% of the total phosphorus is present as product.

‡Other short-path distillation apparatus may be used.

TABLE I

Compound	δ, ppm		J_{PP}, Hz
	PPh_2	PPhR	
$Ph_2PCH_2CH_2PPhMe$	13.0	31.4	26
$Ph_2PCH_2CH_2PPhEt$	12.9	15.9	27
$Ph_2PCH_2CH_2PPh$ (*sec*-Bu)	12.8	6.4	30

References

1. S. O. Grim, R. P. Molenda, and R. L. Keiter, *Chem. Ind.* (*Lond.*), **1970,** 1378.
2. R. B. King, *Acc. Chem. Res.,* **5,** 177 (1972).
3. A. J. Gordon and R. A. Ford, "The Chemist's Companion," p. 439, John Wiley & Sons, Inc., New York, 1972.

52. [(PHENYLISOPROPYLPHOSPHINO)METHYL]-DIPHENYLPHOSPHINE SULFIDE

$$(C_6H_5)_3P(S) + CH_3Li \xrightarrow[Et_2O]{THF} (C_6H_5)_2P(S)CH_2Li + C_6H_6$$

$$(C_6H_5)_2P(S)CH_2Li + (i\text{-}C_3H_7)(C_6H_5)PCl \longrightarrow (C_6H_5)_2P(S)CH_2P(C_6H_5)(i\text{-}C_3H_7) + LiCl$$

Submitted by J. D. MITCHELL* and S. O. GRIM*
Checked by L. SACCONI†

Phosphine-phosphine sulfides are a new class[1] of compounds which can behave as ligands themselves, be oxidized with oxygen, sulfur, or selenium to produce diphosphine chalcogenides, which also are useful ligands,[2] or be reduced with $LiAlH_4$,[3] Si_2Cl_6,[4] etc., to produce unsymmetrical (in the cases below) bistertiary phosphine ligands.[5] For the syntheses below, the compounds have a methylene bridge between the phosphorus atoms, whereas other preparations result in ethylene bridges[3,5] between the phosphorus atoms.

Procedure

Triphenylphosphine sulfide (29.4 g, 0.100 mole) is placed in a 500-ml three-necked flask equipped with a nitrogen inlet and a magnetic stirrer. Approximately 100 ml of freshly distilled tetrahydrofuran and ca. 75 ml of freshly distilled diethyl ether are added to the flask. The slurry is deoxygenated several times with a water aspirator, and nitrogen is admitted each time. As the slurry is stirred, 0.100 mole (62.5 ml of 1.60*M*) of standardized methyllithium in diethyl ether is added dropwise under nitrogen over a 30-min period. A deep-red homogeneous solution of $(C_6H_5)_2P(S)CH_2Li$ is formed,[1] and the solution is stirred for 1 hr more after addition of the methyllithium.

*Department of Chemistry, University of Maryland, College Park, Md. 20742.
†Istituto di Chimica Generale, Laboratorio C.N.R., Università di Firenze, Firenze 50132, Italy.

A second 500-ml three-necked flask is equipped with a nitrogen inlet, magnetic stirrer, and an addition funnel. Deoxygenated diethyl ether (60 ml) along with 18.7 g (0.100 mole) of $(i\text{-}C_3H_7)(C_6H_5)PCl$* is placed in the flask. The solution of $(C_6H_5)_2P(S)CH_2Li$ is transferred under nitrogen to the addition funnel and is *slowly* added at room temperature to the rapidly stirring ether–phosphinous chloride mixture over at least 1 hr. As the addition progresses, the solution gradually turns yellow, and lithium chloride precipitates. After the addition is complete, the yellow reaction mixture is stirred for ca. 10 hr more under nitrogen.

The reaction mixture is transferred to a 500-ml, one-necked, round-bottomed flask, and the solvents are removed with a rotary evaporator. A thick, reddish-yellow oil results† and is dissolved in approximately 100 ml of dichloromethane. The solution is extracted with three 100-ml portions of deoxygenated distilled water to remove the lithium chloride. The dichloromethane solution is dried over sodium sulfate, and the solvent is removed again. The oil remaining is dissolved in 50 ml of a 1:1 CH_2Cl_2-hexane mixture and chromatographed through a column (2-in. diam. by 10 in. long) of neutral alumina with the same solvent as eluent. The product emerges as an off-white band after about 200 ml, and the next ca. 350 ml of eluate is collected.

After evaporation of the solvents, the light-colored oil is dissolved in hot absolute ethanol. Rapid cooling of the ethanol solution in a Dry Ice–isopropanol bath with simultaneous scratching produces the desired product, $(C_6H_5)_2P(S)CH_2P(i\text{-}C_3H_7)(C_6H_5)$, as white crystals.‡ Several crystallizations by the same treatment produce 17.0 g (45% based on Ph_3PS) of $(C_6H_5)_2P(S)CH_2P(i\text{-}C_3H_7)$. *Anal.* Calcd. for $C_{22}H_{24}P_2S$: C, 69.09; H, 6.33; P, 16.20. Found: C, 69.2; H, 6.28; P, 16.0.

*Chloroalkylphenylphosphines were prepared by the following reaction sequence:

$$PhPCl_2 + 2Et_2NH \longrightarrow Ph(Et_2N)PCl + [Et_2NH_2]Cl$$
$$Ph(Et_2N)PCl + RMgX \longrightarrow Ph(Et_2N)PR + MgX_2$$
$$Ph(Et_2N)PR + HCl(g) \longrightarrow PhRPCl + [Et_2NH_2]Cl$$

Chlorodiisopropylphosphine was prepared by a similar route:

$$PCl_3 + 2Et_2NH \longrightarrow Et_2NPCl_2 + [Et_2NH_2]Cl$$
$$Et_2NPCl_2 + 2i\text{-}PrMgX \longrightarrow Et_2NPi\text{-}Pr_2 + 2MgXCl$$
$$Et_2NPi\text{-}Pr_2 + HCl(g) \longrightarrow i\text{-}Pr_2PCl + [Et_2NH_2]Cl$$

†In all preparations, a ^{31}P nmr spectrum of the crude reaction mixture indicates a fairly clean reaction with formation of a noticeable amount of Ph_2MePS ($\delta = -35$ ppm). The amount of Ph_2MePS can be minimized by rigorous exclusion of moisture in the transfer and use of the $Ph_2P(S)CH_2Li$ intermediate.

‡The solid compounds are fairly air-stable, i.e., they show no noticeable change after several days in air, but in solution they should be handled under nitrogen.

Properties

The compound is a white solid which melts at 81–82°. A ^{31}P nmr spectrum in CH_2Cl_2 shows an upfield doublet centered at +19.3 ppm from H_3PO_4 attributed to $-P(C_6H_5)(i\text{-}C_3H_7)$ and a downfield doublet centered at −40.5 ppm attributed to $-P(S)-(C_6H_5)_2(J_{PCP} = 71$ Hz). An ^1H nmr spectrum in $CDCl_3$ shows a phenyl multiplet (τ2.0–2.8), a multiplet centered at ca. τ7.0 due to the nonequivalent methylene protons, a broad multiplet centered at ca. τ8.0 due to the methine protons, and two sets of doublets of doublets due to the nonequivalent methyl groups of the isopropyl group (τ_a1.20, J_{PCCH} = 13.8 Hz, J_{HCCH} = 6.9 Hz; τ_b0.92, J_{PCCH} = 15.9 Hz, J_{HCCH} = 6.9 Hz).

Other Compounds

The synthetic aspects and other properties of some analogs are given in Table I.

TABLE I Synthetic Aspects and Properties of $Ph_2P(S)CH_2PR_2$

Compound	Yield, %	m.p., °C	δ, ppm		J_{PP}, Hz
			P	P=S	
$Ph_2P(S)CH_2PPh_2$	33	103–105	28.0	−40.1	76
$Ph_2P(S)CH_2PPhMe$	23	91–92	43.3	−39.1	66
$Ph_2P(S)CH_2Pi\text{-}Pr_2$	52	107–109	9.3	−41.6	77
$Ph_2P(S)CH_2PPhEt$	48	59–62*	30.9	−39.9	68

*This compound was crystallized with difficulty. Crystals formed after the product was left for several months at −25°.

References

1. D. Seyferth, D. E. Welch, and J. K. Heeren, *J. Am. Chem. Soc.*, **85,** 642 (1963); D. Seyferth and D. E. Welch, *J. Organomet. Chem.*, **2,** 1 (1964); D. Seyferth, U.S. Pat. 3,426,021, February 1969 [*Chem. Abstr.*, **71,** 319 (1969)]; S. O. Grim and J. D. Mitchell, *Syn. React. Inorg. Metal–org. Chem.*, **4,** 221 (1974).
2. W. E. Slinkard and D. W. Meek, *J. Chem. Soc., Dalton Trans.*, **1973,** 1024.
3. R. B. King, J. C. Cloyd, Jr., and P. K. Hendrick, *J. Am. Chem. Soc.*, **95,** 5083 (1973).
4. G. Zon, K. E. DeBruin, K. Naumann, and K. Mislow, *J. Am. Chem. Soc.*, **91,** 7023 (1969).
5. S. O. Grim, R. P. Molenda, and R. L. Keiter, *Chem. Ind. (Lond.)*, **1970,** 1378; S. O. Grim, J. DelGaudio, R. P. Molenda, C. A. Tolman, and J. P. Jesson, *J. Am. Chem. Soc.*, **96:** 34–16 (1974).

53. *N,N,N′,N′*-TETRAKIS(DIPHENYLPHOSPHINOMETHYL)-ETHYLENEDIAMINE[1]

$$4(C_6H_5)_2PH + 4CH_2O + H_2NCH_2CH_2NH_2 \longrightarrow [(C_6H_5)_2PCH_2]_2NCH_2CH_2N[CH_2P(C_6H_5)_2]_2$$

Submitted by L. J. MATIENZO* and S. O. GRIM*
Checked by R. URIARTE† and D. W. MEEK†

A large variety of polydentate ligands containing nitrogen and phosphorus can be prepared easily by the reaction of ethylenediamine or substituted ethylenediamines, formaldehyde, and secondary phosphines, as illustrated below. Furthermore, the scope of this reaction is very wide. It includes the compounds listed here and it is also general for amines which contain at least one N—H bond and for phosphines which contain at least one P—H bond.

Procedure

Diphenylphosphine‡ (7.5 ml, 43 mmoles) is dissolved in 30 ml of nitrogen-saturated benzene in a 125-ml two-necked flask fitted with a water condenser, topped by a nitrogen inlet, and a magnetic stirrer. Ethylenediamine (0.65 g, 0.72 ml, 11 mmoles, d. = 0.90) and 5.3 ml of 37% formalin solution (65 mmoles of CH_2O, 50% excess) are then added, and the remaining neck is stoppered. The flask is heated by means of an oil bath at 60–65° for 1.5 hr,§ after which the reaction is essentially completed.¶ The mixture is transferred to a separatory funnel and the benzene upper layer is removed, dried over sodium sulfate, and evaporated by means of a rotary evaporator with a water aspirator. The oily residue is dissolved in 10 ml of dichloromethane, and about 70 ml of absolute ethanol is added with stirring.** The white crystalline material which precipitates immediately is isolated by filtration on a glass-fritted funnel, dried *in vacuo* for several hours, and recrystallized from ethanol-dichlo-

*Department of Chemistry, University of Maryland, College Park, Md. 20742.
†Chemistry Department, The Ohio State University, Columbus, Ohio 43210.

‡Prepared from the hydrolysis of lithium diphenylphosphide resulting from the cleavage of triphenylphosphine by lithium in tetrahydrofuran; also available from Strem Chemicals, Inc., Danvers, Mass. 01923.

§The temperature control (not in excess of 65°) is important; otherwise CH_2O is eliminated.

¶A ^{31}P nmr spectrum of an aliquot of the reaction mixture reveals that the characteristic doublet of $(C_6H_5)_2PH$ ($\delta = 41$ ppm, $J_{PH} = 215$ Hz) has completely disappeared and the singlet of the product has appeared.

**The checkers emphasize the importance of stirring in order to prevent the formation of an oil instead of the desired crystalline product.

romethane. The yield is 9.1 g (99%).* *Anal.* Calcd. for $C_{54}H_{52}N_2P_4$: C, 76.04; H, 6.15; N, 3.28; P, 14.53. Found: C, 75.8; H, 6.32; N, 3.04; P, 14.1.

Properties

N,N′,N′-Tetrakis(diphenylphosphinomethyl)ethylenediamine is a white crystalline compound, m.p. 77–78°. It is fairly stable in air when pure and can be stored for an appreciable time in a brown bottle in a desiccator. It is soluble in dichloromethane, benzene, and chloroform but not very soluble in ethanol. The compound in CH_2Cl_2 has a ^{31}P singlet at 28.1(1) ppm vs. H_3PO_4. The ^1H nmr spectrum in $CDCl_3$ has peaks at τ2.65 and 2.74 (phenyl), τ6.49 (J_{PCH} = 3.0 Hz) for the methylene protons, and a singlet at τ7.12 (ethylene).†

Other Compounds

Data for other compounds prepared by this method are listed in Table I on page 200.

Reference

1. S. O. Grim and L. J. Matienzo, *Tetrahedron Lett.*, **1973**, 2951.

54. ETHYLENEBIS(NITRILODIMETHYLENE)TETRAKIS-(PHENYLPHOSPHINIC ACID)

$$H_2NCH_2CH_2NH_2 + 4CH_2O + 4C_6H_5P(O)(OH)H \xrightarrow[\Delta]{HCl,\ H_2O} [\{C_6H_5(OH)(O)PCH_2\}_2NCH_2]_2 + 4H_2O$$

Submitted by A. I. PLAZA‡ and S. O. GRIM‡
Checked by R. J. MOTEKAITIS§ and A. E. MARTELL§

An adaptation of the Mannich reaction can be used to produce an analog of EDTA in which phenylphosphinic acid groups replace the carboxylic acid groups in EDTA.[1] The reaction is very simple, straightforward, and versatile

*The checkers obtained 5.0 g of the product initially and another 1.8 g by evaporating the filtrate for a total yield of 76%.

†In some preparations a small extraneous peak occurs at τ6.62, which is probably formaldehyde.

‡Department of Chemistry, University of Maryland, College Park, Md. 20742.

§Chemistry Department, Texas A. & M. University, College Station, Tex. 77843.

TABLE I

Final compound	Starting material		Reaction time, hr	Yield, %	m.p., °C	δ, ppm[a]
	Compounds	Amount used				
$(Ph_2PCH_2)_2NCH_2CH_2N(CH_3)CH_2PPh_2$[b]	Ph_2PH	86 mmol				
	$H_2NCH_2CH_2NHCH_3$	29 mmol				
	$CH_2O(aq)$	103 mmol	2.0	75	73–74	26.9, 28.4[c]
	C_6H_6	50 ml				
$[Ph_2PCH_2(CH_3)NCH_2]_2$[d]	Ph_2PH	110 mmol				
	$CH_3NHCH_2CH_2NHCH_3$	57 mmol				
	$CH_2O(aq)$	137 mmol	1.5	86	79–80	26.9
	C_6H_6	50 ml				
$(Ph_2PCH_2)_2NCH_2CH_2N(CH_3)_2$[e]	Ph_2PH	85 mmol				
	$H_2NCH_2CH_2N(CH_3)_2$	43 mmol				
	$CH_2O(aq)$	110 mmol	1.5	90	Oil	27.6
	C_6H_6	50 ml				
$Ph_2PCH_2N(CH_3)CH_2CH_2N(CH_3)_2$[f]	Ph_2PH	85 mmol				
	$CH_3NHCH_2CH_2N(CH_3)_2$	85 mmol	1.5	96	Oil	26.6
	$CH_2O(aq)$	100 mmol				

[a]Phosphorus chemical shift in CH_2Cl_2 relative to 85% H_3PO_4.

[b]*Anal.* Calcd. for $C_{42}H_{43}N_2P_3$: C, 75.44; H, 6.48; N, 4.19; P, 13.90. Found: C, 75.4; H, 6.60; N, 3.85; P, 14.2.

[c]The relative intensity of the peaks is 1:2, respectively.

[d]*Anal.* Calcd. for $C_{30}H_{34}N_2P_2$: C, 74.37; H, 7.07; N, 5.78; P, 12.78. Found: C, 74.1; H, 7.22; N, 5.72; P, 12.9.

[e]*Anal.* Calcd. for $C_{30}H_{34}N_2P_2$: C, 74.37; H, 7.07; N, 5.78. Found: C, 73.9; H, 7.40; N, 5.48.

[f]*Anal.* Calcd. for $C_{18}H_{25}N_2P$: C, 71.98; H, 8.39. Found: C, 70.8; H, 8.34.

and provides compounds of moderate chelating ability. Variation of the compounds containing the N—H and P—H bonds can lead to a wide variety of compounds.

Procedure

A 250-ml, three-necked, round-bottomed flask, equipped with a water-cooled reflux condenser and mechanical overhead stirrer, is charged with 2.4 g of ethylenediamine (0.04 mole), 22.7 g of phenylphosphinic acid (0.16 mole), 50 ml of water, and 30 ml of concentrated hydrochloric acid.[1] The resultant slurry is stirred and heated with an oil bath. When the temperature reaches about 70°, the reaction mixture becomes a homogeneous, clear, very slightly yellow solution. At refluxing temperatures, 30 ml of 37% formaldehyde solution (about 0.32 mole CH_2O) is added dropwise over 35 min through a pressure-equalizing addition funnel. About halfway through the addition, a cloudy white precipitate begins to form. When addition of the formaldehyde is complete, the reaction mixture is stirred vigorously and refluxed for 2 hr more. After cooling to room temperature, the precipitate is collected by filtration using a Büchner funnel with a water vacuum and then washed by boiling three times with 100-ml portions of each of the following: water, acetone, and methylene chloride. The precipitate is then dried *in vacuo* at room temperature overnight. Yield is 15.7 g (58%) of a white, crystalline powder which melts at 243–245° with decomposition. *Anal.* Calcd. for $C_{30}H_{36}O_8N_2P_4$: C, 53.26; H, 5.36; N, 4.14; P, 18.31. Found: C, 53.0; H, 5.45; N, 3.96; P, 18.5.

Properties

The nonhygroscopic compound is found to be very stable and very insoluble, even with boiling, in common solvents such as water, acetone, methylene chloride, hexane, and acetonitrile. However, the compound is found to be soluble in a stoichiometric amount of aqueous potassium hydroxide, in which the ^{31}P nmr spectrum shows a sharp singlet at −29.3 ppm relative to 85% phosphoric acid.

Other Compounds

Data for two other compounds prepared in this manner are given in Table I on page 202.

Reference

1. K. Moedritzer and R. R. Irani, *J. Org. Chem.*, **31**, 1603 (1966).

TABLE I

Final compound	Starting material		Yield		m.p., °C
	Compounds	Amount used	Grams	%	
$N[CH_2P(O)(OH)C_6H_5]_3$*	NH_4Cl $C_6H_5P(O)$-(H)OH $CH_2O(aq)$ HCl (conc.) H_2O	0.05 mol 0.15 mol 0.3 mol 50 ml 40 ml	12.2	51.5	233–236
$[C_6H_5P(O)(OH)CH_2N(CH_3)$-$CH_2$—$]_2\cdot 2HCl$†,‡	CH_3NHCH_2-CH_2NHCH_3 $C_6H_5P(O)(H)$-OH $CH_2O(aq)$ HCl (conc.) H_2O	0.1 mol 0.2 mol 0.4 mol 40 ml 100 ml	21.7	46.4	218–222

**Anal.* Calcd. for $C_{21}H_{24}O_6NP_3$: C, 52.62; H, 5.05; N, 2.92; P, 19.39. Found: C, 52.3; H, 5.16; N, 2.87; P, 19.1.

†*Anal.* Calcd. for $C_{18}H_{28}O_4N_2P_2Cl_2$: C, 46.07; H, 6.01; N, 5.97; P, 13.20; Cl, 15.11. Found: C, 45.9; H, 6.21; N, 6.00; P, 13.5; Cl, 14.7.

‡This compound is obtained as the dihydrochloride salt. During the reaction this compound does not precipitate but is obtained by concentrating the reaction mixture with a rotary evaporator and allowing the mixture to stand overnight, whereupon crystallization takes place. The compound is quite soluble in water.

55. [2-(PHENYLPHOSPHINO)ETHYL]DIPHENYLPHOSPHINE

Submitted by J. C. CLOYD, JR.,* and R. B. KING*
Checked by T. L. DUBOIS† and D. W. MEEK†

The base-catalyzed addition of P—H bonds across the carbon—carbon double bonds of vinyl phosphorus compounds readily provides a large number of new polyphosphines,[1–4] which are useful chelating ligands.[5–8] The synthesis of $(C_6H_5)_2PCH_2CH_2P(H)C_6H_5$, one member of a class of these new polyphosphines,‡ is described here. Details for the preparation of the vinyl ester, $CH_2{=}CHP(O)(i\text{-}OC_3H_7)C_6H_5$, are also provided.

[2-(Phenylphosphino)ethyl]diphenylphosphine has been prepared by the reaction of $C_6H_5P(H)Na$ with $ClCH_2CH_2P(C_6H_5)_2$[9] and by the radical addition of phenylphosphine to one molar equivalent of $CH_2{=}CHP(C_6H_5)_2$.[10] The base-

*Chemistry Department, University of Georgia, Athens, Ga. 30602.
†Chemistry Department, The Ohio State University, Columbus, Ohio 43210.
‡See first footnote, p. 203.

catalyzed addition of diphenylphosphine to $CH_2{=}CHP(O)(i\text{-}OC_3H_7)C_6H_5$ followed by $LiAlH_4$ reduction of the intermediate ester provides a convenient source of $(C_6H_5)_2PCH_2CH_2P(H)C_6H_5$ in moderate yields and high purity.*

A. ISOPROPYLPHENYLVINYLPHOSPHINATE

$$C_6H_5P(i\text{-}OC_3H_7)_2 + BrCH_2CH_2Br \longrightarrow BrCH_2CH_2P(O)(i\text{-}OC_3H_7)C_6H_5 + i\text{-}C_3H_7Br$$

$$BrCH_2CH_2P(O)(i\text{-}OC_3H_7)C_6H_5 + (C_2H_5)_3N \longrightarrow CH_2{=}CHP(O)(i\text{-}OC_3H_7)C_6H_5 + (C_2H_5)_3NHBr$$

Procedure

Under an atmosphere of dry nitrogen, $C_6H_5P(i\text{-}OC_3H_7)_2$ [11] (226 g, 1 mole) is added dropwise to an excess of $BrCH_2CH_2Br$† (940 g, 5 moles) maintained at 140–150° with an external oil bath. Magnetic stirring is employed, and the dropping funnel is arranged so that the $C_6H_5P(i\text{-}OC_3H_7)_2$ drops directly into the vortex. The oil-bath temperature should be maintained as high as possible, without collecting a distillate other than $i\text{-}C_3H_7Br$. After approximately 10% of the $C_6H_5P(i\text{-}OC_3H_7)_2$ has been added, $i\text{-}C_3H_7Br$ begins to distill (b.p. 60°) from the reaction solution and is continuously collected through a 30-cm Vigreux column and total-reflux distillation head. After all the $C_6H_5P(i\text{-}OC_3H_7)_2$ has been added (3–4 hr), heating is maintained until the distillation of isopropyl bromide ceases (60–70% of theory is collected). The Vigreux column is taken down and the bulk of the excess ethylene bromide is removed by distillation at atmospheric pressure. The remaining traces of ethylene bromide are removed under vacuum, leaving the water-clear liquid $BrCH_2CH_2P(O)(i\text{-}OC_3H_7)C_6H_5$.‡

Benzene (500 ml) and the isopropylphenyl(2-bromoethyl)phosphinate are placed in a flask equipped with a nitrogen inlet, mechanical stirrer, and reflux condenser. Triethylamine (150 g, 1.5 mole; 50% excess) is added all at once, and the reaction mixture is boiled under reflux for 12 hr. After cooling to room temperature, the benzene solution is filtered as rapidly as possible from the

*In general, the vinyl ester, $CH_2{=}CHP(O)(OR)R'$, may be combined with any P–H compound, $R_{3-n}PH_n$, to give, following reduction, $R_{3nm}P[CH_2CH_2P(H)R]_n$.

†The ethylene bromide should be fractionated at atmospheric pressure before use.

‡Purification of this compound by distillation was unsuccessful. During one attempt, at 0.1 torr and a pot temperature of 170°, a sudden rise in pressure (ca. 1 torr) occurred. Cooled to room temperature, the liquid was noticeably thicker. A much reduced yield of the isopropylphenylvinylphosphinate was obtained in this preparation.

precipitated triethylamine hydrobromide. The salt cake is washed with three portions of benzene. The benzene is removed from the combined benzene solutions by distillation at atmospheric pressure. After the remaining traces of solvent have been removed under vacuum, the product is distilled at 97–100° and 0.10–0.15 torr. The yield (based on $C_6H_5P(i\text{-}OC_3H_7)_2$) is 147 g (70%). *Anal.* Calcd. for $C_{11}H_{15}PO_2$: C, 62.8; H, 7.2. Found: C, 62.7; H, 7.2. Isopropylphenylvinylphosphinate is a water-clear liquid with a slightly sweet odor. The ^1H nmr (neat liquid, TMS internal standard) consists of multiplets at $\tau2.20$ and $\tau2.56$ (phenyl protons), $\tau3.90$ (vinyl protons), $\tau5.48$ (methine protons), and four equivalent resonances centered at $\tau8.76$ (methyl protons).

B. [2-(PHENYLPHOSPHINO)ETHYL]DIPHENYLPHOSPHINE

$$(C_6H_5)_2PH + CH_2{=}CHP(O)(i\text{-}OC_3H_7)C_6H_5 \longrightarrow (C_6H_5)_2PCH_2CH_2P(O)(i\text{-}OC_3H_7)C_6H_5$$

$$4(C_6H_5)_2PCH_2CH_2P(O)(i\text{-}OC_3H_7)C_6H_5 + 2LiAlH_4 \longrightarrow 4(C_6H_5)_2PCH_2CH_2P(H)C_6H_5 + LiAl(OH)_4 + LiAl(i\text{-}OC_3H_7)_4$$

Diphenylphosphine[12] (37.2 g, 0.2 mole) and isopropyl phenylvinylphosphinate (42 g, 0.2 mole) are mixed with 300 ml of tetrahydrofuran (freshly distilled, under nitrogen, from sodium-benzophenone ketyl). Sufficient KO-(*t*-Bu) is added until the solution becomes hot and yellow to orange in color (the catalyst is hydrolyzed fairly rapidly by atmospheric moisture, but the additions can be made using a spatula, provided that the bulk of the catalyst is kept under nitrogen and the transfers are made rapidly). If the reactants are pure and the tetrahydrofuran is scrupulously dry, only 20–30 mg is required to produce the heat and color. This solution is boiled under reflux for 16 hr. During the reflux period, the color of the solution should fade, becoming pale yellow to colorless.

The solvent is removed at 40°/40 torr leaving a solid residue. It is not necessary to purify the intermediate $(C_6H_5)_2PCH_2CH_2P(O)(i\text{-}OC_3H_7)C_6H_5$, before the reduction step. The ^1H nmr ($CDCl_3$ solution, TMS internal standard) consisting of multiplets at $\tau2.8$ (phenyl protons), $\tau5.7$ (methine protons), and $\tau8.0$ (methylene protons), and four equal resonances centered at $\tau8.8$ (methyl protons) is a sufficient criterion of purity.

Enough tetrahydrofuran (200–300 ml) is added to dissolve the $(C_6H_5)_2PCH_2$-$CH_2P(O)(i\text{-}OC_3H_7)C_6H_5$ (68 g, 0.17 mole), and anhydrous diethyl ether is added to bring the volume of the solution to approximately 1 l. To this solution is added, in portions, $LiAlH_4$ (20 g, 0.53 mole). In order to control the exothermicity and vigorous gas evolution caused by the addition of the $LiAlH_4$, the additions should be made in at least five separate portions. The resulting slurry is stirred magnetically at room temperature for 4 days.*

Hydrolysis is accomplished by first chilling the reaction mixture (0°) and adding, cautiously and successively, 20 ml of water, 20 ml of 15% aqueous NaOH solution, and 60 ml of water.[14] This procedure results in an easily handled, finely divided white solid. The solid and the organic solution are transferred through a glass bend (45°) to a closed filtering flask (a three-necked round-bottomed flask to which has been attached at the bottom a coarse-sintered-glass disk, Teflon stopcock, and standard-taper male joint. Placing molecular sieves covered by anhydrous Na_2SO_4 directly over the frit will facilitate the filtering and drying of the organic solution).

A flask (the size of the flask depends on the method of purification chosen) is attached to the bottom of the closed filtering flask, and the organic solution is drawn into it under vacuum, removing the solvent and leaving the liquid phosphine (since the solvent may smell of phosphine, it should be trapped at −78° and disposed of safely at a later time).

The crude product may be purified by vacuum distillation or by crystallization from degassed pentane at 0°.† A single crystallization from pentane yields 31 g (57%) of analytically pure $(C_6H_5)_2PCH_2CH_2P(H)C_6H_5$. *Anal.* Calcd. for $C_{20}H_{20}P_2$: C, 74.5; H, 6.2; P, 19.2. Found: C, 74.3; H, 6.1; P, 19.3.

Properties[2]

[2-(Phenylphosphino)ethyl]diphenylphosphine crystallizes from pentane as white needles melting around room temperature (ca. 30°) and boiling at 201–202°/0.08 torr (literature[10] 210–215°/1 torr). The P—H stretching frequency in the infrared spectrum occurs at 2286 cm^{-1}. The ^1H nmr spectrum ($CDCl_3$ solution,‡ TMS internal standard) consists of a multiplet at $\tau 2.78$ (phenyl protons), broad multiplets at $\tau 7.9$ and $\tau 8.2$ (methylene protons), and a doublet of triplets at $\tau 6.4$ (PH proton; triplet separation ca. 6 Hz). The ^{31}P nmr spectrum ($CDCl_3$ solution) consists of a broad resonance at

*A prolonged reaction time at room temperature is preferred to a short reaction time at reflux, since P—C bond cleavage by $LiAlH_4$ under more forcing conditions has been observed.[13]

†The pure product has been isolated using both low-temperature crystallization (jacketed funnel) and vacuum distillation, with little difference in yields. In the distillation, a forerun is obtained (unidentified mixture).

‡Solutions in $CDCl_3$ must be freshly prepared. The checkers stated that they observed the P—H doublet of triplets in C_6D_6 but not in $CDCl_3$.

14.1 ppm [type $(C_6H_5)_2P$–] and a doublet (J_{PH} = 212 ± 10 Hz) at 46.6 ppm [type $C_6H_5(H)P$–] upfield from 85% H_3PO_4 (external standard).

References

1. R. B. King and P. N. Kapoor, *J. Am. Chem. Soc.*, **93,** 4158 (1971).
2. R. B. King, J. C. Cloyd, Jr., and P. N. Kapoor, *J. Chem. Soc., Perkin Trans.*, **1,** 2226 (1973).
3. R. B. King and J. C. Cloyd, Jr., *Z. Naturforsch.*, **27b,** 1432 (1972).
4. R. B. King, J. C. Cloyd, Jr., and P. K. Hendrick, *J. Am. Chem. Soc.*, **95,** 5083 (1973).
5. R. B. King, *Acc. Chem. Res.*, **5,** 177 (1972).
6. R. B. King, R. N. Kapoor, and P. N. Kapoor, *Inorg. Chem.*, **10,** 1941 (1971).
7. R. B. King, R. N. Kapoor, M. S. Saran, and P. N. Kapoor, *Inorg. Chem.*, **10,** 1851 (1971).
8. R. B. King and M. S. Saran, *Inorg. Chem.*, **10,** 1861 (1971).
9. J. C. Cloyd, Jr., and D. W. Meek, *Inorg. Chim. Acta*, **6,** 607 (1972).
10. K. Issleib and H. Weichmann, *Z. Chem.*, **11,** 188 (1971).
11. A. E. Arbusov, G. Kh. Kamai, and O. N. Belorossova, *J. Gen. Chem. (U.S.S.R.)*, **15,** 766 (1945); T. H. Siddall and C. A. Prohaska, *J. Am. Chem. Soc.*, **84,** 3467 (1962).
12. W. Gee, R. A. Shaw, and B. C. Smith, *Inorganic Syntheses*, **9,** 19 (1967).
13. J. C. Cloyd, Jr., and R. B. King, *J. Am. Chem. Soc.*, **97,** 53 (1975).
14. L. F. Fieser and M. Fieser, "Reagents for Organic Synthesis," p. 584, John Wiley & Sons, Inc., New York, 1968.

56. 1,1,1-TRIFLUORO-4-MERCAPTO-4-(2-THIENYL)-3-BUTEN-2-ONE*

(1,1,1-Trifluoro-3-thiothenoylacetone)

$$(C_4H_3S)\underset{\underset{O}{\|}}{C}CH_2\underset{\underset{O}{\|}}{C}CF_3 + H_2S \xrightarrow{HCl} (C_4H_3S)\underset{\underset{SH}{|}}{C}{=}CH\underset{\underset{O}{\|}}{C}CF_3 + H_2O$$

Submitted by M. DAS† and S. E. LIVINGSTONE†
Checked by B. C. PESTEL‡ and S. C. CUMMINGS‡

1,1,1-Trifluoro-3-thenoylacetone has been extensively used for solvent extraction of heavy metals, in particular zirconium and the actinides.[1,2] It has also

*Tautomer is named 4,4,4-trifluoro-1-(2-thienyl)-1-thio-1,3-butanedione or 1,1,1-trifluoro-4-(2-thienyl)-4-thioxo-3-buten-2-one.

†School of Chemistry, University of New South Wales, Kensington, New South Wales 2033, Australia.

‡Chemistry Department, Wright State University, Dayton, Ohio 45431.

been used for extraction of metals in flame photometry.[3] Its monothio analog, 1,1,1-trifluoro-4-mercapto-4-(2-thienyl)-3-buten-2-one, forms stable complexes with many metal ions[4] and may also prove to be a useful reagent in solvent extraction.

Monothio β-diketones can be prepared by the action of hydrogen sulfide and hydrogen chloride on the appropriate β-diketone in alcohol solution.[5] Nevertheless the conditions are rather critical. At room temperature β-diketones are in tautomeric equilibrium between the diketo form (I) and the chelated hydrogen-bonded form (II), and in polar solvents the concentration of the diketo form (I) is increased.[6] Reaction with hydrogen sulfide occurs only with the diketo tautomer (I). Consequently, higher concentrations of hydrogen chloride are required for those β-diketones which exist predominantly in the

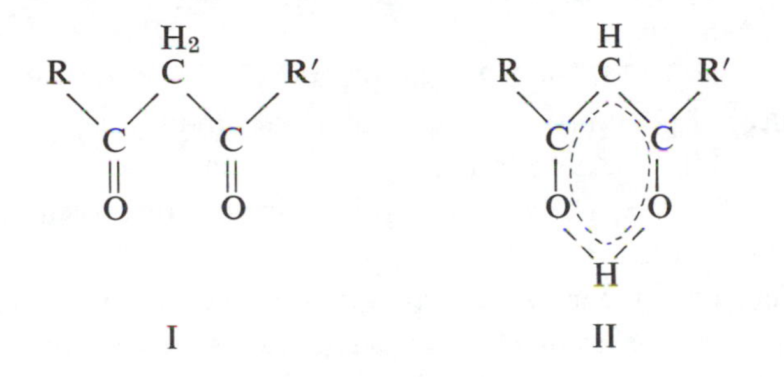

enol form in order to provide the more polar conditions necessary to shift the equilibrium in favor of the diketo form.[5,6] The reaction must be carried out in dilute solution to avoid formation of oligomers.

The reaction can be represented as follows:[6,7]

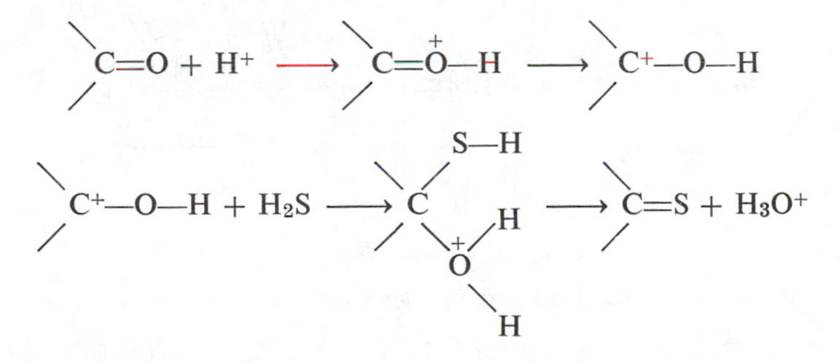

In the case of a β-diketone (I) where $R \neq R'$, it has been shown by mass spectrometry that the nucleophilic attack by hydrogen sulfide takes place at the ketonic group attached to R if the electron-withdrawing power of R is less than that of R′.[5,7] Consequently, 4,4,4-trifluoro-1-(2-thienyl)-1,3-butanedione yields the monothio β-diketone (III) as the only isomer.

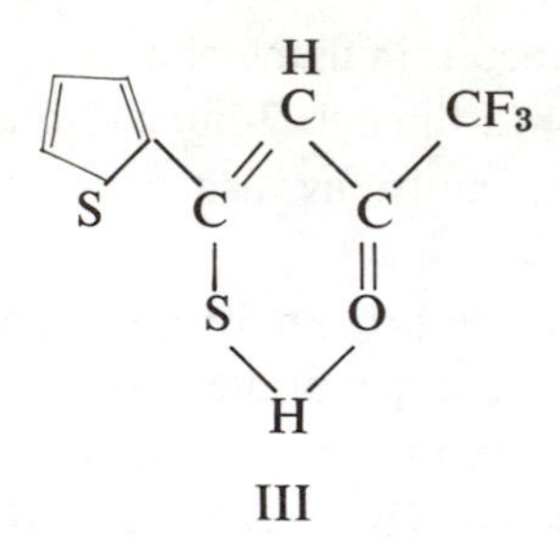

Procedure

■ **Caution.** *The whole preparation and recrystallization should be carried out in a fume hood, since hydrogen sulfide is poisonous and the crude product has an unpleasant smell.*

4,4,4-Trifluoro-1-(2-thienyl)-1,3-butanedione* (5.0 g) is dissolved in dry absolute alcohol (200 ml)† to give a colorless solution which is placed in a 1-l. two-necked flask *D,* fitted with an inlet tube and a calcium chloride guard tube *E* (see Fig. 16). The flask and its contents are cooled to −70° in an alcohol–Dry Ice bath *F.*

Hydrogen sulfide is generated in a Kipp apparatus *A* by the reaction of 5*M* hydrochloric acid on iron(II) sulfide. The gas is washed by allowing it to bubble through water contained in the washing tower *B.* The gas is then passed through three drying towers *C,* packed with granulated calcium chloride.

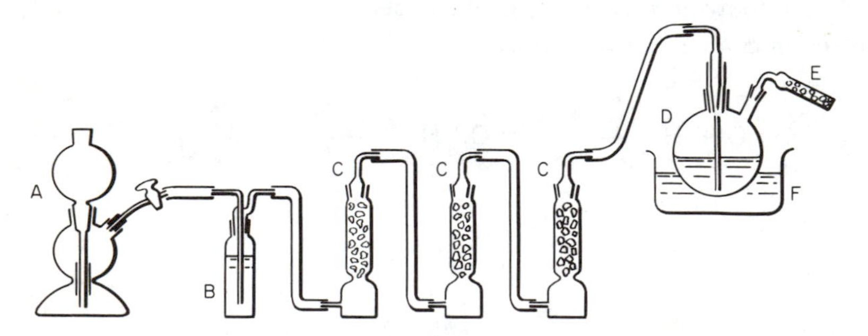

Fig. 16. Apparatus for the passage of hydrogen sulfide.

*Purchased from Aldrich Chemical Co., Milwaukee, under the name thenoyltrifluoroacetone.

†It is essential that *very dry* absolute alcohol be used. Anhydrous alcohol can be prepared by adding metallic sodium (5 g) to commercial reagent grade ethanol (99%; 500 ml), heating the mixture under reflux for 5 hr, and then distilling the alcohol.[8] If the alcohol is not treated in this way, it should be heated at the boiling point for 15 min to remove dissolved oxygen.

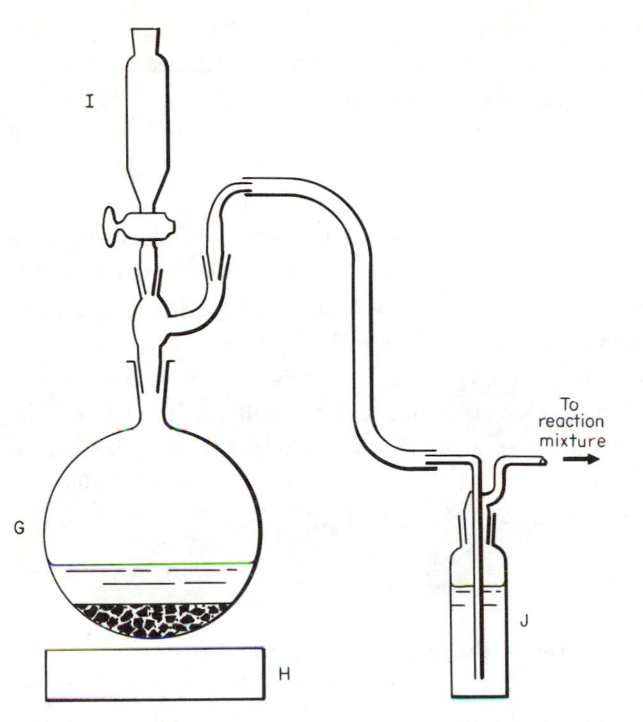

Fig. 17. Apparatus for the generation of hydrogen chloride.

The dried hydrogen sulfide is now passed into the reaction flask *D* at a *very rapid* rate.*

After 30 min the passage of hydrogen sulfide is discontinued, and hydrogen chloride is passed into the reaction mixture, also at a *very rapid* rate, for 15 min. The hydrogen chloride is generated as shown in Fig. 17. Sodium chloride (250 g) and 10 *M* hydrochloric acid (250 ml) are placed in a 2-l. round-bottomed flask *G*, fitted with a two-necked adapter. The NaCl-HCl slurry is warmed to ca. 60° over a hot plate *H*. Concentrated sulfuric acid is added dropwise from the dropping funnel *I* to give a *rapid* but steady stream of hydrogen chloride, which is dried by bubbling through concentrated sulfuric acid in the washing tower *J*.

After the passage of hydrogen chloride, the reaction flask is removed from the cooling bath, and the inlet tube is replaced by a stopper. As the tempera-

*It is important that the gases (H_2S and HCl) be passed *very rapidly* through the solution. Hydrogen sulfide from a pressurized cylinder can be used instead of a Kipp generator, enabling the gas to be passed at a very rapid rate. However, in this case, it is advisable to use a three-necked, round-bottomed flask fitted with two drying tubes to serve as pressure releases, since there is a pressure buildup as the temperature of the solution rises.

ture of the solution rises, the reaction starts and the solution changes color through yellow and orange to red. After 3 hr at room temperature the reaction flask is again placed in the alcohol–Dry Ice bath to cool to −70°. Hydrogen sulfide is again passed rapidly into the flask for 30 min, followed by addition of hydrogen chloride for 15 min. The mixture is then removed from the cooling bath and allowed to stand at room temperature overnight.

The resulting deep-red solution is poured slowly into ice-cold water (400 ml) with constant stirring, whereupon a red solid separates. The mixture is allowed to stand in an ice bath for 30 min with occasional stirring.* The red product is then collected on a filter, washed with water, and dried over silica gel. The yield of crude material is 4.3 g. This crude product is dissolved in boiling petroleum ether (b.p. 40–60°) (40 ml), and the solution is filtered. The filtrate is concentrated to 20 ml and cooled in ice. The red crystals are filtered off and dried *in vacuo* over silica gel. Yield: 3.1 g (58%).† *Anal.* Calcd. for $C_8H_5F_3OS_2$: C, 40.3; H, 2.1; S, 26.9. Found: C, 39.9; H, 2.1; S, 26.9.

Properties

1,1,1-Trifluoro-4-mercapto-4-(2-thienyl)-3-buten-2-one forms odorless, deep-red needles (m.p. 73–74°), insoluble in water, readily soluble in acetone, benzene, and petroleum ether, but less so in alcohol. If it is kept in a brown bottle away from light, the compound is stable virtually indefinitely. The principal bands in its infrared spectrum occur at 1612, 1570 (sh), 1260, and 817 cm^{-1}.[5] The dissociation constant pK_D is 7.05 in 74.5 vol % dioxane-water solution at 30°.[9] The dipole moment is 4.2 D.[10]

The complexes with nickel(II), palladium(II), platinum(II), zinc(II), cadmium(II), mercury(II), and lead(II) and their adducts with nitrogen heterocycles have been described.[4] The magnetic moment of the iron(III) complex is temperature-dependent owing to a spin-paired spin-free equilibrium.[11]

The monothio derivative of dibenzoylmethane, namely, 3-mercapto-1,3-diphenyl-2-propen-1-one, can be prepared from dibenzoylmethane by the same method. Yield: 3.4 g (64%). The red crystals melt at 84°. *Anal.* Calcd. for $C_{15}H_{12}OS$: C, 75.0; H, 5.0; S, 13.3. Found: C, 75.1; H, 5.0; S, 13.4.

References

1. G. H. Morrison and H. Freiser, "Solvent Extraction in Analytical Chemistry," John Wiley & Sons, Inc., New York, 1957.

*This occasional stirring is essential, otherwise the subsequent filtration is very slow.
†The checkers obtained a yield of 3.9 g (78%).

2. F. J. Welcher, "Standard Methods of Chemical Analysis," 6th ed., Part A, p. 163, D. Van Nostrand Company, Inc., Princeton, N.J., 1963.
3. H. A. Flaschka and A. J. Barnard, "Chelates in Analytical Chemistry," Vol. 4, p. 246, Marcel Dekker, Inc., New York, 1972.
4. R. K. Y. Ho, S. E. Livingstone, and T. N. Lockyer, *Aust. J. Chem.,* **21,** 103 (1968).
5. S. H. H. Chaston, S. E. Livingstone, T. N. Lockyer, V. A. Pickles, and J. S. Shannon, *Aust. J. Chem.,* **18,** 673 (1965).
6. S. E. Livingstone, *Coord. Chem. Rev.,* **7,** 59 (1971).
7. J. P. Guemas, doctoral thesis, Université de Nantes, France, 1970.
8. T. Honjyo and T. Kiba, *Bull. Chem. Soc. Jap.,* **45,** 185 (1972).
9. S. H. H. Chaston and S. E. Livingstone, *Aust. J. Chem.,* **19,** 2035 (1966).
10. M. Das and S. E. Livingstone, unpublished results.
11. R. K. Y. Ho and S. E. Livingstone, *Aust. J. Chem.,* **21,** 1987 (1968).

Chapter Seven

COMPOUNDS OF BIOLOGICAL INTEREST

57. METALLOPORPHYRINS

Submitted by A. D. ADLER,* F. R. LONGO,† and V. VÁRADI‡
Checked by R. G. LITTLE§

Metalloporphyrins form a general class of planar tetradentate ligands, and their complexes are of physical as well as of biological interest.[1,2] Reviews on the synthesis and preparation of the metalloderivatives,[3,4] their mechanism of chelation,[5] and their coordination chemistry[6] have been reported in some detail.

The most general and convenient method for the metallation of porphyrins is the procedure employing *N,N*-dimethylformamide (DMF) as the reaction medium.[7] Previous methods[7] employed either bases such as pyridine or organic acids such as acetic acid for the reaction medium. These methods are thermodynamically inefficient, as the base competes with the porphyrin for the metal in the former procedure and the protons of the solvent compete with the metal for the porphyrin in the latter.[3,7] The DMF method has been applied to the synthesis of a wide variety of materials with a considerable variation in both the chelating structure and the metal chelated.[3,4,7]

The most useful purification procedure employed is a variation[7] of the dry-column chromatographic technique[8] which usually works well for all but the

*Department of Chemistry, Western Connecticut State College, Danbury, Conn. 06810.
†Chemistry Department, Drexel University, Philadelphia, Pa. 19104.
‡Bell Telephone Laboratories, Murray Hill, N. J. 07974.
§Department of Chemistry, University of Maryland Baltimore County, Baltimore, Md. 21228.

highly ionized structures. The mechanism of the separation depends upon the formation of a reversible complex between the minority and majority porphyrinic structures present in solution. Concentrated conditions are necessary to form these complexes and to keep them in the complexed form. The impurity complex chromatographs slowly as a tight band, while the majority species will overload the column and pass through rapidly as a broad smear. For these reasons complete conversion of reactants, concentrated loading solutions, and dry-column beds maintaining concentrated conditions and rapid passage all favor better separation and recovery. However, a repetition of the chromatography is sometimes necessary. The finer the precipitate the easier it is to obtain a concentrated loading solution, giving a better and purer yield. Suffering the inconvenience of slow filtration of fine precipitates is therefore recommended over enduring the difficulties encountered in the redissolution of more readily filtered crystalline crude products.

A. *meso*-TETRAPHENYLPORPHYRINCOPPER(II)

[*5,10,15,20-Tetraphenylporphyrinatocopper(II)*]

$$Cu(CH_3COO)_2 + C_{44}H_{30}N_4 \longrightarrow CuC_{44}H_{28}N_4 + 2CH_3COOH$$

or

$$Cu^{2+} + TPPH_2 \longrightarrow CuTPP + 2H^+$$

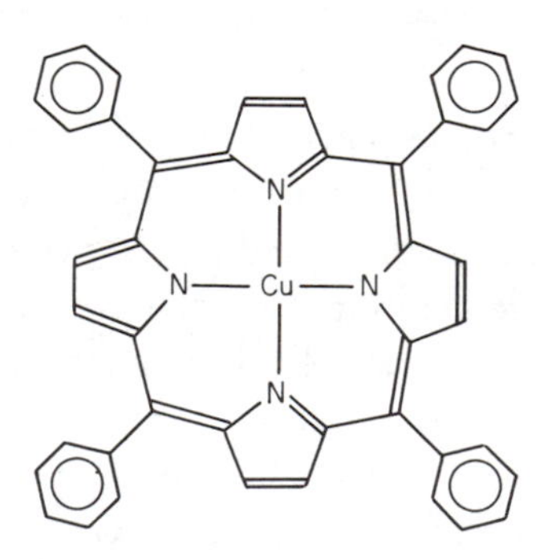

Procedure

The 5,10,15,20-tetraphenyl-21H,23H-porphyrin, or *meso*-tetraphenylporphyrin ($TPPH_2$), employed is prepared by a general method for condensing *meso*-porphyrins,[9] and its properties are described elsewhere[10] (it can now be obtained cheaply from various chemical suppliers, e.g., Strem and Aldrich). If the starting material is not chlorin-free, the copper-chlorin complex will be removed in the chromatographic purification; therefore purification of the porphyrin itself may be neglected at this stage. However, if removal of the chlorin is necessary, it can be effected by the dry-column chromatographic procedure[10] described below, using alumina and chloroform. Purity with respect to the chlorin can be determined by the absence of a complex band for a microchromatographic column using this technique or by visible solution spectrophotometry.[11-14]

DMF (1 l. in a 2-l. flask) is brought to gentle reflux on a hot plate and magnetically stirred. Reagent-grade DMF may be used without further purification or drying. The use of Teflon boiling chips is recommended for this solvent and method. $TPPH_2$ (10 g, 0.016 mole) is added, allowed to dissolve for a few minutes, and then followed by the addition of 4.0 g (0.02 mole) of cupric acetate, $Cu(CH_3COO)_2 \cdot H_2O$. Reaction is allowed to proceed for about 5 min. Complete conversion to the copper complex is then confirmed by observing the replacement of the four-banded TPP spectrum by the two-banded CuTPP spectrum[14] or by checking for the complete quenching of the porphyrin fluorescence under long-wavelength ultraviolet light,[7] as this complex is nonfluorescent. If conversion is not complete, another addition of a few 100 mg of cupric salt will accomplish it. (Cupric chloride may be readily substituted throughout for the acetate.) The reaction mixture is then allowed to cool in an ice-water bath for about 15 min.

After the reaction mixture has cooled, an equal volume of distilled water is stirred into the reaction mixture to precipitate the porphyrinic material. The solution is cooled again and then Büchner-filtered with Whatman paper no. 1, or its equivalent, until the filtrate is clear.* The reddish-brown material on the filter is then washed with water and air-dried. Recovery of crude material at this point should be about 11.0 g ($>$99%) or more, if some inorganic material carries over or an appreciable amount of DMF solvate is formed.

This crude material may now be further purified, 2 g at a time, by *dry-column* chromatography[7] on chromatographic-grade alumina (Fisher A-540 is recommended) with reagent-grade $CHCl_3$ on a 1 ft by 4-in.-diam column. The material is first dissolved in about 200 ml of $CHCl_3$, and the solution is poured carefully on the top of the dry-packed column and allowed to run until just dry. The materials are then eluted by further continuous additions of $CHCl_3$. Insoluble residues, if present, will remain on the top of the column and are thus removed. A sharp band forms near the top and moves slowly down the column. This is the complex formed between the porphyrinic impurities (TPP and CuTPC, where TPC = *meso*-tetraphenylchlorin) and the major product, CuTPP. As noted above, this complex separates well only on *dry* columns and is the secret of successful chromatography of these materials. The purified

*The checker recommends the use of a sintered-glass funnel.

CuTPP moves rapidly off the column as a broad smear and can be collected continuously until the sharp complex band is about $1\frac{1}{2}$–2 in. from the bottom. Inorganic and other impurities remain near the top. The resulting eluate is allowed to evaporate slowly at room temperature in a hood, and the crystalline CuTPP is collected.

Final yield is about 1.9 g (95% based on initial $TPPH_2$). *Anal.* Calcd. for $C_{44}H_{28}N_4Cu$: C, 78.14; H, 4.17; N, 8.29. Found: C, 78.0; H, 4.20; N, 8.27. Purity can also be confirmed by various spectrophotometric methods[11–14] as the molar extinction coefficient in benzene is 20,600 at 538 nm. If the material is required to be absolutely solvent-free, it must be further purified by vacuum sublimation.[7,9] General problems in the analysis of these materials, mainly involving adsorbed solvents, are discussed elsewhere.[10]

Properties

The metalloporphyrin CuTPP is soluble in halogenated hydrocarbons and ligating solvents such as pyridine. It is less soluble in benzene, hydrocarbons, and ethers. It is insoluble in water and only slightly soluble in alcohols. Although it is stable in the solid form, dilute solutions in strong light can undergo photochemical decomposition.[2] The crystal structure[15] and several other spectroscopic properties have been previously reported.[14,16] The characteristic absorption spectrum in benzene has the following molar extinction coefficients in the visible region: 2240 (575 nm), 20,600 (538 nm), and 3790 (505 nm).[11] The material is an organic semiconductor.[17]

B. HEMATOPORPHYRIN IX DIMETHYLESTERCHLOROIRON(III)

[*7,12-Di(1-hydroxyethyl)-3,8,13,17-tetramethyl-2,18-dipropionate-porphyrinatochloroiron(III) dimethyl ester*]

$$FeCl_2 + C_{36}H_{42}N_4O_6 \longrightarrow FeC_{36}H_{40}N_4O_6 + 2HCl$$

or

$$Fe^{2+} + HPDMH_2 \longrightarrow [Fe(II)HPDM] + 2H^+$$

and

$$2[Fe(II)HPDM] + \tfrac{1}{2}O_2 + 2HCl \longrightarrow 2[Fe(III)(HPDM)Cl] + H_2O$$

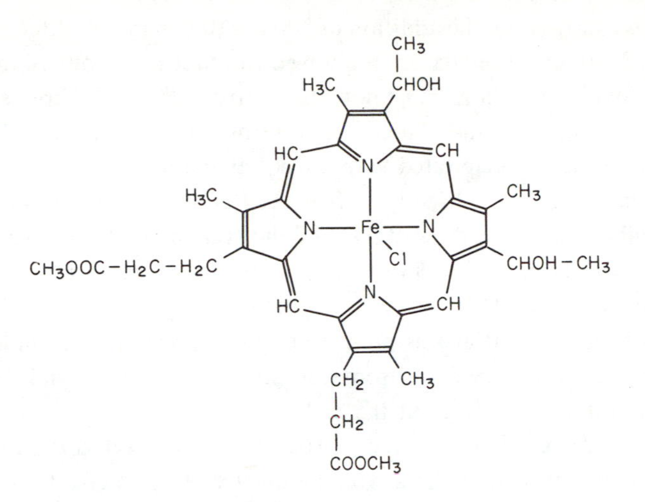

Procedure

The porphyrin may be prepared[10] or purchased from several of the biochemical suppliers (Sigma, Calbiochem, etc.). The sample employed was obtained as the hematoporphyrin IX dihydrochloride from L. Light and Co., converted to the dimethyl ester,* and purified and characterized by literature methods.[10,18,19] The crude diester can also be successfully chromatographed as the free base by the dry-column method described in Sec. A.

It should be noted here that there is a very wide latitude in the quality of the commercially available materials. Recently reported experiences with ferric hematoporphyrin derivatives[20,21] are quite typical. In general, esters are better than the unesterified bases. For the blood-derived materials the usual order of quality goes deutero, hemato, and meso, with protoporphyrin the worst. For a well-defined preparation of natural porphyrins, it is always best to purify the starting material and to characterize it[10] satisfactorily before carrying out any chemical modifications.

The metallation generally follows the previously given procedure.[7] On a hot plate, 100 ml of DMF in a 250-ml flask is brought to gentle reflux and

*The checker esterified hematoporphyrin by refluxing it in methanol with trifluoroacetic acid for 24 hr. The crude ester was then chromatographed on dry alumina and eluted with 1:1 $CHCl_3$-methanol.

magnetically stirred. $HPDMH_2$ (1.0 g, 0.0016 mole) is added and dissolved; this step is followed by the addition of 800 mg (0.004 mole) of ferrous chloride ($FeCl_2 \cdot 4H_2O$), and the mixture is allowed to react for about 10 min. Ferrous salts with oxidizing anions must not be substituted here. Complete conversion is checked as in Sec. A and, if incomplete, completed as in Sec. A. The reaction mixture is cooled, precipitated with an equal volume of distilled water, recooled, filtered, and air-dried as in Sec. A. Recovery of the crude ferric porphyrin will again be in excess of 99% of the starting material except that the product at this point will be a mixture of the hemin, i.e., ferric chloride, form and the μ-oxo form.[10,22]

It is most convenient at this point to convert everything completely to the μ-oxo dimer, which species chromatographs most easily, and then convert this back to the hemin form at the end.*

The crude material is dissolved in 100 ml of $CHCl_3$ and then shaken in a separatory funnel with an equal volume of *N* NaOH to effect the μ-oxo conversion. The chloroform layer is separated, washed once with distilled water, separated again, and then chromatographed directly. The chloroform solution is poured carefully on the top of a dry-packed 1 ft by 2-in.-diam. alumina column and eluted with reagent-grade $CHCl_3$, as in Sec. A. Any unconverted free base present comes off with the solvent front, which is followed shortly by the greenish-brown μ-oxo smear. If any unconverted hemin form is present, it comes down as a very broad, dark overlapping tail on the μ-oxo band. Material can be collected until the sharp, dark-green band containing chlorin or bile pigment, which advances slowly, reaches $1\frac{1}{2}$–2 in. from the bottom. Inorganic impurities, etc., remain on the top. The pooled ferric porphyrin (μ-oxo form plus hemin form) is now completely converted to the ferric chloride form by bubbling for 1 min with anhydrous gaseous HCl and then flushing for 5 min with dry N_2. The solution is allowed to evaporate slowly at room temperature in the hood, and the dried product, Fe(III)(HPDM)Cl, is collected. The final yield is about 1.10 g (95% based on initial $HPDMH_2$). *Anal.* Calcd. for $C_{36}H_{40}N_4O_6FeCl$: C, 60.38; H, 5.63; N, 7.82; Fe, 7.80. Found: C, 60.3; H, 5.75; N, 7.78; Fe, 7.76. Purity can also be confirmed by spectrophotometric and chromatographic methods.[20,21] The pyridine hemochromagen (reduced form in pyridine) has its major absorption bands at 547, 517, and 408 nm.[21]

The ferrous form goes spontaneously to the ferric under the reaction condi-

*The checker notes that the technique given here for purification of the Fe(HPDM)Cl is unsatisfactory for the purification of Fe(TPP)Cl, since free TPP and the μ-oxo dimer do not separate sharply. If, however, the crude reaction mixture is followed on the column by the use of a long-wave ultraviolet source under darkened conditions, the free-base TPP front can be sharply separated from the trailing μ-oxo smear and thus satisfactorily separated.

tions and this is the method of choice, since the ferric ion does not readily chelate directly. Ferrous forms can be prepared by running the reaction under N_2. However, they are readily generated in solution with various reducing agents such as ascorbate, dithionite, etc. This same method can be applied to the synthesis of the unesterified forms of these porphyrins. However, no simple chromatographic purification method has been developed for these forms at this scale and they must be purified by recrystallization[10] from pyridine-chloroform, etc.

Properties

This esterified hemin is soluble in ligating solvents such as pyridine and also halogenated hydrocarbons, ethers such as tetrahydrofuran, and benzene. It is slightly soluble in alcohols and insoluble in water. In the presence of traces of water in a solution it will slowly convert to the μ-oxo form. It can also dehydrate to the proto forms[21] under specific conditions. Other chemical, physical, and biological properties are similar to those for the protoporphyrin IX complex.[10] Extinction coefficients for the various characteristic spectra of this material are not well defined, as the solutions are generally unstable with time, going to mixtures of the various ligated and μ-oxo forms.

References

1. A. D. Adler (ed.), The Chemical and Physical Behavior of Porphyrin Compounds and Related Structures, *Ann. N.Y. Acad. Sci.,* **206** (1973).
2. D. Dolphin and R. H. Felton, *Acc. Chem. Res.,* **7,** 26 (1974).
3. D. Ostfeld and M. Tsutsui, *Acc. Chem. Res.,* **7,** 52 (1974).
4. J. W. Buchler, L. Puppe, K. Rohbock, and H. H. Schneehage, *Ann. N.Y. Acad. Sci.,* **206,** 116 (1973).
5. F. R. Longo, E. M. Brown, D. J. Quimby, A. D. Adler, and M. Meot-Ner, *Ann. N.Y. Acad. Sci.,* **206,** 420 (1973).
6. P. Hambright, *Coord. Chem. Rev.,* **6,** 247 (1971).
7. A. D. Adler, F. R. Longo, F. Kampas, and J. Kim, *J. Inorg. Nucl. Chem.,* **32,** 2443 (1970).
8. B. Loev and M. M. Goodman, *Progr. Sep. Purif.,* **3,** 73 (1970).
9. A. D. Adler, F. R. Longo, J. D. Finarelli, J. Goldmacher, J. Assour, and L. Korsakoff, *J. Org. Chem.,* **32,** 476 (1967).
10. W. S. Caughey, A. Adler, B. F. Burnham, D. Dolphin, S. F. MacDonald, D. C. Mauzerall, and Z. J. Petryka, Porphyrins and Related Compounds, pp. 185ff., in "Specifications and Criteria for Biochemical Compounds," 3d ed., National Academy of Sciences, Washington, D.C., 1972.
11. J. A. Mullins, A. D. Adler, and R. M. Hochstrasser, *J. Chem. Phys.,* **43,** 2548 (1965).
12. G. M. Badger, R. Alan Jones, and R. L. Laslett, *Aust. J. Chem.,* **17,** 1028 (1964).
13. G. D. Dorough and F. M. Huennekens, *J. Am. Chem. Soc.,* **74,** 3974 (1952).

14. G. D. Dorough, J. R. Muller, and F. M. Huennekens, *J. Am. Chem. Soc.*, **73**, 4315 (1951).
15. E. B. Fleischer, *Acc. Chem. Res.*, **3**, 105 (1970).
16. A. D. Adler, J. H. Green, and M. Mautner, *Org. Mass Spectrosc.*, **3**, 995 (1970).
17. A. D. Adler, *J. Polymer Sci. (C)*, **29**, 73 (1970).
18. W. S. Caughey, J. C. Alben, W. Y. Fujimoto, and J. L. York, *J. Org. Chem.*, **31**, 2631 (1966).
19. J. E. Falk, "Porphyrins and Metalloporphyrins," pp. 124ff., Elsevier Publishing Company, Amsterdam, 1964.
20. D. W. Lamson and T. Yonetani, *Anal. Biochem.*, **52**, 647 (1973).
21. D. W. Lamson, A. F. W. Coulson, and T. Yonetani, *Anal. Chem.*, **45**, 2273 (1973).
22. N. Sadasivan, H. J. Eberspaecher, W. H. Fuchsman, and W. S. Caughey, *Biochemistry*, **8**, 534 (1969).

58. (1,4,8,11-TETRAAZACYCLOTETRADECANE)NICKEL(II) PERCHLORATE AND 1,4,8,11-TETRAAZACYCLOTETRADECANE

Submitted* by E. K. BAREFIELD,† F. WAGNER,† A. W. HERLINGER,‡ and A. R. DAHL§
Checked by S. HOLT¶

A large number of complexes of tetraaza macrocyclic ligands is known. These ligands have from 12 to 16 atoms in the ring, and the most common structure is a 14-membered ring. The simplest of the ligands is 1,4,8,11-tetraazacyclotetradecane:

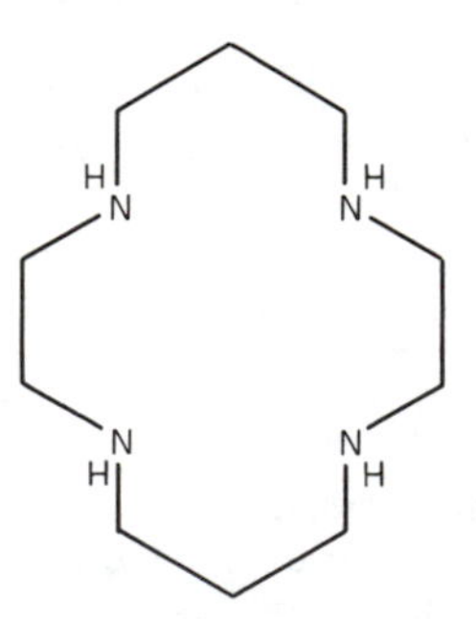

*Financial support by the Research Corporation (Grants 2753–63 and 6703–AK), Petroleum Research Fund (Grant 2013–G3), and by the National Science Foundation (Grant GP 33530) are gratefully acknowledged.

†Department of Chemistry, University of Illinois, Urbana, Ill. 61801.
‡Department of Chemistry, Loyola University, Chicago, Ill. 60626.
§Department of Chemistry, Northwestern University, Evanston, Ill. 60201.
¶Department of Chemistry, University of Wyoming, Laramie, Wyo. 82070.

A variety of metal complexes of this ligand have been prepared, including the metals Co,[1] Ni,[2] and Rh.[3] Coordination is generally in a planar fashion, although cis coordination to Co(III)[4] has been reported. These complexes were first prepared by direct interaction of the ligand with the metal salt. Unfortunately, this ligand, unlike others of its class,[5] has previously been available only through laborious organic syntheses.[1,6] Recently an *in situ* preparation of the nickel(II) complex was reported.[7] The ligand can easily be removed from the nickel(II) and used in the preparation of other metal complexes.

Given here is an improved version of this *in situ* synthesis which provides high yields of the nickel(II) perchlorate complex. Procedures for removing the ligand from the metal and for its purification are also given.

A. (1,4,8,11-TETRAAZACYCLOTETRADECANE)NICKEL(II) PERCHLORATE

([*Ni*(*cyclam*)][ClO_4]$_2$)

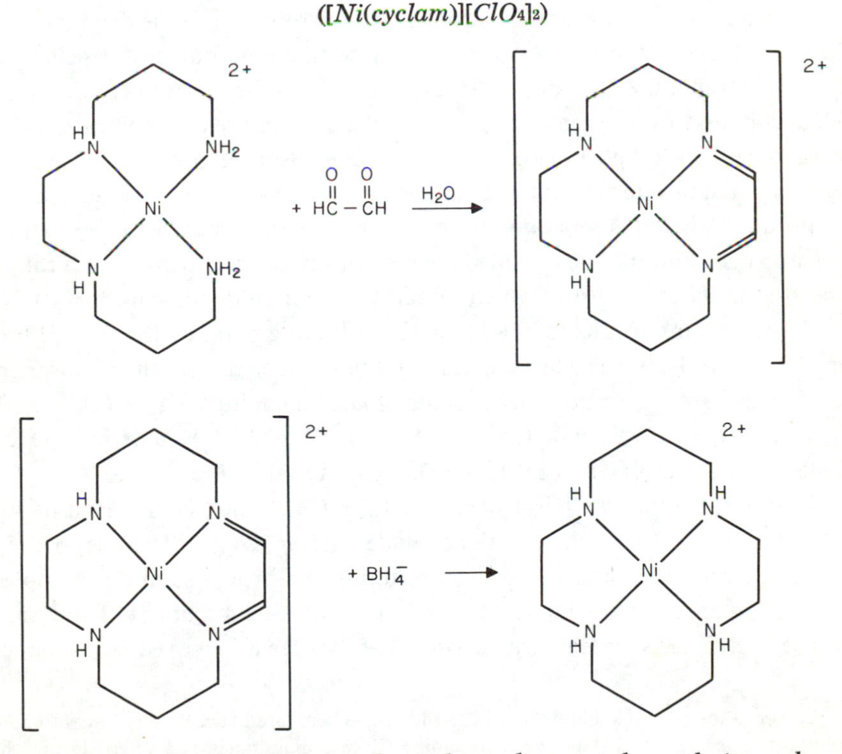

■ **Caution.** *Perchlorate salts of metal complexes can be explosive and must be handled with care. It is very important that such compounds not be heated as solids.*

Procedure

The linear tetramine 1,5,8,12-tetraazadodecane required as starting material is prepared by the following procedure:

$$2H_2N(CH_2)_3NH_2 + BrCH_2CH_2Br \longrightarrow H_2N(CH)_3\overset{+}{N}H_2CH_2CH_2\overset{+}{N}H_2\text{-}$$
$$(CH_2)_3NH_2 + 2Br^- \xrightarrow{KOH} H_2N(CH_2)_3NHCH_2CH_2NH\text{-}$$
$$(CH_2)_3NH_2 + 2H_2O + 2KBr$$

1,3-Diaminopropane, 890 g (12 moles), is placed in a 3-l. three-necked flask equipped with a mechanical stirrer, addition funnel, and thermometer. The diamine is cooled to ice temperature, and 188 g (1 mole) of 1,2-dibromoethane is added dropwise with vigorous stirring.* After the addition is complete (ca. 1 hr), the reaction mixture is heated on a steam bath for 1 hr and then concentrated to one-third its original volume by removal of excess 1,3-diaminopropane on a rotary evaporator. The concentrate is returned to the original reaction vessel, and 150 g of potassium hydroxide (flake or pulverized pellets) is added. This mixture is heated with efficient stirring for 2 hr on a steam bath. After cooling to room temperature, the solids are removed by filtration and washed with several portions of ether to remove adsorbed product.† The ether washings and filtrate are combined and evaporated to a viscous oil on a rotary evaporator. The oil is separated from any solid which forms by ether extraction and decantation. The decanted ether solution is again stripped on a rotary evaporator. The viscous oil which remains is vacuum-distilled using a 10–20-cm Vigreux column. Yield: 100–110 g with a boiling point of 138–148°/2 torr or 115–125°/0.1 torr. The tetramine should be protected from the atmosphere as it forms a solid hydrate with moist air. *Anal.* Calcd. for $C_8H_{22}N_4$: C, 55.10; H, 12.73; N, 32.16. Found: C, 55.0; H, 12.59; N, 32.27. H nmr in D_2O vs. TMS: τ7.43; complex multiplet, $I = 3$, τ8.43, quintet, $I = 1$.

Nickel(II) perchlorate hexahydrate,‡ 54.7 g (0.15 mole), is dissolved in 400 ml of water in a 2-1. beaker. With adequate stirring, 26 g (0.15 mole) of 1,5,-8,12-tetraazadodecane is added. The resulting red-brown solution is cooled to ca. 5° in an ice bath, and 30 ml of 30% (an equivalent amount of 40% may be used) glyoxal is added with stirring. The beaker is removed from the ice

*Lower ratios of amine halide to alkyl halide have been used successfully but with concomitant lower yields of the desired tetramine. Excess diamine is recovered during the work-up procedure and may be used for subsequent preparations.

†Large losses may occur if care is not taken to separate all liquid from the solid residue.

‡This can be prepared easily and inexpensively in solution from $NiCO_3$ and $HClO_4$.

bath and allowed to stand at room temperature for 4 hr.* The solution is again cooled to 5° and treated with 11 g (0.3 mole) of sodium tetrahydroborate in small portions over a 1-hr period.† As the addition proceeds, orange crystals of the product as well as some metallic nickel may form. After complete addition of the tetrahydroborate the solution is removed from the ice bath and heated to 90° on a hot plate or steam bath and held there for 15–20 min. The hot solution is filtered rapidly through a large medium frit. The filtrate is acidified with concentrated $HClO_4$ and allowed to cool. The orange crystalline product which forms is collected by suction and washed well with ethanol. It is finally washed with ether and allowed to air-dry. A second crop of crystals may be obtained by evaporating the filtrate. The yield is 40–45 g (occasionally as high as 50 g), or 65%. The product can be purified, if necessary, by recrystallization from water. *Anal.* Calcd. for $NiC_{10}H_{24}N_4Cl_2O_8$:‡ Ni, 12.82; C, 26.21; H, 5.28; N, 12.24. Found: Ni, 12.8; C, 26.5; H, 5.24; N, 12.2.

Properties

The salt $[Ni(cyclam)](ClO_4)_2$ is diamagnetic and exhibits a single absorption at 450 nm in its electronic absorption spectrum. The infrared spectrum of the perchlorate salt has a single strong N—H stretching absorption at 3220 cm^{-1} (Nujol mull). It is somewhat soluble in most polar solvents. Paramagnetic, six-coordinate, tetragonal complexes can be obtained by treating the perchlorate salt with sodium halides in methanol. A dark-green nickel(III) complex, $[Ni(cyclam)(NCCH_3)_2]^{3+}$, can be prepared by oxidation of the nickel(II) complex with $NOClO_4$ in CH_3CN.[8]

B. 1,4,8,11-TETRAAZACYCLOTETRADECANE

(*Cyclam*)

Procedure

Cyclam is often desired as a ligand for other metal ions. The free base may be obtained in good yields by the following procedure without isolating the nickel complex.

*Reaction times as long as 15 hr have no effect on the yield.

†This addition must be made slowly to avoid severe frothing of the solution.

‡Note that this is a perchlorate salt and as such presents a possible hazard. In 15–20 preparations, however, no problems have been encountered.

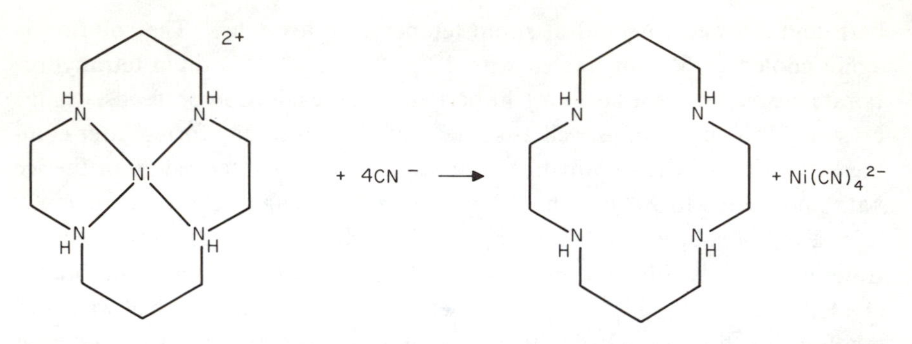

Transfer the solution obtained in Sec. A after reduction with sodium tetrahydroborate (but before addition of perchloric acid) to a 1-l. single-necked flask with 29 g (0.6 mole) of sodium cyanide. Fit the flask with a condenser and reflux the solution for 2 hr. Cool the solution to room temperature, add 15 g of sodium hydroxide, and evaporate on a rotary evaporator until a semisolid remains. Chloroform, 100 ml, is added to the flask, and the mixture is transferred to a large frit. The liquid is drawn off, and the solid is washed twice with 100-ml portions of chloroform. The aqueous layer is then separated from the filtrate and washed five to seven times with 50-ml portions of chloroform. All the chloroform extracts are dried over sodium sulfate and then evaporated to dryness. The yellowish solid is recrystallized from 800 ml of chlorobenzene to yield white needles which are collected by suction and washed with 50 ml of ethyl ether. The product is then air-dried. A second crop obtained by evaporating the filtrate is treated similarly. Yield: 19 g or 63%. *Anal.* Calcd. for $C_{10}H_{24}N_4$: C, 59.99; H, 12.09; N, 28.01. Found: C, 60.1; H, 11.8; N, 28.2. Alternatively, the free base may be recovered from the isolated nickel complex by the above procedure.

Properties

Cyclam is obtained as a white fibrous solid after recrystallization from chlorobenzene and melts at 185–186°. The free base is very soluble in ethanol and chloroform but only slightly soluble in water. The infrared spectrum of the base contains strong N—H stretching absorptions at 3280 and 3200 cm^{-1} (Nujol mull). pK_a values (at 20°) for the protonated base are 10.76, 10.18, 3.54, and 2.67.[9]

References

1. B. Bosnich, C. K. Poon, and M. L. Tobe, *Inorg. Chem.*, **4**, 1102 (1965).
2. B. Bosnich, M. L. Tobe, and G. A. Webb, *Inorg. Chem.*, **4**, 1109 (1965).

3. E. J. Bounsall and S. R. Koprich, *Can. J. Chem.*, **48,** 1481 (1970).
4. C. K. Poon and M. L. Tobe, *J. Chem. Soc.*, **1968,** 1549.
5. L. F. Lindoy and D. H. Busch, *Prep. Inorg. React.*, **6,** 1 (1971).
6. H. Stetter and K. H. Mayer, *Chem. Ber.*, **94,** 1410 (1961).
7. E. K. Barefield, *Inorg. Chem.*, **11,** 2273 (1972).
8. E. K. Barefield and D. H. Busch, *Chem. Commun.*, **1970,** 522.
9. K. H. Mayer, dissertation, München, 1960, cited by D. K. Cabbiness, Ph.D. dissertation, Purdue University, Lafayette, Ind., 1970.

59. *N,N′*-ETHYLENEBIS(MONOTHIOACETYLACETONIMINATO)COBALT(II) AND RELATED METAL COMPLEXES

Submitted by P. R. BLUM,* R. M. C. WEI,* and S. C. CUMMINGS*
Checked by M. DAS† and S. E. LIVINGSTONE†

N,N′-Ethylenebis(monothioacetylacetoniminato)cobalt(II), Co[(sacac)$_2$en] (I),

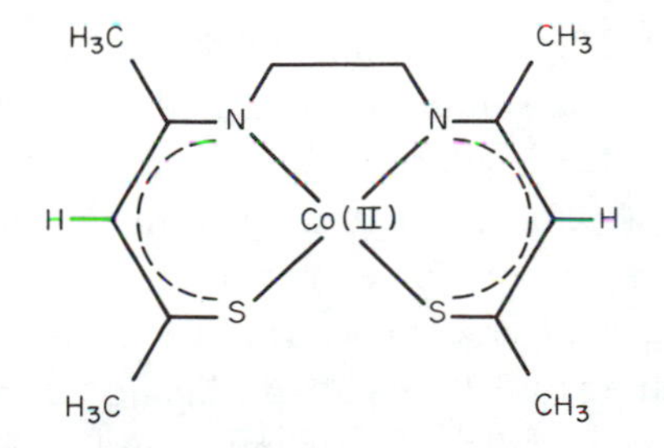

is of considerable interest as a model oxygen carrier since it is the first thioiminato-metal complex which has been shown to combine reversibly with molecular oxygen[1] and because it is the sulfur analog of *N,N′*-ethylenebis(acetylacetoniminato)cobalt(II), an oxygen carrier which has been extensively studied in order to determine those factors which favor reversible oxygenation.[2–4] Until now, synthesis of the thioiminato-metal complexes was impossible, as there was no method available for obtaining the ligands. The simple and convenient reaction scheme presented here for the synthesis of *N,N′*-ethylenebis(monothioacetylacetonimine), H$_2$[(sacac)$_2$en], provides a general

*Chemistry Department, Wright State University, Dayton, Ohio 45431.
†School of Chemistry, University of New South Wales, Kensington, New South Wales 2033, Australia.

route to multidentate thioimine ligands and their metal complexes. The procedure involves a nucleophilic substitution by hydrogen sulfide ion and is an extension of the method reported by Gerlach and Holm.[5]

A. *N,N′*-ETHYLENEBIS(MONOTHIOACETYLACETONIMINE)[6]

$$H_2[(acac)_2en] + 2Et_3OBF_4 \xrightarrow[N_2]{CH_2Cl_2} \text{alkylated ketoamine salt}$$

$$\xrightarrow[EtOH]{NaHS} H_2[(sacac)_2en] + 2NaBF_4 + 2EtOH$$

Procedure

A solution of 11.20 g (0.05 mole) of *N,N′*-ethylenebis(acetylacetonimine)* in 100 ml of dichloromethane is placed in a 500-ml, three-necked, round-bottomed flask fitted with a dropping funnel, condenser, and nitrogen-inlet tube. A solution of 19.00 g (0.10 mole) of triethyloxonium tetrafluoroborate† in 50 ml of dichloromethane is added to the above over a period of 15 min with constant stirring under a stream of dry nitrogen. After the addition is complete, the mixture is stirred at room temperature for an additional 30 min. At the end of this period, a fine suspension of 5.60 g (0.10 mole) of sodium hydrogen sulfide‡ in 100 ml of absolute ethanol is added. The $NaBF_4$ is removed by filtration, and the filtrate is evaporated to dryness in the filter flask under vacuum with an air stream. Concentration of the filtrate via other methods, e.g., with a rotary evaporator, sometimes produces an oil instead of fibrous crystals and therefore is not recommended. The crude ligand is recrystallized from methanol and dried over P_4O_{10}. As the ligand is heat-sensitive, all operations should be carried out at room temperature or lower. Yield of the purified ligand, which is isolated as fibrous yellow needles, ranged from 50–60%. *Anal.* Calcd. for $C_{12}H_{20}N_2S_2$: C, 56.20; H, 7.86; N, 10.92; S, 25.00; mol. wt., 256.5. Found: C. 56.3; H, 7.4; N, 11.0; S, 25.2; mol. wt. (chloroform), 256.

*$H_2[(acac)_2en]$ is prepared by the method of Martell, Belford, and Calvin, *J. Inorg. Nucl. Chem.*, **5,** 170 (1958), by condensation of ethylenediamine with two equivalents of acetylacetone in absolute ethanol.

†Et_3OBF_4 should be freshly prepared by the method of H. Meerwein, *Org. Synth.*, **46,** 113 (1966).

‡Technical grade $NaHS \cdot H_2O$ purchased from Matheson, Coleman and Bell was ground to a powder and dried *in vacuo* for 24–36 hr before use. Anhydrous NaHS is necessary, as the presence of water in the reaction greatly reduces the yield of ligand. The checkers suggest that $NaHS \cdot H_2O$ can also be prepared [R. E. Eibeck, *Inorganic Syntheses,* **7,** 128 (1963)].

B. *N,N'*-ETHYLENEBIS(MONOTHIOACETYLACETONIMINATO)-COBALT(II)[6]

$$H_2[(sacac)_2en] + Co(OAc)_2 \cdot 4H_2O \xrightarrow[N_2]{MeOH} Co[(sacac)_2en]$$

Procedure

A solution of $Co(OAc)_2 \cdot 4H_2O$ (2.50 g, 0.10 mole) in 100 ml of deaerated methanol is prepared in a 250-ml, three-necked, round-bottomed flask fitted with a condenser and nitrogen-inlet tube. The solution is stirred under dry nitrogen while 2.56 g (0.10 mole) of $H_2[(sacac)_2en]$ is added through the third neck, which is then stoppered. The reaction mixture is heated and stirred for 30 min, during which time the solution turns deep red and a bright-red crystalline compound forms. The flask containing the mixture is stoppered and allowed to cool overnight under nitrogen in the refrigerator. The crude product is isolated by filtration under nitrogen, recrystallized in a glove bag from deaerated acetone, and then dried *in vacuo* over P_4O_{10}. Yield of the purified complex is 55%. *Anal.* Calcd. for $CoC_{12}H_{18}N_2S_2$: C, 46.00; H, 5.79; N, 8.94; S, 20.47; Co, 18.81; mol. wt., 313.3. Found: C, 46.1; H, 5.3; N, 9.1; S, 20.5; Co, 19.0; mol. wt. (chloroform), 321.

Properties

N,N'-Ethylenebis(monothioacetylacetonimine) forms as yellow needles (m.p. 140–141°) which are soluble in polar organic solvents. Methanol solutions of the ligand exhibit an intense $\pi \rightarrow \pi^*$ electronic transition at ca. 385 nm (ε = ca. 38,000). Major bands in the infrared spectrum of the ligand occur at 1585, 1525, 1275, 1117, and 719 cm^{-1}. Resonances in the ^1H nmr spectrum ($CHCl_3$) have been assigned as follows: τ7.94, singlet, —CH_3 protons adjacent to N; τ7.53, singlet, —CH_3 protons adjacent to S; τ3.93, singlet, =CH protons; τ4.16, singlet, —SH protons; τ6.30, multiplet, —CH_2CH_2 protons.

The compound $Co[(sacac)_2en]$ is square-planar with a magnetic moment of 2.19 B.M. The red crystals dissolve readily in organic solvents such as dichloromethane, dichloroethane, tetrachloroethylene, chloroform, carbon tetrachloride, ether, acetone, and the lower-molecular-weight alcohols, but they are insoluble in water. The visible spectrum in $C_2H_4Cl_2$ exhibits a weak absorption at 980 nm (ε = ca. 38) with more intense bands at 529, 481, 413, 346, 333, 300, and 255 nm (ε = ca. 10^3–10^4). In addition, a lower energy band typical of square-planar cobalt(II) complexes occurs at 1870 nm in tetrachloroethylene solutions. Dried samples of Co[$(sacac)_2$en] are stable in air for several months, but the complex is air-oxidized in solution at room temperature within a few hours. In solvents, such as toluene at very low temperatures, the complex will bind reversibly with molecular oxygen in the presence of Lewis bases to form 1:1 adducts of the type Co[$(sacac)_2$en]·base·O_2.[1]

Other Compounds

The procedures reported here are readily adapted to the synthesis of other thioimine ligands and other metal complexes.[6–8] Four structurally similar ligands, *N,N'*-propylenebis(monothioacetylacetonimine), H_2[$(sacac)_2$pn], *cis*- and *trans*-*N,N'*-cyclohexylbis(monothioacetylacetonimine), *cis*- and *trans*-H_2[$(sacac)_2$chxn], *N,N'*-trimethylenebis(monothioacetylacetonimine), H_2-[$(sacac)_2$tn], and *N,N'*-ethylenebis(monothiobenzoylacetonimine), H_2[$(bzsac)_2$-en], have been prepared from the corresponding bis(keto) amines. Cobalt(II), nickel(II), copper(II), palladium(II), zinc(II), and cadmium(II) complexes have been synthesized from almost all these ligands. Only the cobalt(II) complexes require inert-atmosphere conditions for their isolation.

References

1. M. E. Koehler and S. C. Cummings, *Abstr. 165th Natl. Meet. Am. Chem. Soc., Dallas, Tex., April 1973;* M. E. Koehler, M.S. thesis, Wright State University, Dayton, Ohio, 1973.
2. A. L. Crumbliss and F. Basolo, *Science,* **164,** 1168 (1969); *J. Am. Chem. Soc.,* **92,** 55 (1970).
3. B. M. Hoffman, D. L. Diemente, and F. Basolo, *J. Am. Chem. Soc.,* **92,** 61 (1970).
4. H. C. Stynes and J. A. Ibers, *J. Am. Chem. Soc.,* **94,** 1559 (1972).
5. D. H. Gerlach and R. H. Holm, *J. Am. Chem. Soc.,* **91,** 3457 (1969).
6. R. M. C. Wei and S. C. Cummings, *Inorg. Nucl. Chem. Lett.,* **9,** 43 (1973).
7. P. R. Blum, R. M. C. Wei, and S. C. Cummings, *Inorg. Chem.,* **13,** 450 (1974).
8. D. R. Treter, M.S. thesis, Wright State University, Dayton, Ohio, 1973.

60. METALLATRANES*

$$nC_2H_5GeCl_3 + 1.5nH_2O \xrightarrow{NaOH} (C_2H_5GeO_{1.5})_n + 3nHCl$$

$$\frac{1}{n}(C_2H_5GeO_{1.5})_n + (HOCH_2CH_2)_3N \longrightarrow C_2H_5\overset{\downarrow}{Ge}(OCH_2CH_2)_3N + 1.5H_2O$$

Submitted by J. OCHS† and M. ZELDIN† (Secs. A and B) and M. ZELDIN† and R. GSELL† (Sec. C)
Checked by L. HILL‡ and C. D. SCHMULBACH‡ (Sec. A), J. BRANCA‡ and C. D. SCHMULBACH‡ (Sec. B), and R. SCOBELL,‡ J. L. DARIN,‡ and T. BALLANTINE‡ (Sec. C)

Transesterification reactions provide an applicable synthetic route to the preparation of organometallic compounds with unusual structures. Tricyclic organometallic compounds, which have the general formula $R—\overset{\downarrow}{M}(OCH_2CH_2)_3N$, can be prepared in high yield by the transesterification of an organometallic triester or organometalloxane polymer with triethanolamine.[1-3] Where M has an additional available coordination site, a monomeric compound is formed in which an intramolecular transannular dative bond between nitrogen and M is formed. A number of these compounds, particularly where M is a main group IV element, have been the subject of considerable study owing to their surprising physiological activity.[3,4] (■ **Caution.** *The metallatranes may be neurotoxins and should be handled with extreme care.*)

A. 1-ETHYL-2,8,9-TRIOXA-5-AZA-GERMATRICYCLO[3.3.3.0]-UNDECANE

Procedure

Trichloroethylgermane§ (2.25 g, 0.011 mole) is added to a 250-ml, two-necked, round-bottomed flask fitted with a 20-cm Vigreux column-condenser

*This class of compound has had the following nomenclature: organometallatrane; organo-*triptych*-metalloxazolidine; organo-(2,2′,2″-nitrilotriethoxy)-metallane; 1-organo-2,8,9-trioxa-5-aza-metallabicyclo[3.3.3.]undecane or more correctly 1-organo-2,8,9-trioxa-5-aza-metallatricyclo[3.3.3.0]undecane. The stannous derivatives prepared in Sec. C have been named as *N*-substituted bicyclic derivatives.

†Department of Chemistry, Polytechnic Institute of New York, Brooklyn, N.Y. 11201.
‡Department of Chemistry and Biochemistry, Southern Illinois University, Carbondale, Ill. 62901.

§Research Organic/Inorganic Corp., Sun Valley, Calif. 91352, used without further purification.

unit, dropping funnel, and magnetic stirring bar. A 25% aqueous NaOH solution is added dropwise with stirring until the resulting mixture is alkaline to litmus. Triethanolamine (3.06 g, 0.021 mole) (Fisher, vacuum-distilled twice before use) is added with 100 ml of toluene (Fisher). The mixture is refluxed with vigorous stirring, and the toluene-water azeotrope (b.p. 85°) is removed continually until the reflux temperature remains constant at 110°. The contents of the flask are filtered hot through a medium-porosity sintered-glass funnel. The clear filtrate is evaporated to dryness at 40° (1 torr) to give 2.0 g of crude white solid (74% yield based on starting trichloroethylgermane). The product is purified by two recrystallizations from heptane followed by sublimation at 35° (0.1 torr). *Anal.* Calcd. for $C_8H_{17}O_3NGe$: C, 38.77; H, 6.92; N, 5.65; Ge, 29.29; mol. wt., 247.6. Found: C, 38.8; H, 7.07; N, 5.87; Ge, 29.6; $m/e(^{74}Ge)$, 249; mol. wt. (dichloroethane), 241.3.*

Properties

1-Ethyl-2,8,9-trioxa-5-aza-germatricyclo[3.3.3.0]undecane is a white needle crystalline solid which melts at 144–145° after purification. The compound is soluble in polar solvents such as CH_2Cl_2 and $CHCl_3$, slightly soluble in aromatic solvents and insoluble in alkanes. The compound hydrolyzes slowly in water and reacts slowly with methyl iodide in acetonitrile to form the quaternary salt, $C_2H_5Ge(OCH_2CH_2)_3NCH_3{}^+I^-$. ^1H nmr (TMS, dichloroethane, 60 MHz): $\tau 6.30$, triplet, OCH_2; $\tau 7.22$, triplet, NCH_2; $\tau 8.7–9.3$ multiplet, C_2H_5Ge.

B. 1-ETHYL-2,8,9-TRIOXA-5-AZA-STANNATRICYCLO[3.3.3.0]-UNDECANE

$$C_4H_9Li + HN(C_2H_5)_2 \longrightarrow LiN(C_2H_5)_2 + C_4H_{10}$$

$$3LiN(C_2H_5)_2 + C_2H_5SnCl_3 \longrightarrow C_2H_5Sn[N(C_2H_5)_2]_3 + 3LiCl$$

$$C_2H_5Sn[N(C_2H_5)_2]_3 + 3C_2H_5OH \longrightarrow C_2H_5Sn(OC_2H_5)_3 + 3HN(C_2H_5)_2$$

$$C_2H_5Sn(OC_2H_5)_3 + (HOCH_2CH_2)_3N \longrightarrow C_2H_5\overset{\leftarrow}{Sn}(OCH_2CH_2)_3N + 3C_2H_5OH$$

Procedure

Into a dried 1-l. three-necked flask fitted with a mechanical stirrer and a

*The checkers found a molecular weight of 247 by vapor-pressure osmometry in $CHCl_3$.

125-ml dropping funnel is placed 318 ml of a 1.5*M* (0.48 mole) pentane solution of C_4H_9Li prepared from C_4H_9Cl and Li by established methods[5] [alternatively, C_4H_9Li (1.67*M*) in pentane solution, purchased from Matheson, Coleman and Bell, can be used]. Ethyl ether (100 ml, Mallinckrodt anhydrous, used without further purification or drying) is added. After cooling the flask to 0°, 35 g (0.48 mole) of $HN(C_2H_5)_2$ is added slowly with vigorous stirring. Stirring is continued for at least 15 min after addition is complete. To the solution of $LiN(C_2H_5)_2$ at 0° is added 40.7 g (0.16 mole) of $C_2H_5SnCl_3$ (prepared by the method of Neumann and Burkhardt)[6] in 50 ml of benzene (dried over and distilled from P_4O_{10}) dropwise with stirring. Stirring is continued for 30 min after addition. The LiCl precipitate is allowed to settle overnight, and the yellow supernatant liquid is decanted and filtered through a medium-porosity sintered-glass funnel. The clear filtrate is evaporated to dryness at reduced pressure to give $C_2H_5Sn[N(C_2H_5)_2]_3$, a yellow oil, which is purified by vacuum distillation (yield: 41 g, 70%; b.p. 90°/0.5 torr).[7] The pure compound is extremely sensitive to moisture and should be handled under dry nitrogen.

Then 19.5 g (0.054 mole) of $C_2H_5Sn[N(C_2H_5)_2]_3$ is placed in a 250-ml, two-necked, round-bottomed flask containing a magnetic stirring bar and fitted with a 125-ml dropping funnel, and 75 ml of pentane (technical grade, dried over and distilled from P_4O_{10}) is added. A solution of 14.6 g (0.318 mole) absolute ethanol [dried over $Mg(OC_2H_5)_2$] in 75 ml of pentane is added dropwise with stirring. When addition is complete, the pentane and excess ethanol are removed under reduced pressure at room temperature. The residual yellow oil, $C_2H_5Sn(OC_2H_5)_3$, is purified by vacuum distillation (yield: 9.03 g, 60%; b.p. 130°/0.1 torr).[7]

To a 500-ml round-bottomed flask containing a magnetic stirring bar and fitted with a condensing head and collection flask is added 5.90 g (0.021 mole) of $C_2H_5Sn(OC_2H_5)_3$ and 250 ml of benzene (Fisher, dried over and distilled from P_4O_{10}); then 3.09 g (0.021 mole) of triethanolamine (Mallinckrodt, vacuum-distilled twice before use) is added. The reaction mixture is refluxed at 110° with vigorous stirring. When evolution of ethanol in the distillate is complete, as shown by gas-liquid phase chromatography or by a negative iodoform test for ethanol, the reaction mixture is cooled to room temperature and filtered.

As solvent is removed under vacuum, the product precipitates from solution; 6.2 g (100%) of crude product is isolated by filtration and washed several times with small portions of anhydrous benzene. Purification may be effected by sublimation at 75° (10^{-4} torr) or by recrystallization from hot xylene. *Anal.* Calcd. for $C_8H_{17}O_3NSn$: C, 32.69; H, 5.83; N, 4.77; mol. wt. 293.9.

Found: C, 32.7; H, 5.80; N, 4.76; $m/e(^{120}Sn)$, 295; mol. wt. (dichloromethane), 296.4.

Properties

1-Ethyl-2,8,9-trioxa-5-aza-stannatricyclo[3.3.3.0]undecane is a white solid melting at 179.5–181°. It is soluble in CH_2Cl_2, slightly soluble in aromatic solvents and insoluble in alkanes. ^1H nmr (TMS, CH_2Cl_2, 60 MHz): $\tau 6.17$, broad, O—CH_2; $\tau 7.11$, broad, N—CH_2; $\tau 8.74$, singlet, C_2H_5Sn.

C. 2,8-DIOXA-5-AZA-1-STANNOBICYCLO[2.2.0]OCTANE

(Stannoatrane)

$$SnCl_2 + 2CH_3OH + 2(C_2H_5)_3N \longrightarrow Sn(OCH_3)_2 + 2(C_2H_5)_3NHCl$$

$$Sn(OCH_3)_2 + 2C_4H_9OH \longrightarrow Sn(OC_4H_9)_2 + 2CH_3OH$$

$$Sn(OC_4H_9)_2 + (HOCH_2CH_2)_2NR \longrightarrow \overset{\downarrow\!\!-\!\!-\!\!-\!\!-\!\!-\!\!-\!\!-\!\!|}{Sn(OCH_2CH_2)_2NR} + 2C_4H_9OH$$

The development of the chemistry of tin(II) compounds has been exceedingly slow. The neglect of this field has no doubt been due to the difficulty of preparing and analyzing classes of compounds that are extremely sensitive to moisture and oxygen. With the advent of better techniques for handling air-sensitive materials and the recognition that anhydrous $SnCl_2$ is soluble in a variety of nonaqueous solvents, new synthetic methods have been developed for the preparation of reactive tin(II) intermediates such as the stannous alkoxides.[8] Transesterification of the alkoxides with alcohols and diols permits the synthesis in high yields of a wide variety of new and interesting molecules with unusual bonding properties.[9] The title compound is a cyclic diester derivative of tin(II) in which an intramolecular transannular dative bond exists between nitrogen and tin.

Procedure

After 100 g of $SnCl_2 \cdot 2H_2O$ and 500 ml of acetic anhydride are combined in a 1-l. round-bottomed flask, the flask is stoppered and allowed to stand for several hours with occasional shaking. The solid anhydrous $SnCl_2$ is filtered in an oxygen- and moisture-free, nitrogen-filled glove bag and washed, first with several portions of fresh acetic anhydride, then with 250 ml of anhydrous diethyl ether. The $SnCl_2$ thus obtained is dried and stored under vacuum until further use. The yield is quantitative.

In an inert atmosphere, 50 g of anhydrous $SnCl_2$ is dissolved in about 1 l. of absolute methanol (dried by refluxing over magnesium metal and distilled before use). To the resulting clear solution is added enough dry triethylamine to produce a permanent precipitate. The precipitate is filtered and washed with 300 ml of dry methanol and then with 500 ml of anhydrous ether. The product, $Sn(OCH_3)_2$, is dried and stored under vacuum. Triethylamine hydrochloride can be recovered from the filtrate by addition of a precipitating solvent such as diethyl ether. The yield of tin(II) methoxide is near 100%.[8]

Tin(II) methoxide (11.2 g, 0.062 mole) in 700 ml of dry toluene is added to a 1-l., two-necked, round-bottomed flask containing an excess of dry 1-butanol (11.1 g, 0.15 mole) and equipped with a distilling column and water condenser head, a dropping funnel, and magnetic stirring bar. The entire preparation is carried out under a dry nitrogen atmosphere. The contents of the flask (slurry) are then refluxed with stirring until all the tin(II) methoxide has reacted, as is indicated by the disappearance of the methoxide slurry. The methanol-toluene (64°) and excess butanol-toluene (106°) azeotropes are then removed by distillation. To the resulting solution of tin(II) butoxide in boiling toluene is added dropwise a small excess of freshly distilled diethanolamine (7.4 g, 0.070 mole) in 30 ml dry *n*-butanol. As the diethanolamine is added, a white solid precipitates from solution. After all the diethanolamine is added, the contents of the flask are refluxed with stirring for a few minutes, and the 1-butanol–toluene and excess diethanolamine-toluene (ca. 108°) azeotropes are removed by distillation. The contents of the flask are then cooled and filtered through a medium-porosity sintered-glass funnel under a dry nitrogen atmosphere using Schlenk-tube techniques. The residual white solids are washed several times with small portions of dry chloroform and dried under vacuum. The yield is 12.4 g (90%). *Anal.* Calcd. for $C_4H_9NO_2Sn$: C, 21.66; H, 4.09; N, 6.31; Sn, 53.52. Found: C, 21.9; H, 4.32; N, 6.24; Sn, 54.0.

Properties

The 2,8-dioxa-5-aza-stannobicyclo[2.2.0]octane is a white air-sensitive solid that decomposes without melting above 300°. The material is generally insoluble in polar and nonpolar solvents. Although the product is analytically pure as obtained, it can be sublimed at 150–175° at 10^{-4} torr. Formulation of the product as $Sn(OCH_2CH_2)_2NH$ (with an N→Sn bond) is supported by elemental analysis and mass and infrared spectroscopy. The mass spectrum contains parent ions which show a tin cluster at 219, 220, 221, 222, 223, 225, 227 *m/e* with the predicted percentage isotopic composition. The infrared spectrum has the following characteristic absorptions: 3097 (m) cm^{-1} (N—H); 1500–1350 cm^{-1} (C—H region); 1090 (m), 1060 (s), 1020 (s) cm^{-1} (C—O region); and 612 (s), 555 (s), 508 (s), 464 (s) cm^{-1} (Sn—O region).

In an exactly analogous manner $Sn(OCH_2CH_2)_2NCH_3$ can be obtained from the reaction of tin(II) butoxide and *N*-methyldiethanolamine in dry toluene. However, since the product is slightly soluble in hot toluene, the volume of the final reaction solution must be reduced to about 100 ml to facilitate precipitation of the product. The material is then filtered and dried *in vacuo* to give 13.1 g (90%) of product.

The product is a white, crystalline, air-sensitive solid which can be purified by recrystallization from toluene or sublimation at 80° and 10^{-4} torr. The pure material has a melting point of 188–190°. *Anal.* Calcd. for $C_5H_{11}O_2NSn$; C, 25.47; H, 4.70; N, 5.94; Sn, 50.34. Found: C, 25.3; H, 4.79; N, 5.91; Sn, 50.0. The mass spectrum contains parent ions which show the expected tin cluster at 233, 234, 235, 236, 237, 239, and 241 *m/e*. The infrared spectrum has the following characteristic absorptions: 2950(s), 1455–1360(s) cm^{-1} (C—H region); 1090–1065 cm^{-1} (C—O region); 620–450 cm^{-1} (Sn—O region); ^1H nmr (TMS, dichloroethane, 60 MHz): τ6.05, triplet, OCH_2; τ7.34, triplet, NCH_2; τ7.55, singlet, NCH_3.

References

1. H. C. Brown and E. A. Fletcher, *J. Am. Chem. Soc.*, **73,** 2808 (1961).
2. C. L. Frye, G. E. Vogel, and T. A. Hall, *J. Am. Chem. Soc.*, **83,** 996 (1961).
3. M. Voronkov, *Pure Appl. Chem.*, **13,** 39 (1966).
4. M. G. Voronkov and E. Lukevics, *Russ. Chem. Rev.* (*Engl.*), **38**(12), 975 (1969).
5. L. F. Fieser and M. Fieser, "Reagents for Organic Synthesis," p. 95, John Wiley & Sons, Inc., New York, 1967.
6. W. Neumann and G. Burkhardt, *Ann.*, **663,** 11 (1963).
7. J. Loberth and M. Kula, *Chem. Ber.*, **97,** 3444 (1964).
8. J. Morrison and H. Haendler, *J. Inorg. Nucl. Chem.*, **29,** 393 (1967).
9. R. Gsell and M. Zeldin, *J. Inorg. Nucl. Chem.*, **37,** 1133 (1975).

61. *cis*-BIS(DIETHYLDITHIOCARBAMATO)-DINITROSYLMOLYBDENUM

$$Mo(CO)_6 + Br_2 \longrightarrow MoBr_2(CO)_4 + 2CO$$

$$MoBr_2(CO)_4 + 2NaS_2CN(C_2H_5)_2 \longrightarrow Mo(CO)_2[S_2CN(C_2H_5)_2]_2 + 2CO + 2NaBr$$

$$Mo(CO)_2[S_2CN(C_2H_5)_2]_2 + 2NO \longrightarrow Mo(NO)_2[S_2CN(C_2H_5)_2]_2 + 2CO$$

Submitted by J. A. BROOMHEAD,* J. BUDGE,* and W. GRUMLEY*
Checked by J. W. McDONALD†

Molybdenum complexes with sulfur donor atoms play an important role in the nitrogenase enzyme system, and the study of their chemistry is of continuing interest. The present synthesis concerns *cis*-dinitrosylbis(*N,N*-diethyldithiocarbamato)molybdenum, originally described by Johnson et al.[1] Their method of synthesis involves the conversion of molybdenum hexacarbonyl to the unstable $MoBr_2(NO)_2$ using NOBr, followed by reaction with the *N,N*-diethyldithiocarbamate $[S_2CN(C_2H_5)_2]$ ligand. Yields are 50–60% based on molybdenum hexacarbonyl. Both NOBr and $MoBr_2(NO)_2$ are unstable and must be freshly prepared. The method given here avoids these intermediates and makes use of carbonyl complexes instead. Molybdenum hexacarbonyl is treated with bromine followed by sodium *N,N*-diethyldithiocarbamate to yield $Mo(CO)_2[S_2CN(C_2H_5)_2]_2$. Subsequent reaction with NO affords *cis*-$Mo(NO)_2[S_2CN(C_2H_5)_2]_2$ in 75% yield.

Procedure

■ **Caution.** *Fume hood must be used.*

Molybdenum hexacarbonyl (5 g, 0.019 mole) is ground to a fine powder and placed in a 250-ml, two-necked, round-bottomed flask fitted with a gas inlet tube, dropping funnel, and magnetic stirrer. The reaction vessel is cooled to −10° in an acetone–Dry Ice bath. Dichloromethane (100 ml) is added, and a stream of dry nitrogen is maintained throughout the subsequent reaction procedure. Bromine (3 g, 0.019 mole) is added dropwise and gas evolution is evident. When the addition of bromine is complete, the cooling bath is removed,

*Chemistry Department, Faculty of Science, Australian National University, Canberra, A.C.T. 2600, Australia. The authors wish to thank Dr. B. Tomkins for helpful discussions and the Australian Research Grants Committee for supporting the work.
†Kettering Scientific Research Laboratory, Yellow Springs, Ohio 45387.

and the reaction mixture is allowed to evaporate to dryness. The effluent gas is tested with aqueous silver nitrate–nitric acid solution until there is complete removal of bromine. This takes about $\frac{3}{4}$ hr. Methanol (50 ml, deoxygenated) is then added followed 10 min later by a solution of sodium *N,N*-diethyldithiocarbamate (8.55 g, 0.038 mole) in methanol (50 ml). A mixture of the orange tricarbonyl complex and purple dicarbonyl complex precipitates.[3,*] After a further 5 min, dichloromethane (100 ml) is added to the stirred suspension to dissolve the molybdenum-containing products (NaBr formed does not dissolve), and nitrogen oxide is passed through the solution for 30 min. The color changes to a deep brown during this time.[†] At this stage and during subsequent purification the reaction vessel is covered with aluminum foil to minimize exposure to light. The product complex is photosensitive, especially in solution, and gives the seven-coordinate tris(*N,N*-diethyldithiocarbamato)-nitrosylmolybdenum complex along with other unidentified products.[2] The dark-brown solution is next evaporated to dryness with a rotary evaporator, and the residue is extracted with chloroform (50 ml). Further purification is best effected by column chromatography. It is not necessary to exclude air, but photochemical reactions should be avoided by wrapping the column with dark paper or aluminum foil. A 50- by 3-cm column is prepared by suspending silica gel powder in 1:1 *n*-hexane-chloroform. One-half of the chloroform solution of the crude product is placed on the column and eluted with 1:1 *n*-hexane-chloroform (ca. 500 ml). The complex readily elutes as a brown band and is recovered by evaporation of the solvent. The remainder of the crude material is chromatographed similarly, and the combined products are recrystallized from hot methanol (200 ml). The yield is 6.5 g (0.014 mole). *Anal.* Calcd. for $C_{10}H_{20}N_4S_4O_2Mo$: C, 26.55; H, 4.66; N, 12.38; S, 28.36. Found: C, 26.52; H, 4.54; N, 12.29; S, 28.70.

Properties

cis-Bis(diethyldithiocarbamato)dinitrosylmolybdenum is a brown crystalline solid stable in air but undergoing slow photochemical decomposition in solution.[2] It is soluble in organic solvents such as acetone, chloroform, and

*The checker reports a 3.2-g yield of carbonyl complexes here based on 2.0 g of $Mo(CO)_6$. The mixture of carbonyl complexes can be filtered off and stored under nitrogen and is useful for the synthesis of other bis(dithiocarbamato)molybdenum compounds.

†Thin-layer chromatography using a drop of the reaction mixture spotted onto Bakerflex silica gel IB-F strips (20 by 10 cm) and eluted with dichloromethane (in the air) readily monitors the disappearance of the purple intermediate complexes. Brown *cis*-$Mo(NO)_2[S_2CN(C_2H_5)_2]_2$ moves with the solvent front while the purple complexes have R_F values of ca. 0.5.

methanol. Solutions are unstable with respect to atmospheric oxidation although the reaction is slow. The nmr and infrared spectra of the compound have been reported, and the cis geometry was assigned on the basis of these measurements.[1] For evaporated chloroform films and KBr pellets, the characteristic strong infrared bands are at 1755, 1642 (γ N≡O), and 1505 cm^{-1} (γ C=N). The mass spectrum gives a strong parent molecular ion and fragmentation consistent with loss of NO and breakdown of the diethyldithiocarbamate ligand.[1] The stoichiometry of the present preparation remains obscure.

References

1. B. F. G. Johnson, K. H. Al-Obaidi, and J. A. McCleverty, *J. Chem. Soc. (A)*, **1969,** 1668.
2. J. A. Broomhead and W. Grumley, unpublished work.
3. R. Colton, G. R. Scollary, and I. B. Tomkins, *Aust. J. Chem.*, **21,** 15 (1968).

62. SOME η-CYCLOPENTADIENYL COMPLEXES OF TITANIUM(III)

Submitted by C. R. LUCAS* and M. L. H. GREEN†
Checked by B. H. TAYLOR‡ and F. N. TEBBE‡

The bis-η-cyclopentadienyltitanium system has been studied[1] fairly extensively in comparison with other organotitanium species. Bearing in mind some of the challenging and useful aspects of the chemistry of the bis-η-cyclopentadienyltitanium system,[2-4] it is of interest to include related systems such as the mono-η-cyclopentadienyltitanium one in the expanding examination of organotitanium chemistry.

Particularly in the field of homogeneous catalysis, but in other fields as well,[5-7] some of the more interesting and unusual chemistry of titanium is associated with lower-valent species.

Preparations of $[(\eta\text{-}C_5H_5)TiCl_2]_n$ have been described,[8,9] but they are less satisfactory than the procedure given here. Relatively large quantities (>10 g) of η-cyclopentadienyltitanium dichloride can be readily obtained by this method, and from that compound a mononuclear phosphine complex, $(\eta\text{-}C_5H_5)$-

*Department of Chemistry, Memorial University of Newfoundland, St. John's, Newfoundland A1C 5S7, Canada.

†Inorganic Chemistry Laboratory, South Parks Road, Oxford, OX1 3QR, England.

‡Central Research Department, E. I. du Pont de Nemours & Co., Wilmington, Del. 19898.

$TiCl_2(PR_3)_2$, is also readily obtained. When standard Schlenk-tube techniques are used, both compounds can be prepared in 1 day or less in quantities suitable for starting materials for further research.

A. DICHLORO(η-CYCLOPENTADIENYL)TITANIUM(III) POLYMER

$$2n(\eta\text{-}C_5H_5)TiCl_3 + nZn \longrightarrow 2[(\eta\text{-}C_5H_5)TiCl_2]_n + nZnCl_2$$

Procedure

The $(\eta\text{-}C_5H_5)_2TiCl_2$ used in this preparation was obtained from Alfa Inorganics, whereas the $TiCl_4$ and zinc dust came from British Drug Houses. All reagents were used as supplied without further purification.

Orange $(\eta\text{-}C_5H_5)TiCl_3$* (6.0 g, 27 mmoles) and zinc dust (1.0 g, 15 mg-atom) are suspended at room temperature under an atmosphere of dry nitrogen in tetrahydrofuran (THF) (350 ml) (freshly distilled from calcium hydride and freed of dissolved oxygen) and stirred for 3 hr. The green solution is filtered, and its volume is reduced *in vacuo* on a hot-water bath to 80 ml. Upon cooling, green crystals of $(\eta\text{-}C_5H_5)TiCl_2(THF)_2$ form which are separated and heated *in vacuo* at 100° for 2 hr. The removal of tetrahydrofuran from the complex during this time is accompanied by a color change to dark purple. Yield is 3.5 g† (70% based on titanium). *Anal.* Calcd. for $C_5H_5Cl_2Ti$: C, 32.7; H, 2.7; Cl, 38.6; Ti, 26.1. Found: C, 32.6; H, 2.9; Cl, 38.2; Ti, 26.3.

Properties

The purple solid is pyrophoric in air but is stable indefinitely when stored under dry nitrogen at room temperature. Because it oxidizes so readily, the solid should always be washed before further use with either toluene or ether until the washings are colorless; this treatment removes small amounts of yellow oxidized material. The compound is soluble only in solvents with which it reacts to form a complex, for example, $(\eta\text{-}C_5H_5)TiCl_2(L)_2$ (L = THF or py).

*Orange $(\eta\text{-}C_5H_5)TiCl_3$ can be prepared by heating a mixture of $(\eta\text{-}C_5H_5)_2TiCl_2$ (11.4 g, 46 mmoles) and $TiCl_4$ (10 ml, 91 mmoles) which has been sealed *in vacuo* in a dry 50-ml glass tube. After 60 hr at 120°, the tube is cooled and opened under an inert atmosphere. The contents are removed, washed with hexane, and sublimed *in vacuo* at 150° into an air-cooled glass tube. The yield of orange sublimate is 11.8 g. Its infrared spectrum (Nujol) is 3110 (w), 1881 (w), 1802 (w), 1020 (m), 937 (w), 835 (s) cm^{-1}. The compound should be stored and handled under an inert atmosphere.

†The scale of the reaction may be increased threefold without difficulty.

It reacts with hydroxylic solvents to give colored solutions which slowly deposit Ti_2O_3.

The structure of $[(\eta\text{-}C_5H_5)TiCl_2]_n$ is unknown, but its insolubility suggests that it may be polymeric. The compound is paramagnetic, and its infrared spectrum (Nujol) is 3110 (m), 1830 (w), 1740 (w), 1650 (w), 1440 (s), 1016 (s), 822 (vs), 395 (s), 360 (s), 330 (vs), and 300 (vs) cm^{-1}.

B. DICHLORO(η-CYCLOPENTADIENYL)BIS-(DIMETHYLPHENYLPHOSPHINE)-TITANIUM(III)*

$$[(\eta\text{-}C_5H_5)TiCl_2]_n + 2n(CH_3)_2(C_6H_5)P \longrightarrow n(\eta\text{-}C_5H_5)TiCl_2[(CH_3)_2(C_6H_5)P]_2$$

*Procedure**

Purple $[(\eta\text{-}C_5H_5)TiCl_2]_n$ (1.1 g, 6.0 mmoles) is washed with dry toluene† under an atmosphere of dry nitrogen until the washings are colorless and then suspended in dry toluene† (100 ml) at room temperature. Dimethylphenylphosphine (1.6 ml, 12.0 mmoles) (obtained from Strem Chemicals and used without purification) is added under an atmosphere of dry nitrogen, and the mixture is stirred magnetically overnight. Volatile substances are then removed *in vacuo* at 50°, and the residue is washed four times with dry petroleum ether.† The green solid is recrystallized from dry ether.† Yield is typically about 85%. *Anal.* Calcd. for $C_{21}H_{27}Cl_2P_2Ti$: C, 54.8; H, 5.9; Cl, 15.4; P, 13.5; Ti, 10.4. Found: C, 54.6; H, 6.1; Cl, 15.8; P, 13.7; Ti, 10.8.

Properties

The compound is sensitive to air and should be stored or handled under an inert atmosphere. In solution, it is rather unstable thermally, and temperatures must be kept below ca. 50°. It is soluble in ethers and aromatic hydrocarbons and insoluble in aliphatic hydrocarbons. It reacts with hydroxylic solvents to give solutions which slowly deposit Ti_2O_3. The compound is paramagnetic, and its infrared spectrum (Nujol) is 3105 (w), 3070 (w), 1478 (m), 1435 (s), 1417 (m), 1278 (s), 1018 (m), 1009 (m), 999 (m), 945 (s), 904 (s), 868 (m), 804 (s), 750 (vs), 733 (s), 709 (m), 699 (s), 492 (s), and 350 (s) cm^{-1}.

*This procedure can also be followed using the ligands methyldiphenylphosphine, ethylenebis(dimethylphosphine) and ethylenebis(diphenylphosphine). Thermal stability and solubility of the products decrease with ligand size.[10]

†Freshly distilled from calcium hydride and free from dissolved oxygen.

References

1. R. S. P. Coutts and P. C. Wailes, *Adv. Organomet. Chem.*, **9,** 136 (1970).
2. J. J. Salzmann and P. Mosimann, *Helv. Chim. Acta,* **50,** 1831 (1967).
3. J. E. Bercaw, R. H. Marvich, L. G. Bell, and H. H. Brintzinger, *J. Am. Chem. Soc.*, **94,** 1219 (1972).
4. G. Henrici-Olive and S. Olive, *J. Polymer Sci.* (B), **8,** 271 (1970).
5. Yu. G. Borodko, I. N. Iwleva, L. M. Kachapina, E. F. Kvashina, A. K. Shilova, and A. E. Shilov, *Chem. Commun.*, **1973,** 169.
6. T. Saito, *Chem. Commun.*, **1971,** 1422.
7. C. R. Lucas, M. L. H. Green, R. A. Forder, and C. K. Prout, *Chem. Commun.*, **1973,** 97.
8. R. S. P. Coutts, R. L. Martin, and P. C. Wailes, *Aust. J. Chem.*, **24,** 1533 (1971).
9. P. D. Bartlett and B. Seidel, *J. Am. Chem. Soc.*, **83,** 581 (1961).
10. M. L. H. Green and C. R. Lucas, *J. Chem. Soc., Dalton Trans.*, **1972,** 1000.

Correction

We have detected an error in Volume XIV.

14. DIANIONOBIS(ETHYLENEDIAMINE) COBALT(III) COMPLEXES

Submitted by J. SPRINGBØRG† and C. E. SCHÄFFER†
Checked by JOHN M. PRESTON‡ and BODIE DOUGLAS‡

E. *cis*-BIS(ETHYLENEDIAMINE)DINITROCOBALT(III) NITRITE

$$[Co(en)_2(CO_3)]^+ + 2H^+ + H_2O \rightarrow cis\text{-}[Co(en)_2(H_2O)_2]^{3+} + CO_2$$
$$cis\text{-}[Co(en)_2(H_2O)_2]^{3+} + 3NO_2^- \rightarrow cis\text{-}[Co(en)_2(NO_2)_2]NO_2 + 2H_2O$$

Lines 8 through 12 from the bottom of page 73 should read as follows:

Preparation from Cobalt(II) Chloride

One hundred fifty milliliters of 12 *M* hydrochloric acid is added to *solution A* (0.82 mole of cobalt(II) salt (see preparation A, p. 66) with stirring and cooling in an ice bath.

†Chemistry Department I, The H. C. Ørsted Institute, University of Copenhagen, DK-2100 Denmark.
‡University of Pittsburgh, Pittsburgh, Pa. 15213.

INDEX OF CONTRIBUTORS

Adler, A. D., 213
Aitken, G. B., 83
Anand, S. P., 24
Armor, J. N., 75
Ashby, E. C., 137
Attig, T. G., 192
Atwood, J. L., 137

Ballantine, T., 229
Barefield, E. K., 220
Barth, R. C., 188
Bauer, D. P., 63
Bennett, M. A., 161, 164
Bianco, V. D., 155, 161, 164
Block, B. P., 89
Blum, P. R., 225
Bottomley, F., 9, 13, 75
Branca, J., 229
Bray, J., 24
Broomhead, J. A., 235
Brown, T. M., 78, 131
Budge, J., 235
Burmeister, J. L., 131
Butler, I. S., 53

Caglio, G., 49
Carroll, P., 87
Caulton, K. G., 16
Cenini, S., 47, 51
Centofanti, L. F., 166
Chan, J., 161
Chang, J. C., 80
Chia, P. S. K., 168
Clark, H. C., 155
Clark, R. J. H., 120, 166
Cloyd, J. C., Jr., 202
Coles, M. A., 120
Coville, N. J., 53
Cummings, S. C., 206, 225

Dahl, A. R., 220
Darin, J. L., 229
Das, M., 206, 225
Davison, A., 68
De Jongh, R. O., 127
Del Gaudio, J., 192
Dieck, R. L., 131
Dietz, E. A., Jr., 153
Dolcetti, G., 29, 32, 35, 39
Domingos, A. J. P., 103
Doronzo, S., 155, 161, 164
Douglas, B. E., 93
Douglas, W. M., 63
Downie, S., 164
DuBois, T. L., 202
Dyer, G., 168

Eady, C., 45
Eisenberg, R., 21
Ellis, J. E., 68

Feltham, R. D., 5, 16, 29, 32
Fenster, A. E., 53
Fiess, P., 155
Fitzgerald, R. J., 41
Flagg, E. E., 89
Forster, A., 39, 45
Fredette, M. C., 35

Galliart, A., 131
Gandolfi, O., 32
Ghedini, M., 32
Gillman, H. D., 89
Goddard, R., 80
Granchi, M. P., 93
Green, M. L. H., 237
Grice, N., 103
Grim, S. O., 181, 184, 188, 192, 195, 198, 199
Grumley, W., 235
Gsell, R., 229
Gwost, D., 16

Hagen, A. P., 139
Halstead, G. W., 147
Hamilton, J. B., 78
Harris, R. O., 9
Haymore, B. L., 41
Heintz, R. M., 51
Herlinger, A. W., 220
Hill, L., 229
Hoffman, N. W., 32
Holt, S., 220
Hota, N. K., 9
Howell, J. A. S., 103
Hsieh, A. T. T., 61
Hurst, H. J., 143

Ileperuma, O. A., 5, 16, 32
Ittel, S. D., 113

James, B. R., 45, 47, 49
Johnson, B. F. G., 39, 45, 103
Johnson, E. D., 5, 21
Johnson, E. H., 117

Kauffman, G. B., 93
King, R. B., 24, 202
Kirksey, K., 78
Kita, W. G., 24
Klein, A., 87
Kolich, C. H., 97
Kristoff, J. S., 53, 61

Labinger, J. A., 107
Levason, W., 174, 184, 188
Lewis, J., 103
Lin, H.-M. W., 41
Lin, S. W., 13
Lindley, E. V., Jr., 93
Lines, L., 166
Little, R. G., 213
Littlecott, G. W., 113
Livingstone, S. E., 168, 206, 225
Lock, C. J. L., 35
Longo, F. R., 213
Lucas, C. R., 107, 237

McAmis, L. L., 139
McAuliffe, C. A., 174, 184, 188
McCleverty, J. A., 24
MacDiarmid, A. G., 63
McDonald, J. W., 235
McQuillan, G. P., 83
McQuillin, F. J., 113
McVicker, G. B., 56
Mangat, M., 13
Mantovani, A., 47, 51
Markham, R. T., 153
Marks, T. J., 56, 147
Martell, A. E., 199
Martin, D. R., 153
Matienzo, L. J., 198
Mays, M. J., 61
Meek, D. W., 168, 180, 198, 202
Mitchell, J. D., 195
Mitra, G., 87
Moeller, T., 131
Molin, M., 127
Morassi, R., 174
Morris, D. E., 51
Motekaitis, R. J., 199

Nannelli, P., 89
Nappier, J. R., 16
Nappier, T. E., 29, 32
Nibert, J. H., 87
Norton, J. R., 35, 39

Ochs, J., 229
O'Donnell, T. A., 143

Parshall, G. W., 68
Partenheimer, W., 117
Pestel, B. C., 206

Peterson, E. J., 131
Peterson, J. L., 180
Pettit, R., 103
Pierpont, C. G., 21
Plackett, D. V., 47
Plaza, A. I., 184, 199
Powell, K. G., 113
Prasad, H. S., 137

Raymond, K. N., 147
Reed, C. A., 29
Reed, J., 21
Reimer, K. J., 155
Rempel, G. L., 45, 49
Ruckenstein, A., 131
Ruff, J. K., 63
Russo, P. J., 63

Sacconi, L., 174, 195
Satija, S. K., 1
Schäfer, H., 63
Schmulbach, C. D., 97, 229
Schmutzler, R., 153
Schram, E. P., 97
Schreiner, A. F., 13
Schwartz, J., 107
Scobell, R., 229
Seddon, D., 24
Seebach, G. L., 83
Selbeck, H., 127
Selbin, J., 87
Sepelak, D. J., 139
Seyam, A. M., 56, 147
Shah, D. P., 180
Shaver, A. G., 155
Sherrill, H. J., 87
Shriver, D. F., 61
Smith, J. N., 78
Smith, K. D., 137
Smith, P. W., 120
Stelzer, O., 153
Strouse, C. E., 1
Swanson, B. I., 1

Tayim, H. A., 117
Taylor, B. H., 237
Tebbe, F. N., 237
Teo, W. K., 45, 49
Tong, S. B., 9
Towarnicky, J., 97

Uriarte, R., 198

Van der Linde, R., 127
Van Dyke, C. H., 139
Váradi, V., 213
Visser, A., 127

Wachter, W. A., 147
Wagner, F., 220
Wasson, J. R., 83
Wei, R. M. C., 225
Wilke, G., 127
Wilson, P. W., 143
Wonchoba, E. R., 68
Workman, M. O., 168
Worrell, J. H., 80

Zeldin, M., 229

SUBJECT INDEX

Names used in this Subject Index for Volume XVI, as well as in the text, are based for the most part upon the "Definitive Rules for Nomenclature of Inorganic Chemistry," 1957 Report of the Commission on the Nomenclature of Inorganic Chemistry of the International Union of Pure and Applied Chemistry, Butterworths Scientific Publications, London, 1959; American version, *J. Am. Chem. Soc.,* **82,** 5523–5544 (1960); and the latest revisions [Second Edition (1970) of the Definitive Rules for Nomenclature of Inorganic Chemistry]; also on the Tentative Rules of Organic Chemistry—Section D; and "The Nomenclature of Boron Compounds" [Committee on Inorganic Nomenclature, Division of Inorganic Chemistry, American Chemical Society, published in *Inorganic Chemistry,* **7,** 1945 (1968) as tentative rules following approval by the Council of the ACS]. All of these rules have been approved by the ACS Committee on Nomenclature. Conformity with approved organic usage is also one of the aims of the nomenclature used here.

In line, to some extent, with *Chemical Abstracts* practice, more or less inverted forms are used for many entries, with the substituents or ligands given in alphabetical order (even though they may not be in the text); for example, derivatives of arsine, phosphine, silane, germane, and the like; organic compounds; metal alkyls, aryls, 1,3-diketone and other derivatives and relatively simple specific coordination complexes: *Iron, cyclopentadienyl-* (also as *Ferrocene*); *Cobalt(II), bis*(2,4-*pentanedionato*)- [instead of *Cobalt(II) acetylacetonate*]. In this way, or by the use of formulas, many entries beginning with numerical prefixes are avoided; thus *Vanadate(III), tetrachloro-*. Numerical and some other prefixes are also avoided by restricting entries to group headings where possible: *Sulfur imides,* with formulas; *Molybdenum carbonyl,* $Mo(CO)_6$; both *Perxenate,* $HXeO_6^{3-}$, and *Xenate(VIII)*, $HXeO_6^{3-}$. In cases where the cation (or anion) is of little or no significance in comparison with the emphasis given to the anion (or cation), one ion has been omitted; e.g., also with less well-known complex anions (or cations): $CsB_{10}H_{12}CH$ is entered only as *Carbaundecaborate*(1−), *tridecahydro-* (and as $B_{10}CH_{13}-$ in the Formula Index).

Under general headings such as *Cobalt(III) complexes* and *Ammines,* used for grouping coordination complexes of similar types having names considered unsuitable for individual headings, formulas or names of specific compounds are not usually given. Hence it is imperative to consult the Formula Index for entries for specific complexes.

As in *Chemical Abstracts* indexes, headings that are phrases are alphabetized straight through, letter by letter, not word by word, whereas inverted headings are alphabetized first as far as the comma and then by the inverted part of the name. Stock Roman numerals and Ewens-Bassett Arabic numbers with charges are ignored in alphabetizing unless two or more names are otherwise the same. Footnotes are indicated by *n.* following the page number.

Amine, tris-[2-(diphenylarsino)ethyl]-, 177
———, tris-[2-(diphenylphosphino)ethyl]-, 176
Ammines:
of cobalt(III), 93–96
of osmium, 9–12
of ruthenium, 13–15, 75
Ammonium, tetraethyl-, cyanate, 132
———, tetraethyl-, cyanide, 133
Ammonium cyanate, 136
Aniline, *o*-methylthio-, 169
Arsine, (*o*-bromophenyl)dimethyl-, 185
———, dimethyl(pentafluorophenyl)-, 183
———, tris *o*-(dimethylarsino)phenyl-, 186
Arsonium, tetraphenyl-, cyanate, 134
———, tetraphenyl-, cyanide, 135
———, tetraphenyl-, hexacarbonylniobate, 72
———, tetraphenyl-, hexacarbonyltantalate, 71
5-Aza-2,8,9-trioxa-germatricyclo(3.3.3.0)-undecane, 1-ethyl-, 229
5-Aza-2,8,9-trioxa-stannatricyclo(3.3.3.0)-undecane, 1-ethyl-, 230

Benzamine (*see* Aniline)
Benzene-d_5, bromo-, 164
Bromobenzene-d_5, 164
o-Bromophenyl methyl sulfide, 169
o-Bromothioanisole, 169
3-Buten-2-one, 1,1,1-trifluoro-4-mercapto-4-(2-thienyl)-, 206

Chromate(III), diaquabis(malonato)-, potassium, trihydrate, 81
———, tris(malonato)-, potassium, trihydrate, 80
Chromium, tetranitrosyl-, 2
Chromium(III), aquahydroxobis(phosphinato)-, 90
———, hydroxobis(phosphinato)-, 91
———, malonato complexes, 80–82
Chromium(III) bis(dioctylphosphinate), aquahydroxo-, poly-, 90
———, hydroxo-, poly-, 91
Chromium(III) bis(diphenylphosphinate), aquahydroxo-, poly-, 90
———, hydroxo-, poly-, 91
Chromium(III) bis(methylphenylphosphinate), aquahydroxo-, poly-, 90
———, hydroxo-, poly-, 91
Chromium(III) bis(phosphinates), polymers, complexes, 89–92
Chromium nitrosyl, $Cr(NO)_4$, 2
Cobalt, bis(dimethylcarbamodithioato-*S,S'*)nitrosyl-, 7
———, bis(methyldiphenylphosphine)-mato)nitrosyl-, 7
———, bis(methyldiphenylphosphine)-dichloronitrosyl-, 29
———, dinitrosyl[1,2-bis(diphenylphosphino)ethane]-, tetraphenylborate, 19
———, dinitrosylbis(triphenylphosphine)-, tetraphenylborate, 18
———, dinitrosyl[1,2-ethanediylbis(diphenylphosphine)]-, tetraphenylborate, 19
———, dinitrosyl[ethylenebis(diphenylphosphine)]-, tetraphenylborate, 19
———, dinitrosyl-(*N,N,N',N'*-tetramethylethylenediamine), tetraphenylborate, 17
———, nitrosyltris(triphenylphosphine)-, 33
Cobalt(II), *N,N'*-ethylenebis(thioacetylacetoniminato)-, 227
———, sulfatotris(selenourea)-, 85
———, tetrakis(selenourea)-, diperchlorate, 84
———, tris(selenourea)-, sulfate, 85
Cobalt(III), amminebromobis(ethylenediamine)-, α-bromocamphor-π-sulfonate, 93
———, amminebromobis(ethylenediamine)-, chloride, 93, 95, 96
———, amminebromobis(ethylenediamine)-, dithionate, 94
———, amminebromobis(ethylenediamine)-, halides, 93
———, amminebromobis(ethylenediamine)-, nitrate, 93
Cobalt complexes, amminebromobis(ethylenediamine)-, 93–96
———, dinitrosyl-, 16–19
———, with selenourea, 83–85
Copper(II), *meso*-tetraphenylporphyrin-, 214
Copper complexes, with tetraphenylporphyrin, 214

Cyanate, ammonium, 136
———, tetraethylammonium, 132
———, tetraphenylarsonium, 134
Cyanide, tetraethylammonium, 133
———, tetraphenylarsonium, 135
Cyclam (*see* 1,4,8,11-Tetraazacyclotetradecane)

Ethanamine, 2-(diphenylphosphino)-*N,N*-diethyl-, 160
1,2-Ethanediamine (*see* Ethylenediamine)
Ethylenediamine, *N,N*-bis(diphenylphosphinomethyl)-*N′,N′*-dimethyl-, 199
———, *N,N′*-bis(diphenylphosphinomethyl)-*N,N′*-dimethyl-, 199
———, *N*-(diphenylphosphinomethyl)-*N,N′,N′*-trimethyl-, 199
———, *N,N,N′,N′*-tetrakis(diphenylphosphinomethyl)-, 198
———, *N,N,N′*-tris(diphenylphosphinomethyl)-*N′*-methyl-, 199

Iridium, bromodinitrogenbis(triphenylphosphine)-, 42
———, bromonitrosylbis(triphenylphosphine)-, tetrafluoroborate, *trans*-, 42
———, chlorodinitrogenbis(triphenylphosphine)-, 42
———, chloronitrosylbis(triphenylphosphine)-, tetrafluoroborate, *trans*-, 41
Iron, (benzylideneacetone)tricarbonyl-, 104
———, bis(diethylcarbamodithioato-*S,S′*)nitrosyl-, 5
———, bis(*N,N*-diethyldithiocarbamato)nitrosyl-, 5
———, (chlorodifluorophosphine)tetracarbonyl-, 66
———, [(diethylamino)difluorophosphine]-tetracarbonyl-, 64
———, tetracarbonyl[(diethylamino)difluorophosphine]-, 64
———, tetracarbonyl(trifluorophosphine)-, 67
———, tricarbonyl-η-diene-, complexes, 103
Iron(III), chlorohematoporphyrin(IX)dimethylester-, 216
Iron carbonyl, $Fe(CO)_4$, complexes with difluorophosphines, 63–67
Iron complexes, with hematoporphyrin, 216

Lithium, [2-(methylthio)phenyl]-, 170

Magnesium, bis(dicarbonyl-η-cyclopentadienyliron)bis(tetrahydrofuran)-, 56
———, bis[dicarbonyl-η-cyclopentadienyl-(tributylphosphine)molybdenum]-tetrakis(tetrahydrofuran)-, 59
———, bis(tetracarbonylcobalt)tetrakis-(pyridine)-, 58
———, bis[tricarbonyl(tributylphosphine)-cobalt]tetrakis(tetrahydrofuran)-, 58
———, diindenyl-, 137
———, tetrakis(pyridine)bis(tetracarbonyl-cobalt)-, 58
———, transition metal carbonyl derivatives, 56–60
Manganese, (carbonothioyl)dicarbonyl(η-cyclopentadienyl)-, 53
———, carbonyltrinitrosyl-, 4
———, dicarbonyl(η-cyclopentadienyl)(thiocarbonyl)-, 53
Mercury(II), bis(selenourea)-, dihalides, 85
———, di-η-chlorodichlorobis(selenourea)di-, 86
Mercury(II) complexes, with selenourea, 85–86
Mercury(II) halides, complexes with selenourea, 85
Metallatranes, 229–234
Methyl difluorophosphite, 166
Molybdenum, bis(diethyldithiocarbamato)-dinitrosyl-, 235
———, η-cyclopentadienyldiiodonitrosyl-, bis-, 28
———, dibromo-η-cyclopentadienylnitrosyl-, bis-, 27
———, dicarbonyl-η-cyclopentadienylnitrosyl-, 24
———, dichloro-η-cyclopentadienylnitrosyl-, bis-, 26
———, diiodo-η-cyclopentadienylnitrosyl-, bis-, 28

Nickel(II), (1,4,8,11-tetraazacyclotetradecane)-, perchlorate, 221
Nickel tetrafluorooxovanadate(IV) heptahydrate, 87
Niobate, hexacarbonyl-, tetraphenylarsonium, 72

Niobate, hexacarbonyl-, tris[bis(2-methoxyethyl)ether]potassium, 69
Niobium, bis(cyclopentadienyl)-, complexes, 107–113
Niobium(III), bis(cyclopentadienyl)(dimethylphenylphosphine)hydrido-, 110
———, bis(cyclopentadienyl)(tetrahydroborate)-, 109
———, bromobis(cyclopentadienyl)(dimethylphenylphosphine)dihydrido-, 112
Niobium(IV), bis(2,2′-bipyridine)tetrakis-(isothiocyanato)-, 78
———, dichlorobis(cyclopentadienyl)-, 107
Niobium(V), bis(cyclopentadienyl)(dimethylphenylphosphine)dihydrido-, hexafluorophosphate, 111
  tetrafluoroborate, 111

Organometallatranes, 229–234
Osmium, pentaamminenitrosyl-, trihalide, monohydrate, 11
———, tetraamminehalonitrosyl-, dihalide, 12
———, tetraamminehydroxonitrosyl-, dihalide, 11
Osmium(II), pentaammine(dinitrogen)-, diiodide, 9
Osmium(III), hexaammine-, triiodide, 10
———, pentaammineiodo-, diiodide, 10
Osmium complexes, ammines, 9–12

Palladium, ethylenebis(tricyclohexylphosphine)-, 129
———, ethylenebis(triphenylphosphine)-, 127
———, ethylenebis(tri-*o*-tolylphosphite)-, 129
2-Pentanethione, 4,4′-(alkanediyldinitrilo)bis-, 228
  transition metal complexes, 228
———, 4,4′-(1,2-ethanediyldinitrilo)bis-, 226
  cobalt(II) complex, 227
Phenylphosphinic acid, [ethylenebis-(methylnitrilomethylene)]bis-, dihydrochloride, 201
———, [ethylenebis(nitrilodimethylene)]-tetrakis-, 199
———, trimethylaminetris-, 201
Phosphinate, isopropylphenylvinyl-, 203
Phosphinates, chromium(III) bis-, polymers, complexes, 89–92
Phosphine, benzyldiphenyl-, 159
———, bis[*o*-(methylthio)phenyl]phenyl-, 172
———, butyldiphenyl-, 158
———, cyclohexyldiphenyl-, 159
———, dimethyl(pentafluorophenyl)-, 181
———, diphenyl-, 161
———, [2-(diphenylarsino)ethyl]diphenyl-, 191
———, [2-(diphenylarsino)vinyl]diphenyl-, 189
———, diphenyl[2-(phenylphosphino)ethyl]-, 202
———, ethyldiphenyl-, 158
———, [2-(isopropylphenylphosphino)ethyl]diphenyl-, 192
———, methyldiphenyl-, 157
———, [*o*-(methylthio)phenyl]diphenyl-, 171
———, trimethyl-, 153
———, tri(phenyl-d_5)-, 163
———, tris[*o*-(methylthio)phenyl]-, 173
Phosphine sulfide, diphenyl[(phenylisopropylphosphino)methyl]-, 195
Phosphite, difluoro-, methyl, 166
Phosphorus sulfur ligands, 168–173
Platinum, (aryl-1,3-propanediyl)dichloro-, 114
———, (aryl-1,3-propanediyl)dichlorobis-(pyridine)-, 115
———, dichloro(hexyl-1,3-propanediyl)-, 114
———, dichloro(hexyl-1,3-propanediyl)bis(pyridine)-, 115
———, dichloro(1,3-propanediyl)- derivatives, 114
———, dichloro(1,3-propanediyl)bis(pyridine)- derivatives, 115
Porphyrin complexes:
  with copper, 214
  with iron, 216
Potassium, tris[bis(2-methoxyethyl)ether]-, hexacarbonyl niobate, 69
———, tris[bis(2-methoxyethyl)ether]-, hexacarbonyl tantalate, 71
Potassium diaquabis(malonato)chromate(III), trihydrate, 81
Potassium tris(malonato)chromate(III), trihydrate, 80

Rhenium, carbonylchloronitrosyl compounds, 35–38
———, octacarbonyldi-μ-chlorodi-, 35
———, pentacarbonyltri-μ-chloronitrosyldi-, 36
———, tetracarbonyldi-μ-chlorodinitrosyldi-, 37
Rhodium, hexadecacarbonylhexa-, 49
———, nitrosyltris(triphenylphosphine)-, 33
Rhodium carbonyl, $Rh_6(CO)_{16}$, 49
Ruthenium, bis(triphenylphosphine)chlorodinitrosyl-, tetrafluoroborate, 21
———, chlorodinitrosylbis(triphenylphosphine)-, tetrafluoroborate, 21
———, (1,5-cyclooctadiene)tricarbonyl-, 105
———, decacarbonyldi-μ-nitrosyltri-, 39
———, dodecacarbonyltri-, 45, 47
———, hexacarbonyldi-μ-chlorodichlorodi-, 51
———, tetraammineacetatonitrosyl-, diperchlorate, 14
———, tetraammineacidonitrosyl-, diperchlorate, 14
———, tetraamminechloronitrosyl-, dichloride, 13
———, tetraamminecyanatonitrosyl-, diperchlorate, 15
———, tricarbonyl(1,5-cyclooctadiene)-, 105
———, tricarbonyl-η-diene complexes, 103
Ruthenium(II), pentaammine(dinitrogen oxide)-, dihalides, 75
Ruthenium carbonyl, $Ru_3(CO)_{12}$, 45, 47
Ruthenium complexes, amminenitrosyl-, 13–15

Selenourea:
cobalt(II) complexes, 83–85
mercury halide complexes, 85–86
Silane, difluorodimethyl-, 141
———, fluoro(methyl)- derivatives, 139–142
———, methyltrifluoro-, 139
———, trifluoro(methyl)-, 139
Silver(I), cycloalkene(fluorodiketonato)-compounds, 117–119
———, cycloheptene(hexafluoropentanedionato)-, 118
Silver (I), cyclohexene(hexafluoropentanedionato)-, 118
———, cyclooctadiene(hexafluoropentanedionato)-, 117
———, cyclooctadiene(trifluoropentanedionato)-, 118
———, cyclooctatetraene(hexafluoropentanedionato)-, 118
———, cyclooctatetraene(trifluoropentanedionato)-, 118
———, cyclooctene(hexafluoropentanedionato)-, 118
———, olefin(fluorodiketonato)-, compounds, 117–119
Stibine, tris[o-(dimethylarsino)phenyl]-, 186
Sulfide, o-bromophenyl methyl, 169
———, o-(diphenylphosphino)phenyl methyl, 171

Tantalate, hexacarbonyl-, tetraphenylarsonium, 71
———, hexacarbonyl-, tris[bis(2-methoxyethyl)ether]potassium, 71
1,4,8,11-Tetraazacyclotetradecane, 223
1,4,8,11-Tetraazacyclotetradecanenickel perchlorate, 221
Thallium(III), tris(pentacarbonylmanganese)-, 61
Thioacetylacetonimine, N,N'-ethylenebis-, 226
cobalt(II) complex, 227
Thioanisole, o-bromo-, 169
Thorium(IV), chlorotris(cyclopentadienyl)-, 149
Titanium, methyl-, trihalides, 120–126
———, methyltribromo-, 124
———, methyltrichloro-, 122
———, tribromo(methyl)-, 124
———, trichlorobis(dimethylphosphine)-, 100
———, trichlorobis(methylphosphine)-, 98
———, trichlorobis(triethylphosphine)-, 101
———, trichlorobis(trimethylphosphine)-, 100
———, trichloro(methyl)-, 122
Titanium(III), bis(alkylphosphine)trichloro-, 97
———, dichloro(cyclopentadienyl)-, polymer, 238

Titanium (III), dichloro(cyclopentadienyl)bis-(dimethylphenylphosphine)-, 239
Triethylamine, 2-(diphenylphosphino)-, 160
——, 2,2′,2″-tris(diphenylarsino)-, 177
——, 2,2′,2″-tris(diphenylphosphino)-, 176

Uranium(IV), chlorotris(cyclopentadienyl)-, 148
Uranium chloride, UCl_6, 143
Uranium hexachloride, 143

Vanadata(IV), tetrafluorooxo-, nickel, heptahydrate, 87

FORMULA INDEX

The chief aim of this index, like that of other formula indexes, is to help in locating specific compounds or ions, or even groups of compounds, that might not be easily found in the Subject Index, or in the case of many coordination complexes are to be found only as general entries in the Subject Index. *All* specific compounds, or in some cases ions, with definite formulas (or even a few less definite) are entered in this index or noted under a related compound, whether entered specifically in the Subject Index or not.

Wherever it seemed best, formulas have been entered in their usual form (i.e., as used in the text) for easy recognition: Si_2H_6, XeO_3, NOBr. However, for the less simple compounds, including coordination complexes, the significant or central atom has been placed first in the formula in order to throw together as many related compounds as possible. This procedure often involves placing the cation last as being of relatively minor interest (e.g., alkali and alkaline earth metals), or dropping it altogether: MnO_4Ba; $Mo(CN)_8K_4 \cdot 2H_2O$; $Co(C_5H_7O_2)_3Na$; $B_{12}H_{12}^{2-}$. Where there may be almost equal interest in two or more parts of a formula, two or more entries have been made: Fe_2O_4Ni and $NiFe_2O_4$; $NH(SO_2F)_2$, $(SO_2F)_2NH$, and $(FSO_2)_2NH$ (halogens other than fluorine are entered only under the other elements or groups in most cases); $(B_{10}CH_{11})_2Ni^{2-}$ and $Ni(B_{10}CH_{11})_2^{2-}$.

Formulas for organic compounds are structural or semistructural so far as feasible: $CH_3COCH(NHCH_3)CH_3$. Consideration has been given to probable interest for inorganic chemists, i.e., any element other than carbon, hydrogen, or oxygen in an organic molecule is given priority in the formula if only one entry is made, or equal rating if more than one entry: only $Co(C_5H_7O_2)_2$, but $AsO(+)\text{-}C_4H_4O_6Na$ and $(+)\text{-}C_4H_4O_6AsONa$. Names are given only where the formula for an organic compound, ligand, or radical may not be self-evident, but not for frequently occurring relatively simple ones like C_5H_5(cyclopentadienyl), $C_5H_7O_2$ (2,4-pentanedionato), C_6H_{11}(cyclohexyl), C_5H_5N(pyridine). A few abbreviations for ligands used in the text are retained here for simplicity and are alphabetized as such, "bipy" for bipyridine, "en" for ethylenediamine or 1,2-ethanediamine, "diphos" for ethylenebis(diphenylphosphine) or 1,2-bis(diphenylphosphino)ethane or 1,2-ethanediylbis(diphenylphosphine), and "tmeda" for N,N,N′,N′-tetramethylethylenediamine or N,N,N′,N′-tetramethyl-1,2-ethanediamine.

Footnotes are indicated by *n.* following the page number.

$Ag(C_5HF_6O_2)(C_6H_{10})$, 118
$Ag(C_5HF_6O_2)(C_7H_{12})$, 118
$Ag(C_5HF_6O_2)(C_8H_8)$, 118
$Ag(C_5HF_6O_2)(C_8H_{12})$, 117
$Ag(C_5HF_6O_2)(C_8H_{14})$, 118
$Ag(C_5H_4F_3O_2)(C_8H_8)$, 118
$Ag(C_5H_4F_3O_2)(C_8H_{12})$, 118
$As(CH_3)_2(C_6F_5)$, 183

$[As(CH_3)_2C_6H_4]_3As$, 186
$[As(CH_3)_2C_6H_4]_3Sb$, 187
$As(CH_3)_2(C_6H_4Br)$, 185
$As[C_6H_4As(CH_3)_2]_3$, 186
$[As(C_6H_5)_2CH{=}CH]P(C_6H_5)_2$, 189
$[As(C_6H_5)_2C_2H_4]P(C_6H_5)_2$, 191
$[As(C_6H_5)_2C_2H_4]_3N$, 177
$[As(C_6H_5)_4](CN)$, 135
$[As(C_6H_5)_4](OCN)$, 134
$[As(C_6H_5)_4][Nb(CO)_6]$, 72
$[As(C_6H_5)_4][Ta(CO)_6]$, 71

$(BF_4)[Nb(C_5H_5)_2H_2\{P(CH_3)_2(C_6H_5)\}]$, 111
$(BH_4)Nb(C_5H_5)_2$, 109
BrC_6D_5, 164
$BrC_6H_4(SCH_3)$, 169

$(CN)[N(C_2H_5)_4]$, 133
$(CO)Mn(NO)_3$, 4
$[(CO)_2Fe(C_5H_5)]_2Mg(C_4H_8O)_2$, 56
$(CO)_2Mn(C_5H_5)(CS)$, 53
$(CO)_2Mo(NO)(C_5H_5)$, 24
$[(CO)_2Mo\{P(C_4H_9)_3\}(C_5H_5)]_2Mg(C_4H_8O)_4$, 59
$[(CO)_2ReCl_2(NO)]_2$, 37
$[(CO)_3Co\{P(CH_3)(C_6H_5)_2\}]_2Mg(tmeda)_2$, 59
$[(CO)_3CoP(C_4H_9)_3]_2Mg(C_4H_8O)_4$, 58
$(CO)_3Fe[C_6H_5CH{=}CHC(O)CH_3]$, 104
$(CO)_3Ru(C_8H_{12})$, 105
$[(CO)_3RuCl_2]_2$, 51
$[(CO)_4Co]_2Mg(C_5H_5N)_4$, 58
$(CO)_4Fe(PClF_2)$, 66
$(CO)_4Fe[PF_2N(C_2H_5)_2]$, 64
$(CO)_4Fe(PF_3)$, 67
$[(CO)_4ReCl]_2$, 35
$[(CO)_5Mn]_3Tl$, 61
$(CO)_5ReCl_3(NO)$, 36
$[(CO)_6Nb][As(C_6H_5)_4]$, 72
$[(CO)_6Nb][K\{(CH_3OCH_2CH_2)_2O\}_3]$, 69
$(CO)_6Ru_2Cl_4$, 51
$[(CO)_6Ta][As(C_6H_5)_4]$, 71
$[(CO)_6Ta][K\{(CH_3OCH_2CH_2)_2O\}_3]$, 71
$(CO)_{10}Os_3(NO)_2$, 40
$(CO)_{10}Ru_3(NO)_2$, 39
$(CO)_{12}Ru_3$, 45, 47
$(CO)_{16}Rh_6$, 49
$(CS)Mn(C_5H_5)(CO)_2$, 53
$(C_4H_3S)C(SH){=}CHC(O)CF_3$, 206
$C_6H_4(Br)(SCH_3)$, 169
$C_{10}H_{24}N_4$, 223
$[CoBr(en)_2(NH_3)]Br_2$, 93
$[CoBr(en)_2(NH_3)]Cl_2$, 93, 95, 96
$[CoBr(en)_2(NH_3)](NO_3)_2$, 93
$[CoBr(en)_2(NH_3)](O_3SOC_{10}H_{14}Br)_2$, 93
$[CoBr(en)_2(NH_3)](S_2O_6)$, 94
$[Co(CO)_3\{P(CH_3)(C_6H_5)_2\}]_2Mg(tmeda)$, 59
$[Co(CO)_3\{P(C_4H_9)_3\}]_2Mg(C_4H_8O)_4$, 58
$[Co(CO)_4]_2Mg(C_5H_5N)_4$, 58
$Co(C_2H_2N_2)[C(CH_3)CH_2C(S)CH_3]_2$, 227
$CoCl_2(NO)[P(CH_3)(C_6H_5)_2]_2$, 29
$[Co\{(NH_2)_2CSe\}_3](ClO_4)_2$, 84
$[Co\{(NH_2)_2CSe\}_3](SO_4)$, 85
$Co(NO)[P(C_6H_5)_3]_3$, 33
$Co(NO)[S_2CN(CH_3)_2]_2$, 7
$[Co(NO)_2(diphos)][B(C_6H_5)_4]$, 19
$[Co(NO)_2\{P(C_6H_5)_3\}_2][B(C_6H_5)_4]$, 18
$[Co(NO)_2(tmeda)][B(C_6H_5)_4]$, 17
$Co(N_2C_2H_2)[C(CH_3)CH_2C(S)CH_3]_2$, 227
$[Cr(C_3H_2O_4)_2(H_2O)_2]K \cdot 3H_2O$, 81
$[Cr(C_3H_2O_4)_3]K_3 \cdot 3H_2O$, 80
$[Cr(H_2O)(OH)\{OP(CH_3)(C_6H_5)(O)\}_2]_x$, 90
$[Cr(H_2O)(OH)\{OP(C_6H_5)_2(O)\}_2]_x$, 90
$[Cr(H_2O)(OH)\{OP(C_8H_{17})_2(O)\}_2]_x$, 90
$Cr(NO)_4$, 2
$[Cr(OH)\{OP(CH_3)(C_6H_5)(O)\}_2]_x$, 91
$[Cr(OH)\{OP(C_6H_5)_2(O)\}_2]_x$, 91
$[Cr(OH)\{OP(C_8H_{17})_2(O)\}_2]_x$, 91
$CuC_{44}H_{28}N_4$, 214

$F_2PO(CH_3)$, 166
$[Fe(C_5H_5)(CO)_2]_2Mg(C_4H_8O)_2$, 56
$Fe(CO)_3(C_6H_5CH{=}CHCOCH_3)$, 104
$Fe(CO)_4(PClF_2)$, 66
$Fe(CO)_4[PF_2N(C_2H_5)_2]$, 64
$Fe(CO)_4(PF_3)$, 67
$FeCl(C_{36}H_{40}N_4O_6)$, 216
$Fe(NO)[S_2CN(C_2H_5)_2]_2$, 5

$Ge(C_2H_5)(OCH_2CH_2)_3N$, 229
$GeC_8H_{17}O_3N$, 229

$[Hg\{(NH_2)_2CSe\}Cl_2]_2$, 86
$Hg[(NH_2)_2CSe]_2Br_2$, 86
$Hg[(NH_2)_2CSe]_2Cl_2$, 85

$[IrBr(NO)\{P(C_6H_5)_3\}_2](BF_4)$, 42
$IrBr(N_2)[P(C_6H_5)_3]_2$, 42
$[IrCl(NO)\{P(C_6H_5)_3\}_2](BF_4)$, 41
$IrCl(N_2)[P(C_6H_5)_3]_2$, 42

$[K\{(CH_3OCH_2CH_2)_2O\}_3]Nb(CO)_6$, 69
$[K\{(CH_3OCH_2CH_2)_2O\}_3]Ta(CO)_6$, 71
$K[Cr(C_3H_2O_4)_2(H_2O)_2]\cdot 3H_2O$, 81
$K_3[Cr(C_3H_2O_4)_3]\cdot 3H_2O$, 80

$LiC_6H_4S(CH_3)$, 170

$Mg(C_4H_8O)_2[Fe(C_5H_5)(CO)_2]_2$, 56
$Mg(C_4H_8O)_4[Co(CO)_3P(C_4H_9)_3]_2$, 58
$Mg(C_4H_8O)_4[Mo(CO)_2P(C_4H_9)_3(C_5H_5)]_2$, 59
$Mg(C_5H_5N)_4[Co(CO)_4]_2$, 58
$Mg(C_9H_7)_2$, 137
$Mg[Co(CO)_3\{P(CH_3)(C_6H_5)_2\}]_2(tmeda)_2$, 59
$Mg[Co(CO)_3P(C_4H_9)_3]_2(C_4H_8O)_4$, 58
$Mg[Co(CO)_4]_2(C_5H_5N)_4$, 58
$Mg[Fe(C_5H_5)(CO)_2]_2(C_4H_8O)_2$, 56
$Mn(CO)(NO)_3$, 4
$Mn(CO)_2(CS)(C_5H_5)$, 53
$[Mn(CO)_5]_3Tl$, 61
$Mn(C_5H_5)(CO)_2(CS)$, 53
$[MoBr_2(NO)(C_5H_5)]_2$, 27
$Mo(CO)_2(NO)(C_5H_5)$, 24
$[Mo(CO)_2\{P(C_4H_9)_3\}(C_5H_5)]_2Mg(C_4H_8O)_4$, 59
$[Mo(C_5H_5)Br_2(NO)]_2$, 27
$Mo(C_5H_5)(CO)_2(NO)$, 24
$[Mo(C_5H_5)Cl_2(NO)]_2$, 26
$[Mo(C_5H_5)I_2(NO)]_2$, 28
$[MoCl_2(NO)(C_5H_5)]_2$, 26
$[MoI_2(NO)(C_5H_5)]_2$, 28
$Mo(NO)_2[S_2CN(C_2H_5)_2]_2$, 235

$[N\{CH_2P(C_6H_5)_2\}(CH_3)CH_2]_2$, 199
$N[CH_2P(C_6H_5)_2](CH_3)C_2H_4N(CH_3)_2$, 199
$[N\{CH_2P(C_6H_5)_2\}_2CH_2]_2$, 198
$N[CH_2P(C_6H_5)_2]_2C_2H_4N[CH_2P(C_6H_5)_2](CH_3)$, 199
$N[CH_2P(C_6H_5)_2]_2C_2H_4N(CH_3)_2$, 199
$[N\{CH_2P(O)(OH)(C_6H_5)\}_2CH_2]_2$, 199
$N[CH_2P(O)(OH)(C_6H_5)]_3$, 202
$N[C_2H_4As(C_6H_5)_2]_3$, 177
$N[C_2H_4P(C_6H_5)_2]_3$, 176
$N(C_2H_5)_2[C_2H_4P(C_6H_5)_2]$, 160
$[N(C_2H_5)_4](CN)$, 133
$[N(C_2H_5)_4](OCN)$, 131
$NGe(C_2H_5)(OCH_2CH_2)_3$, 229
$(NH_2)C_6H_4(SCH_3)$, 169
$(NH_4)(OCN)$, 136
$(NO)CoCl_2[P(CH_3)(C_6H_5)_2]_2$, 29
$(NO)Co[P(C_6H_5)_3]_3$, 33
$(NO)Co[S_2CN(CH_3)_2]_2$, 7
$(NO)Fe[S_2CN(C_2H_5)_2]_2$, 5
$[(NO)IrBr\{P(C_6H_5)_3\}_2](BF)_4)$, 42
$[(NO)IrCl\{P(C_6H_5)_3\}_2](BF_4)$, 41
$[(NO)MoBr_2(C_5H_5)]_2$, 27
$(NO)Mo(CO)_2(C_5H_5)$, 24
$[(NO)MoCl_2(C_5H_5)]_2$, 26
$[(NO)MoI_2(C_5H_5)]_2$, 28
$[(NO)OsBr(NH_3)_4]Br_2$, 12
$[(NO)OsCl(NH_3)_4]Cl_2$, 12
$[(NO)OsI(NH_3)_4]I_2$, 12
$[(NO)Os(NH_3)_4]Br_2$, 11
$[(NO)Os(NH_3)_4]Cl_2$, 11
$[(NO)Os(NH_3)_4]I_2$, 11
$[(NO)Os(NH_3)_5]Br_3\cdot H_2O$, 11
$[(NO)Os(NH_3)_5]Cl_3\cdot H_2O$, 11
$[(NO)Os(NH_3)_5]I_3\cdot H_2O$, 11
$[(NO)Re(CO)_2Cl_2]_2$, 37
$(NO)Re_2(CO)_5Cl_3$, 36
$(NO)Rh[P(C_6H_5)_3]_3$, 33
$[(NO)Ru(C_2H_3O_2)(NH_3)_4](ClO_4)$, 14
$[(NO)RuCl(NH_3)_4]Cl_2$, 13
$[(NO)Ru(NCO)(NH_3)_4](ClO_4)$, 15
$[(NO)_2Co(diphos)][B(C_6H_5)_4]$, 19
$[(NO)_2Co\{P(C_6H_5)_3\}_2][B(C_6H_5)_4]$, 18
$[(NO)_2Co(tmeda)][B(C_6H_5)_4]$, 17
$(NO)_2Mo[S_2CN(C_2H_5)_2]_2$, 235
$(NO)_2Os_3(CO)_{10}$, 40
$[(NO)_2RuCl\{P(C_6H_5)_3\}_2](BF_4)$, 21
$(NO)_2Ru_3(CO)_{10}$, 39
$(NO)_3Mn(CO)$, 4
$(NO)_4Cr$, 2
$NSn(C_2H_5)(OCH_2CH_2)_3$, 230
$(N_2C_2H_4)[C(CH_3)CH_2C(S)CH_3]_2$, 226
$N_4C_{10}H_{24}$, 223
$Nb(BH_4)(C_5H_5)_2$, 109
$Nb(bipy)_2(NCS)_4$, 78
$NbBr(C_5H_5)_2[P(CH_3)_2(C_6H_5)]$, 112
$[Nb(CO)_6][As(C_6H_5)_4]$, 72
$[Nb(CO)_6][K\{(CH_3OCH_2CH_2)_2O\}_3]$, 69
$Nb(C_5H_5)_2(BH_4)$, 109
$Nb(C_5H_5)_2Cl_2$, 107
$Nb(C_5H_5)_2(H)[P(CH_3)_2(C_6H_5)]$, 110
$[Nb(C_5H_5)_2(H)_2\{P(CH_3)_2(C_6H_5)\}](BF_4)$, 111
$NbCl_2(C_5H_5)_2$, 107
$Nb(NCS)_4(bipy)_2$, 78
$Ni(C_{10}N_4H_{24})(ClO_4)_2$, 221
$Ni(N_4C_{10}H_{24})(ClO_4)_2$, 221
$Ni(VOF_4)\cdot 7H_2O$, 87

$(OCN)[N(C_2H_5)_4]$, 131
$(OCN)(NH_4)$, 136
$[OsBr(NH_3)_4(NO)]Br_2$, 12
$[OsCl(NH_3)_4(NO)]Cl_2$, 12
$[OsI(NH_3)_4(NO)]I_2$, 12
$[OsI(NH_3)_5]I_2$, 10
$[Os(NH_3)_4(NO)(OH)]Br_2$, 11
$[Os(NH_3)_4(NO)(OH)]Cl_2$, 11
$[Os(NH_3)_4(NO)(OH)]I_2$, 11
$[Os(NH_3)_5(NO)]Br_2 \cdot H_2O$, 11
$[Os(NH_3)_5(NO)]Cl_2 \cdot H_2O$, 11
$[Os(NH_3)_5(NO)]I_2 \cdot H_2O$, 11
$[Os(NH_3)_5(N_2)]I_2$, 9
$[Os(NH_3)_6]I_3$, 10
$Os_3(CO)_{10}(NO)_2$, 40

$P[CH{=}CHAs(C_6H_5)_2](C_6H_5)_2$, 189
$[\{P(CH_3)(C_6H_5)(O)O\}Cr(H_2O)(OH)]_x$, 90
$[\{P(CH_3)(C_6H_5)(O)O\}Cr(OH)]_x$, 91
$P(CH_3)(C_6H_5)_2$, 157
$[\{P(CH_3)(C_6H_5)_2\}Co(CO)_3]Mg(tmeda)_2$, 59
$[P(CH_3)(C_6H_5)_2]CoCl_2(NO)$, 29
$[P(CH_3)H_2]_2TiCl_3$, 98
$P(CH_3)_2(C_6F_5)$, 181
$[P(CH_3)_2(C_6H_5)]NbBr(C_5H_5)_2$, 112
$[P(CH_3)_2(C_6H_5)]Nb(C_5H_5)_2(H)$, 110
$[\{P(CH_3)_2(C_6H_5)\}Nb(C_5H_5)_2(H)_2](BF_4)$, 111
$[\{P(CH_3)_2(C_6H_5)\}Nb(C_5H_5)_2(H)_2](PF_6)$, 111
$[P(CH_3)_2(C_6H_5)]_2Ti(C_5H_5)Cl_2$, 239
$[P(CH_3)_2H]_2TiCl_3$, 100
$P(CH_3)_3$, 153
$[P(CH_3)_3]_2TiCl_3$, 100
$P[C_2H_4As(C_6H_5)_2](C_6H_5)_2$, 191
$P[C_2H_4N(C_2H_5)_2](C_6H_5)_2$, 160
$P[C_2H_4P(C_3H_7)(C_6H_5)](C_6H_5)_2$, 192
$P[C_2H_4P(C_6H_5)(H)](C_6H_5)_2$, 202
$P(C_2H_5)(C_6H_5)_2$, 158
$[P(C_2H_5)_3]_2TiCl_3$, 101
$[P(C_3H_7)(C_6H_5)CH_2]P(S)(C_6H_5)_2$, 195
$[P(C_3H_7)(C_6H_5)C_2H_4]P(C_6H_5)_2$, 192
$P(C_3H_7)(C_6H_5)_2$, 158
$[P(C_4H_9)_3Co(CO)_3]_2Mg(C_4H_8O)_4$, 58
$[\{P(C_4H_9)_3\}Mo(CO)_2(C_5H_5)]Mg(C_4H_8O)_4$, 59
$P(C_6D_5)_3$, 163
$P(C_6H_4CH_3)(C_6H_5)_2$, 159
$P(C_6H_4SCH_3)(C_6H_5)_2$, 171
$P(C_6H_4SCH_3)_2(C_6H_5)$, 172
$P(C_6H_4SCH_3)_3$, 173
$[P(C_6H_5)(H)C_2H_4]P(C_6H_5)_2$, 202
$[\{P(C_6H_5)_2CH_2\}(CH_3)NCH_2]_2$, 199
$[P(C_6H_5)_2CH_2](CH_3)NC_2H_4N(CH_3)_2$, 199
$[\{P(C_6H_5)_2CH_2\}_2NCH_2]_2$, 198
$[P(C_6H_5)_2CH_2]_2NC_2H_4N[CH_2P(C_6H_5)_2]$-$(CH_3)$, 199
$[P(C_6H_5)_2CH_2]_2NC_2H_4N(CH_3)_2$, 199
$[P(C_6H_5)_2C_2H_4]_3N$, 176
$P(C_6H_5)_2(C_6H_{11})$, 159
$P(C_6H_5)_2(C_7H_7)$, 159
$P(C_6H_5)_2H$, 161
$[\{P(C_6H_5)_2(O)O\}Cr(H_2O)(OH)]_x$, 90
$[\{P(C_6H_5)_2(O)O\}Cr(OH)]_x$, 91
$[P(C_6H_5)_3]_2(C_2H_4)Pd$, 127
$[\{P(C_6H_5)_3\}_2IrBr(NO)](BF_4)$, 42
$[P(C_6H_5)_3]_2IrBr(N_2)$, 42
$[\{P(C_6H_5)_3\}_2IrCl(NO)](BF_4)$, 41
$[\{P(C_6H_5)_3\}_2RuCl(NO)_2](BF_4)$, 21
$[P(C_6H_5)_3]_3Co(NO)$, 33
$[P(C_6H_5)_3]_3Rh(NO)$, 33
$[P(C_6H_{11})_3]_2(C_2H_4)Pd$, 129
$[\{P(C_8H_{17})_2(O)O\}Cr(H_2O)(OH)]_x$, 90
$[\{P(C_8H_{17})_2(O)O\}Cr(OH)]_x$, 91
$(PClF_2)Fe(CO)_4$, 66
$[PF_2N(C_2H_5)_2]Fe(CO)_4$, 64
$PF_2O(CH_3)$, 166
$(PF_3)Fe(CO)_4$, 67
$(PF_6)\ Nb(C_5H_5)_2(H)_2\ P(CH_3)_2(C_6H_5)$, 111
$P(O)(CH{=}CH_2)(C_3H_7)(C_6H_5)$, 203
$[\{P(O)(OH)(C_6H_5)CH_2\}_2NCH_2]_2$, 199
$[P(O)(OH)(C_6H_5)CH_2]_3N$, 201
$[P(O)(OH)(C_6H_5)CH_2N(CH_3)CH_2]_2 \cdot 2HCl$, 202
$[P(OC_6H_4CH_3)_3]_2(C_2H_4)Pd$, 129
$POF_2(CH_3)$, 166
$P(S)[CH_2P(C_3H_7)(C_6H_5)](C_6H_5)_2$, 195
$Pd(C_2H_4)[P(C_6H_5)_3]_2$, 127
$Pd(C_2H_4)[P(C_6H_{11})_3]_2$, 129
$Pd(C_2H_4)[P(OC_6H_4CH_3)_3]_2$, 129
$Pt[C_3H_4(CH_3)(C_4H_9)]Cl_2$, 114
$Pt[C_3H_4(C_6H_5)_2](C_5H_5N)_2Cl_2$, 115
$Pt[C_3H_4(C_6H_5)_2]Cl_2$, 114
$Pt[C_3H_5(CH_2C_6H_5)](C_5H_5N)Cl_2$, 115
$Pt[C_3H_5(CH_2C_6H_5)]Cl_2$, 114
$Pt[C_3H_5(C_6H_4CH_3)](C_5H_5N)Cl_2$, 115
$Pt[C_3H_5(C_6H_4CH_3)]Cl_2$, 114
$Pt[C_3H_5(C_6H_4NO_2)](C_5H_5N)Cl_2$, 115
$Pt[C_3H_5(C_6H_4NO_2)]Cl_2$, 114
$Pt[C_3H_5(C_6H_5)](C_5H_5N)Cl_2$, 115
$Pt[C_3H_5(C_6H_5)]Cl_2$, 114
$Pt[C_3H_5(C_6H_{13})](C_5H_5N)Cl_2$, 115
$Pt[C_3H_5(C_6H_{13})]Cl_2$, 114
$Pt(C_3H_6)(C_5H_5N)Cl_2$, 115
$Pt(C_3H_6)Cl_2$, 114

$[Re(CO)_2Cl_2(NO)]_2$, 37
$[Re(CO)_4Cl]_2$, 35
$Re_2(CO)_5Cl_3(NO)$, 36
$Rh(NO)[P(C_6H_5)_3]_3$, 33
$Rh_6(CO)_{16}$, 49
$Ru(CO)_3(C_8H_{12})$, 105
$[Ru(CO)_3Cl_2]_2$, 51
$[Ru(C_2H_3O_2)(NH_3)_4(NO)](ClO_4)_2$, 14
$[RuCl(NH_3)_4(NO)]Cl_2$, 13
$[RuCl(NO)_2\{P(C_6H_5)_3\}_2](BF_4)$, 21
$[Ru(NCO)(NH_3)_4(NO)](ClO_4)_2$, 15
$[Ru(NH_3)_5(N_2O)]Br_2$, 75
$[Ru(NH_3)_5(N_2O)]Cl_2$, 75
$[Ru(NH_3)_5(N_2O)]I_2$, 75
$Ru_2(CO)_6Cl_4$, 51
$Ru_3(CO)_{10}(NO)_2$, 39
$Ru_3(CO)_{12}$, 45, 47

$[(S)C(CH_3)CH_2C(CH_3)]_2(C_2N_2H_2)Co$, 227
$[(S)C(CH_3)CH_2C(CH_3)]_2(C_2N_2H_4)$, 226
$[S(CH_3)C_6H_4]Li$, 170
$[S(CH_3)C_6H_4](NH_2)$, 169
$[S(CH_3)C_6H_4]P(C_6H_5)_2$, 171
$[S(CH_3)C_6H_4]_2P(C_6H_5)$, 172
$[S(CH_3)C_6H_4]_3P$, 173
$S(H)C(C_4H_3S){=}CHC(O)CF_3$, 206
$(S)P[CH_2P(C_3H_7)(C_6H_5)](C_6H_5)_2$, 195
$[S_2CN(CH_3)_2]_2Co(NO)$, 7
$[S_2CN(C_2H_5)_2]Mo(NO)_2$, 235
$[S_2CN(C_2H_5)_2]_2Fe(NO)$, 5

$Sb[C_6H_4As(CH_3)_2]_3$, 187
$[\{SeC(NH_2)_2\}Co](ClO_4)_2$, 84
$[\{SeC(NH_2)_2\}Co](SO_4)$, 85
$[\{SeC(NH_2)_2\}Hg]Br_2$, 86
$[\{SeC(NH_2)_2\}Hg]Cl_2$, 85
$[\{[SeC(NH_2)_2]Hg\}Cl_2]_2$, 86
$Si(CH_3)F_3$, 139
$Si(CH_3)_2F_2$, 141
$Sn(C_2H_5)(OCH_2CH_2)_3N$, 230
$SnC_8H_{17}O_3N$, 230

$[Ta(CO)_6][As(C_6H_5)_4]$, 71
$[Ta(CO)_6][K\{(CH_3OCH_2CH_2)_2O\}_3]$, 71
$Th(C_5H_5)_3Cl$, 149
$Ti(CH_3)Br_3$, 124
$Ti(CH_3)Cl_3$, 122
$[Ti(C_5H_5)Cl_2]_n$, 238
$Ti(C_5H_5)Cl_2[P(CH_3)_2(C_6H_5)]_2$, 239
$TiCl_3[P(CH_3)H_2]_2$, 98
$TiCl_3[P(CH_3)_2H]_2$, 100
$TiCl_3[P(CH_3)_3]_2$, 100
$TiCl_3[P(C_2H_5)_3]_2$, 101
$Tl[Mn(CO)_5]_3$, 61

$U(C_5H_5)_3Cl$, 148
UCl_6, 143

$(VF_4O)Ni{\cdot}7H_2O$, 87